Powertrain

More information about this series at http://www.springer.com/series/7569

Michael Paulweber • Klaus Lebert

Powertrain Instrumentation and Test Systems

Development – Hybridization – Electrification

Springer

Michael Paulweber
AVL List GmbH
Graz
Austria

Klaus Lebert
University of Applied Sciences
Kiel
Germany

ISSN 1613-6349
Powertrain
ISBN 978-3-319-32133-2 ISBN 978-3-319-32135-6 (eBook)
DOI 10.1007/978-3-319-32135-6

Library of Congress Control Number: 2016943115

Translation from the German language edition: Mess- und Prüfstandstechnik. Antriebsstrangentwicklung • Hybridisierung • Elektrifizierung by Michael Paulweber and Klaus Lebert, © Springer Fachmedien Wiesbaden GmbH 2014. All Rights Reserved.

Printed on acid-free paper

This Springer imprint is published by Springer Nature
The registered company is Springer International Publishing AG Switzerland

Foreword to the Series Powertrain

For decades, the series of volumes entitled "Die Verbrennungskraftmaschine" ("The Internal Combustion Engine"), edited by Hans List, served as an essential reference for engineers in their practical work and for students at universities. Given the pace of technology, I decided, in 2002, to develop a new concept for the series and publish it under the title "Powertrain." The new title conveyed the idea that internal combustion engines should increasingly be seen as components of drive systems. From that time on, the intent of the series was given further thought, and it was finally decided this year to continue the series under the same title ("Powertrain"), however, with a new layout and with a newly appointed scientific board. As before, the main intent of the series is still to identify and discuss all interactions between the various individual components of an automotive powertrain. The new idea is to increasingly promote the English versions alongside the German editions.

Starting with the fundamentals that include a description of the required background information, the purpose of the series is also to address the new components of future drive systems and the way they impact each other in a system-level analysis. In addition to the technical contents, the series also deals with the tools, methods, and processes needed for component development. It examines the conditions in different economic areas and discusses the influences these have on the concepts.

The series of volumes is intended not only for students at universities or advanced technical colleges but also as a reference book for those working in the industry. It invites readers wishing to acquire the necessary in-depth knowledge to draw from the authors' wealth of experience.

Special thanks go to the members of the Scientific Board for their assistance in the organization of this very wide-ranging topic and in the choice of authors. The members of the Scientific Board are:

Herbert Kohler, Vice President, Daimler
Jun Li, Vice President, FAW
Rolf D. Reitz, Professor, University of Wisconsin-Madison

I would like to take this opportunity to thank all the authors who expressed their willingness to share their knowledge in this series of books and contributed their time and effort. I also wish to thank Springer-Verlag.

AVL List GmbH, Graz, Austria Helmut List

Preface

In order to master the great challenges society faces today, the automotive industry, too, is required to contribute its part. CO_2 and emission reduction efforts, advancements toward accident-free mobility, especially also for the aging population, or the need to adapt vehicles to local requirements in a global economy are placing totally new demands on the drive system development process. On the one hand, software is becoming more and more dominant; on the other hand, powertrain architecture is no longer the constant it used to be (internal combustion engine—transmission—shafts—wheels).

As a result, there is now also a great need for simulation, a technology that has meanwhile become a firmly established part of engineering work at test beds. The greatest challenge in the area of instrumentation and test bed engineering is to manage the tremendously increased complexity. Failing to do so will result in development costs (and therefore testing costs) skyrocketing even further.

This book is an attempt to provide an overview of the ways in which these trends are impacting the instrumentation and test systems needed to develop advanced powertrains. Due to the breadth of topics covered, the book required the assistance of many experts. The authors would like to express their sincere gratitude to all specialists for their valuable contributions. Our special thanks go to Mrs. Hermine Pirker. Without her tireless work and organizational support, this book would never have been completed. We also owe a big thank you to Sarah Töfferl for the linguistic revision of the manuscript and the preparation of the illustrations as well as Anita Hoffmann and Elisabeth Stossier for the translations into English.

This book is intended for powertrain (component) development engineers, test bed planners, test bed operators and beginners and deals with the increasingly complex test systems for powertrain components and systems. It seeks to convey an overview of the diverse types of test beds for all components of an advanced powertrain. Additionally, the book focuses on specific topics such as instrumentation, control, simulation, hardware-in-the-loop, automation or test facility management.

Graz
Kiel
September 2014

Michael Paulweber
Klaus Lebert

Contents

Further Authors

Rodolph Belleux Emission Test Systems, Neuss, Germany

Alexander Bergmann AVL List GmbH, Graz, Austria

Christopher Christ AVL Deutschland GmbH, Mainz, Germany

Matthieu Clauet AVL List GmbH, Graz, Austria

Michael Conrad AVL List GmbH, Graz, Austria

Michael Cottogni AVL List GmbH, Graz, Austria

Matthias Dank AVL List GmbH, Graz, Austria

Heimo Draschbacher AVL List GmbH, Graz, Austria

Tobias Düser AVL Zöllner GmbH, Bensheim, Germany

Johann Eitzinger AVL List GmbH, Graz, Austria

Kurt Engeljehringer AVL List GmbH, Graz, Austria

Reinhard Glanz AVL List GmbH, Graz, Austria

Roland Greul AVL List GmbH, Graz, Austria

Bernhard Gröchenig AVL List GmbH, Graz, Austria

Thomas Guntschnig AVL List GmbH, Graz, Austria

Horst Hammerer SET Power Systems GmbH, Wangen, Germany

Volker Hennige AVL List GmbH, Graz, Austria

Gerald Hochmann AVL List GmbH, Graz, Austria

Helmut Kokal AVL List GmbH, Graz, Austria

Johannes Kregar AVL List GmbH, Graz, Austria

Christoph Kügele AVL List GmbH, Graz, Austria

Ferdinand Mosbacher AVL List GmbH, Graz, Austria

Gerhard Müller AVL List GmbH, Graz, Austria

Werner Neuwirth AVL List GmbH, Graz, Austria

Harald Nonn AVL Deutschland GmbH, Mainz-Kastel, Germany

Gerhard Papst AVL List GmbH, Graz, Austria

Egon Petschenig AVL List GmbH, Graz, Austria

Klaus Pfeiffer AVL List GmbH, Graz, Austria

Felix Pfister AVL List GmbH, Graz, Austria

Peter Priller AVL List GmbH, Graz, Austria

Kurt Reininger AVL List GmbH, Graz, Austria

Katharina Renner AVL List GmbH, Graz, Austria

Gerald Sammer AVL List GmbH, Graz, Austria

Richard Schauperl AVL List GmbH, Graz, Austria

Bernhard Schick AVL List GmbH, Graz, Austria

Andreas Schochlow AVL List GmbH, Graz, Austria

Nikolas Schuch AVL List GmbH, Graz, Austria

Markus Schwarzl AVL List GmbH, Graz, Austria

Rüdiger Teichmann AVL List GmbH, Graz, Austria

Joachim Vetter AVL List GmbH, Graz, Austria

Marie Vogels AVL List GmbH, Graz, Austria

Christoph Weidinger AVL List GmbH, Graz, Austria

Michael Wiesinger AVL List GmbH, Graz, Austria

Josef Zehetner AVL List GmbH, Graz, Austria

Symbols and Abbreviations

a	Acceleration
a_K	Flow capacity
A	Area
A_{eff}	Effective flow area
A_V	Reference diameter
β	Area ratio
c	Stiffness, torsional stiffness
c(α)	Piston speed
c_m	Mean piston speed
c_p	Specific heat capacity
CA	Crankshaft angle
d	Damping
dQ	Charge volume
dt	Infinitesimal time period
d_v	Inner valve seat diameter
D	Cylinder bore diameter
Dp	Pressure difference
Δh	Specific enthalpy difference
Δl	Change in length
ε	Strain
F	Force
F_k	Bore cross-section
F_x	Rolling resistance on road
F_{xR}	Rolling resistance on roller
η	Ratio of exciting frequency to natural frequency
η_e	Effective efficiency
H_u	Calorific value
i	Gear ratio, vehicle
i_{diff}	Gear ratio, differential
I	Current

I_0	Incident light flux
j	Dominant harmonic of internal combustion engine
κ	Isentropic exponent
l_0	Initial length
L_{min}	Minimum air/fuel ratio
λ	Air/fuel ratio
λ_L	Volumetric efficiency
m	Mass
$\dot{m}$	Mass flow rate
$\mu\sigma$	Flow coefficient
$\mu\sigma\beta$	Bore-related flow coefficient
n_D	Speed
$(n_D/n)_m$	Swirl number
$(n_D/n)_{m,red}$	Reduced swirl number
$(n_D/n)_{red}$	Reduced rotation coefficient
n_T/n	Tumble coefficient
N	Opacity
Ø	Angle at the circumference
p	Boost pressure
p_0	Upstream pressure
p_1	Downstream pressure
P_e	Effective power
p_m	Mean pressure
P	Engine power, power
Q	Internal energy
Θ_D	Inertia of dynamometer
Θ_{diff}	Inertia of differential
$\Theta_{transmission}$	Inertia of transmission
Θ_{ICE}	Inertia of engine
Θ_{PT}	Inertia of powertrain
Θ_{wheel}	Inertia of wheel/tire
r	Tire radius
r_{dyn}	Dynamic tire radius
Rg	Gas constant
R_r	Roller radius
R	Electric resistance, cylinder radius
ρ	Air density
s	Stroke, piston stroke
t	Time
T	Temperature
T_0	Temperature at point in time $t = 0$

T_A	Outlet temperature
T_E	Inlet temperature
T_d	Engine torque
T_z	Torque around cylinder axis
U_{th}	Thermopower
v	Velocity
V	Volume
Vg	Transmission coefficient by taking into account damping
V_h	Piston displacement
V_{st}	Volumetric flow
ω	Engine speed
ω_0	Natural frequency of the oscillating system
x	Cyclic irregularity
z	Number of valves

Introduction 1

1.1 Drivers of Automotive Development

The major challenges society faces in this century are having significant effects on how the automotive industry is evolving (see Fig. 1.1).

Global warming is being widely discussed in the media. The automotive industry is required to contribute its part by reducing vehicle-based CO_2 emissions. Automobile manufacturers (in short: OEMs—Original Equipment Manufacturers) are addressing such requirements by implementing downsizing concepts, electric mobility (particularly in megacities), hybrid powertrain concepts or by employing alternative energy resources (e.g. bio-fuels) or electric vehicles with hydrogen fuel cells.

Growing urbanization is leading to bigger and bigger cities. With space being one of the most valuable resources in rapidly expanding megacities, new concepts are required to ensure the continuation of individual mobility. This is why automakers are working hard on the development of self-parking systems or automatic cruise control technologies including highly automated vehicles. The growing vehicle density, coupled with a rising proportion of aging people, is leading to a hugely increased risk of traffic accidents. Again, the automotive industry is responding by offering innovative ADAS (Advanced Driver Assistance Systems) and, sooner or later, even partially or completely self-driving cars.

A further trend among the emerging generation is that young people expect being able to communicate with others and access global content via Google, Facebook, etc. anywhere and anytime. This aspect raises the demands on vehicle operation, as Google and Apple—to single out two "pioneers of simplicity"—have set new standards in this respect.

Apart from that, young adults today care much less about having a car of their own than the generation born before 1990, so it is becoming imperative for auto manufacturers to

M. Paulweber, K. Lebert, *Powertrain Instrumentation and Test Systems*, Powertrain,
DOI 10.1007/978-3-319-32135-6_1

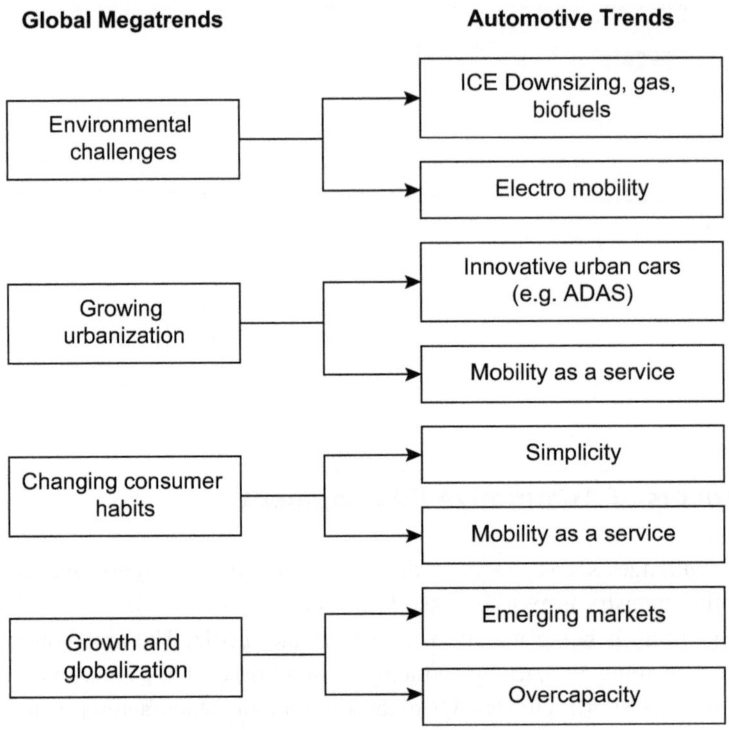

Fig. 1.1 Global megatrends and their implications for the automotive industry [1]

focus on multimedia devices, or new business concepts such as mobility as a service. This is an area where products with a typical life time of just about several months (e.g. cell phones) meet products in the automotive industry with a lifecycle of 10 or more years. The interaction between entertainment electronics and the safety-related vehicle electronics poses new challenges, particularly to validation processes during development.

As shown in Fig. 1.2, nearly all countries worldwide are planning a steady reduction of CO_2 emissions in new vehicles. The only way for this to be accomplished is by employing new powertrain concepts, some of which are either still in development or are already being marketed in initial (small-)series vehicles. The associated buzzwords, such as series hybrid, parallel hybrid, mild hybrid, range extender, electric vehicle or long-range e-mobility (fuel cell electric vehicles), are being widely talked about, but will not be discussed any further in this book.

There is one thing these new concepts have in common: the principal architecture of the automotive powertrain is undergoing its first radical change in almost 100 years. Up until recently, the basic layout always remained the same: the internal combustion engine is connected via a clutch to a transmission, and the output of the transmission is

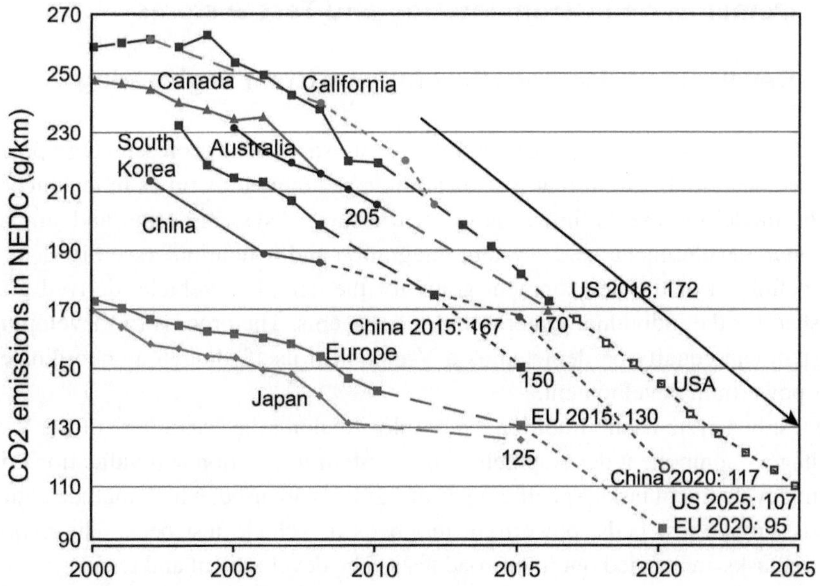

Fig. 1.2 Global trends to rapidly reduce CO_2 emissions [2]

transferred via shafts to the wheels. Architectures in modern hybrid vehicles, however, differ very widely. Such diversity makes it necessary to employ full vehicle simulation in the very first development stages in order to find the type of architecture that solves the demands placed on the vehicle most efficiently. Since OEMs are accustomed to developing the individual components in parallel, the exact requirements of such components and their interfaces have to be specified early on at the beginning of the development process. In the past, the task of translating the requirements of the complete vehicle to component requirements used to be carried out by chief engineers, who had thorough knowledge and understanding of the entire powertrain architecture, combined with vast experience. In light of the changing powertrain architecture and its increasing flexibility in hybrid vehicles, electric vehicles or fuel cell vehicles, automakers clearly lack such long time past experience. The amount of pressure this puts on simulation is in turn causing the close integration of simulation activities into the design and test phase.

This approach requires the utilization of detailed models of the powertrain components in early development stages, resulting, however, in much higher costs for this early phase. To compensate for this cost increase, automakers are making an effort to re-use such models in later stages of the development process. The buzzwords in this respect are "model-based testing," "model-based calibration," etc. The additional benefit is a shortened development time by frontloading, which is described in further detail in the next section.

1.2 Demands on Instrumentation and Test Systems

1.2.1 Development Methodology in Powertrain Engineering

The product creation process in the automotive industry can be represented graphically as a so-called V-Model, a term that has been described numerous times in technical literature. The model represents the sequence of the stages "system design and simulation," "component development" and "system integration and validation" (see Fig. 1.3). Based on the definition of the development goals for the complete vehicle, derived goals are established for the individual systems and sub-systems. The process for developing each sub-system can equally be depicted as a V-model in itself, though a subordinated one, e.g. for powertrain development.

The test bed systems are traditionally employed along the "right leg" of the V-model, in the stages "component development" and "system integration and validation." For the different development tasks, specific types of test beds are used, which include component test beds, engine test beds, powertrain test beds or vehicle test beds. The majority of validation tasks are carried out in on-road tests. The development and testing tasks can be divided into the main groups mechanical development, electrics/electronics development and software development. The validation for the first part is again divided into "mechanics development/endurance strength testing," "drivability calibration," "emissions and fuel consumption optimization" as well as "noise, vibration and harshness testing (NVH)." Typically, the development is staged and results in different prototypes, often called A, B and C prototypes. The requirements for these stages are defined according to the expected degree of maturity of the prototype.

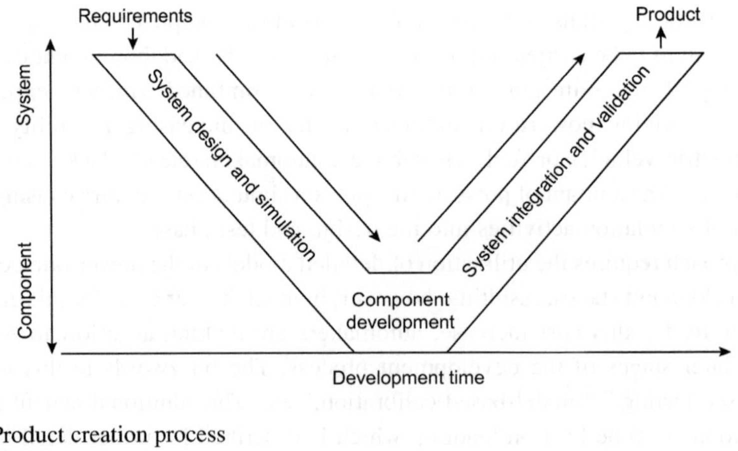

Fig. 1.3 Product creation process

1.2.2 Impact of Development Methodology

The growing pressure to innovate and the demand for shorter development cycles, along with new statutory requirements, require changes in the development methodology. As a result, there are shifts in the demands on test bed systems. The desire for shorter development times stands opposed to the growing complexity needed to satisfy the requirements mentioned in the previous section (see Fig. 1.4).

A core aspect of the evolved work methodology is the tendency to shift development tasks to early phases in the development process. This approach is referred to as frontloading (see Fig. 1.5) and enables an early validation of the assumptions made during the concept and simulation stage.

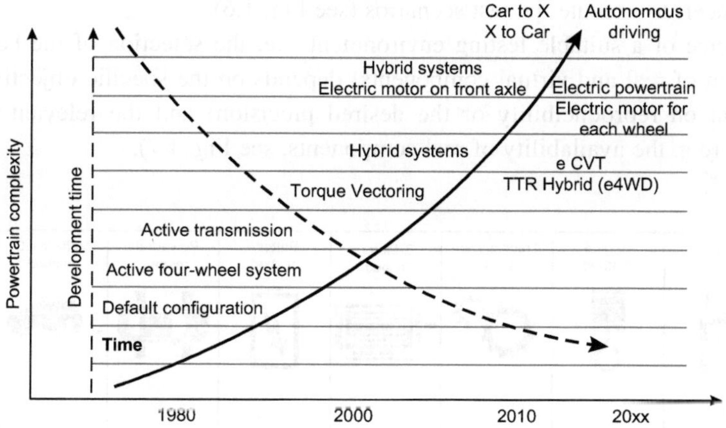

Fig. 1.4 Growing complexity

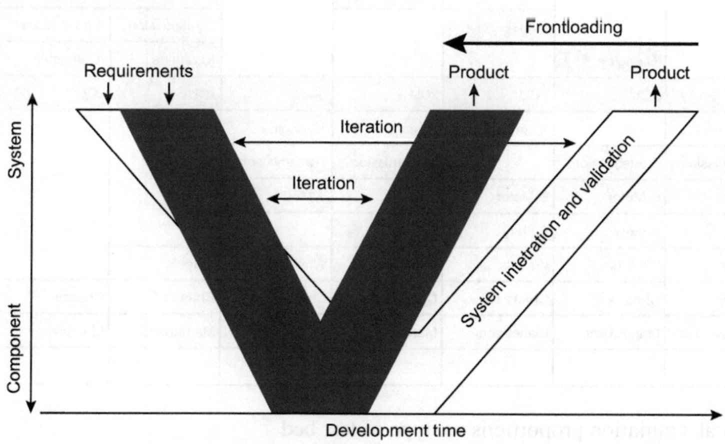

Fig. 1.5 Frontloading in the V-model

A key attribute of frontloading is the simulation of components not physically present at that stage (in office work and in test bed validation). Such simulations must be module-based to enable re-use on different types of test beds. As real and virtual components are combined, the simulation has to be conducted in real-time. We refer to this approach as "X in the loop" or "XiL" test beds.

As a result of the growing system complexity arising from the interaction among intelligent sub-systems, such as engine control unit, transmission control unit or driver assistance systems, it is no longer possible to describe the load scenarios for individual components on the basis of synthetic load profiles. Instead, it is critical to describe the development and testing tasks based on real use scenarios. This means that regardless of the test environments utilized it is always the scenarios that are tested in reality (such as a fleet cycle or a safety-critical maneuver). Real components and simulated components are used in concert to execute the test scenarios (see Fig. 1.6).

The choice of a suitable testing environment (i.e. the selection of the best possible combination of real and virtual components) depends on the specific objective (e.g. the requirement on reproducibility or the desired precision) and the relevant framework conditions (e.g. the availability of real components, see Fig. 1.7).

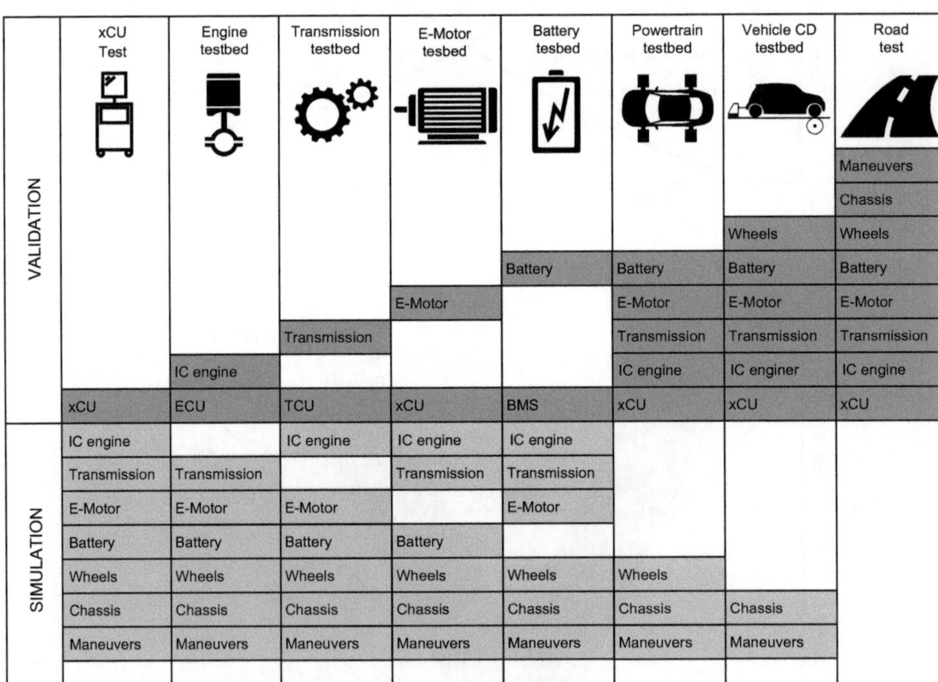

Fig. 1.6 Real/simulation proportions vs. type of test bed

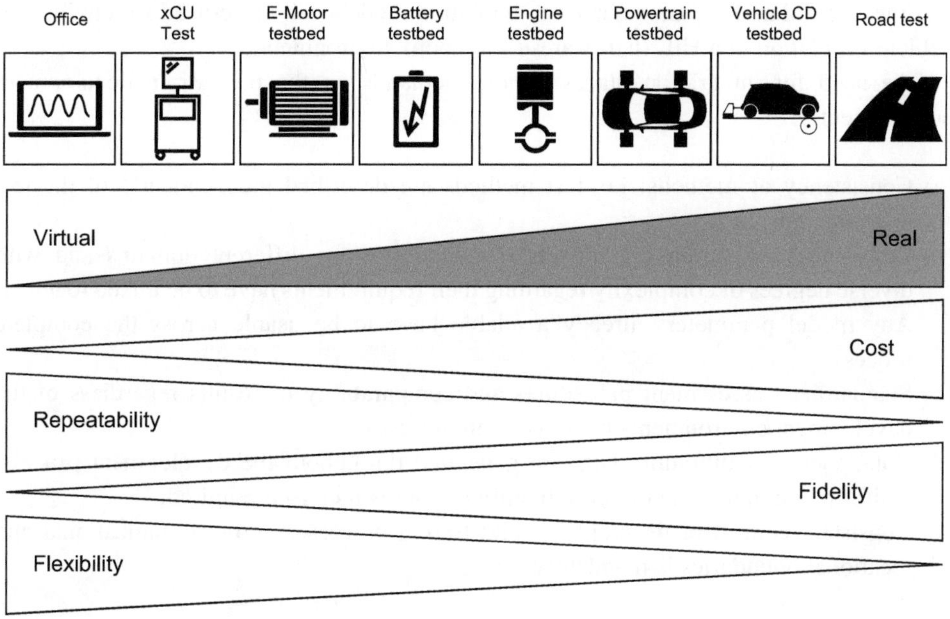

Fig. 1.7 Selection criteria for real and virtual components

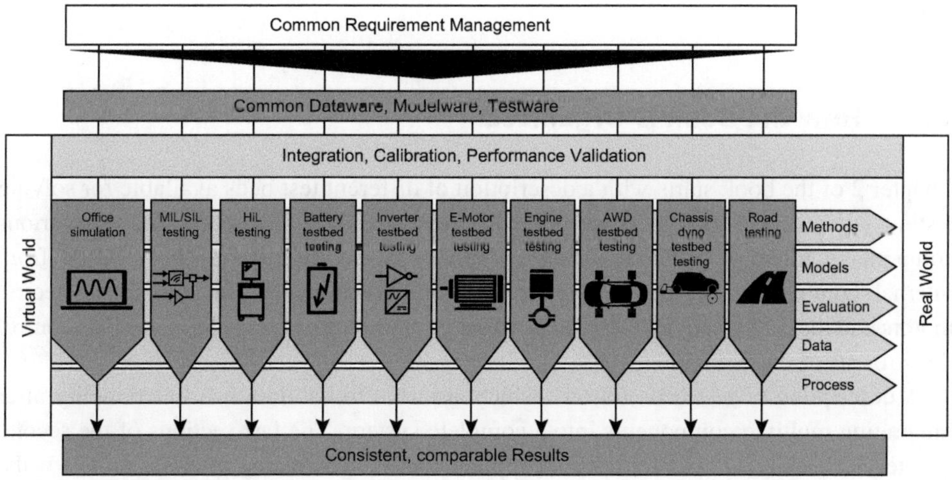

Fig. 1.8 Consistent development platform

1.2.3 Networked Development Environments

Critical to an efficient development process is the integration of different development environments into a single process flow (see Fig. 1.8). Accordingly, it is necessary to make the design data from the CAD systems available to the simulation models. Equally, real measured data for parameterizing simulation models can be re-used for engine and vehicle simulation in a HiL (hardware-in-the-loop) environment.

The need for integration places various demands on the powertrain development environment:

(a) Consistency of methods; i.e. test methods are described independently of the test environment.
(b) Consistency of simulation models; i.e. models from different domains and with diverse degrees of complexity regarding their requirements have to be linked together. Any model parameters already available have to be usable across the complete process.
(c) Standardized assessment procedures and comparability of results regardless of the development environments being used are necessary.
(d) Data, such as calibration data, are generated throughout the development process. Uniform assignment and interpretability of data is therefore required.
(e) Consistent control of development and testing processes across organizational and locational boundaries is mandatory.

Though only partially realized in practice today, networked development processes will most likely rapidly gain in significance. In this regard, it is also crucial to enable networking across heterogeneous system landscapes. A possible architecture is depicted in Fig. 1.8.

1.3 How the Book Is Organized

Chapter 2 of the book starts with a description of different test beds available for solving testing tasks in automotive powertrain development. The test setups for the various components found in modern vehicle powertrains are discussed. These range from internal combustion engines and their auxiliary components, transmissions, electric motors and the related power electronics to battery systems or fuel cells, as well as control units for all of these components.

A description of test beds follows which are used to validate full functionality after integrating multiple components into a complete system. The last sections of the second chapter deal with test beds for specific applications such as racing or emission test beds.

The subsequent sections are organized according to the basic architecture of a test bed, as shown in Fig. 1.9.

This architecture divides the test bed into three levels:

The **hardware layer** comprises the two lower layers above the actual unit under test as shown in Fig. 1.9. These contain among other things the sensors and the data acquisition and actuator modules that are frequently connected to an automation system via bus systems. Chapter 3 presents the individual hardware parts that make up test beds.

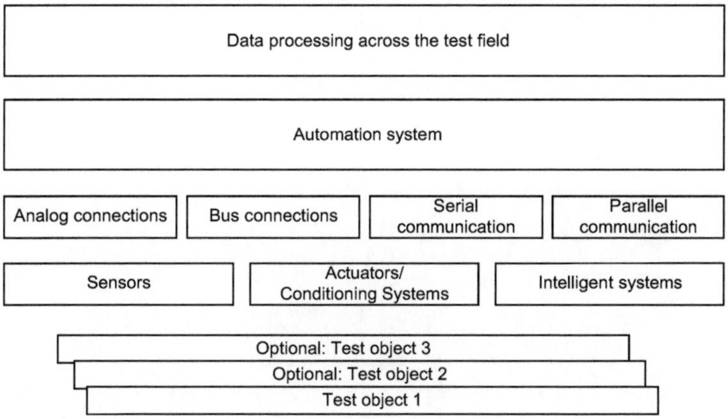

Fig. 1.9 General test bed architecture

On the **automation layer** the data are recorded, processed and stored. Automatic test runs are executed here. Chapter 4 addresses the software aspect of a test bed and describes the individual features of an advanced automation system, covering measured data acquisition, signal processing, data management, control and simulation functions, as well as further automation tasks. The necessary safety features are dealt with in the last section of this chapter.

Multiple test beds are combined to larger development centers. This level requires **data storage**, **data integration** and **data processing across a test facility**. Chapter 5 addresses possible ways of implementing the necessary requirements with regard to this layer.

References

1. KPMG International Cooperative, *KPMG's Global automotive executive survey 2013* (KPMG International Cooperative, 2013)
2. ICCT, European CO_2 emission performance standards for passenger cars and light commercial vehicles, 12 July 2012. [Online]. http://www.theicct.org/sites/default/files/publications/ICCT%20Policy%20Update%20EU%20PV_LCV%20CO2%20July2012final.pdf. Accessed 28 Apr 2014

Types of Test Beds

2

This chapter presents different test bed configurations that are used in the development of advanced powertrains in automotive engineering. First, the discussion focuses on the internal combustion engine test bed, the internal combustion engine still being the most important drive unit available. Its further development accounts for a substantial amount of development effort in present-day vehicles.

This is followed by a description of test beds that are used to test other powertrain components. These test beds additionally deliver valuable data for validating and calibrating simulation models of such single components.

An ever-increasing share of the development budget for modern vehicles is being invested in the research & development of a great variety of control units. Such control units contribute substantially toward making our vehicles more environmentally friendly, comfortable and safe. The section on control unit test beds (often called hardware-in-the-loop or HiL test beds) introduces the respective verification and validation test facilities.

A description of test beds follows which are used to validate full functionality after integrating multiple components into a complete system. This task is often accomplished either on powertrain test beds or on chassis dynamometer test systems. The final sections of this chapter address test beds for specific applications such as racing or emissions certification.

2.1 Combustion Engine Test Beds

2.1.1 Scope of Application

Dependent on the objective of their use, test beds for internal combustion engines are generally distinguished by the following areas and types of application:

M. Paulweber, K. Lebert, *Powertrain Instrumentation and Test Systems*, Powertrain,
DOI 10.1007/978-3-319-32135-6_2

- Research:
  - Single-cylinder engine test bed
  - Flow test bed (see also Sect. 3.3.15)
- Development:
  - Performance test bed
  - Function test bed
  - Endurance test bed
  - Calibration test bed
  - Emission certification test bed
- Production:
  - End-of-line break-in test bed
  - Quality assurance test bed

Single-cylinder engine test beds primarily serve research purposes and help to assess the combustion process optically. **Flow test beds** allow the examination of the charge motion that also has an essential impact on combustion. For further details, please refer to Sect. 2.1.5.

A **performance test bed** is used to determine the engine power P output by the unit under test over the entire operating range, under the test conditions that represent the scheduled unit-under-test utilization, and based on the captured engine speed ω (see also Sect. 3.3.7) and the measured engine torque T_d (see also Sect. 3.3.6). Engine torque T_d is the effective torque which the internal combustion engine outputs to the drive shaft.

$$P = T_d * \omega \tag{2.1}$$

Equation 2.1 Power

A **function test bed** serves to optimize, verify and secure engine-related overall system features. In addition to engine power, functions such as fuel consumption (see Sect. 3.3.8) and emissions behavior (see Sects. 3.3.11 and 3.3.14) downstream of the exhaust manifold and downstream of the emission aftertreatment system (if any) are analyzed. A further critical examination involves the unit under test's response behavior to changing load levels.

To investigate and ensure the durability and long-term stability of the internal combustion engine and the related components planned for series production, the engine is subjected to a thorough testing procedure on an **endurance test bed**. Such tests may take up to several hundred operating hours and therefore need reliable monitoring systems to ensure automatic, unmanned test bed operation.

Dedicated **calibration test beds** are used in series development to impress a specific and optimal engine behavior on engine control units, such as the exhaust emission values output by the engine when a specifically defined sequence of operating points is followed. To name a few examples, this might include the setting of variable injection-system parameters (e.g. diesel engine: injection amount(s), time, pressure).

Internal combustion engines, which are employed in heavy-duty vehicles (trucks, buses, etc.) and in vehicles or drive systems for off-highway applications (construction machinery, agricultural or forestry equipment, stationary diesel generators), are subjected to a final homologation test on the **exhaust emission certification engine test bed** (see Sect. 2.7) to ensure compliance with the required exhaust emission limits. By contrast, in the case of passenger cars (or two- and three-wheelers with up to 3500 kg gross vehicle weight), exhaust emission tests are performed on a chassis dynamometer (see Sect. 2.5).

The **end-of-line production engine test bed** (see Fig. 2.1) is an integral and key element in the production of internal combustion engines. This category of test beds has the purpose of assessing the engine quality in production according to stringent, predefined criteria. The employed instrumentation and test systems as well as the test methods are part of the complete engine-specific production process.

The growing technological complexity of internal combustion engines is leading to hot tests (these are tests performed on a fired unit under test) increasingly being run during production in a loaded state, i.e. under certain load cycles and therefore using a load unit. Only by doing so can the engine's functionality and quality be adequately ensured.

In specific cases, cold tests are required; this means that the engine is tested unfired (without combustion). Test systems for assessing conformity of production (COP) complement the range of random production-testing solutions.

As they do not belong to the development test beds, which are the subject of this book, production test beds will not be discussed any further in the remaining chapters.

Fig. 2.1 Production engine test bed with pallet transportation system

2.1.2 Setup of a Test Bed for Internal Combustion Engines

To meet the objective of permitting as broad a range of use as possible, test beds intended for the development of internal combustion engines typically have the following main components (see Fig. 2.2):

- Dynamometer
- Test bed mechanics
- Engine media conditioning
- Consumption measurement systems for fuel, combustion air and urea

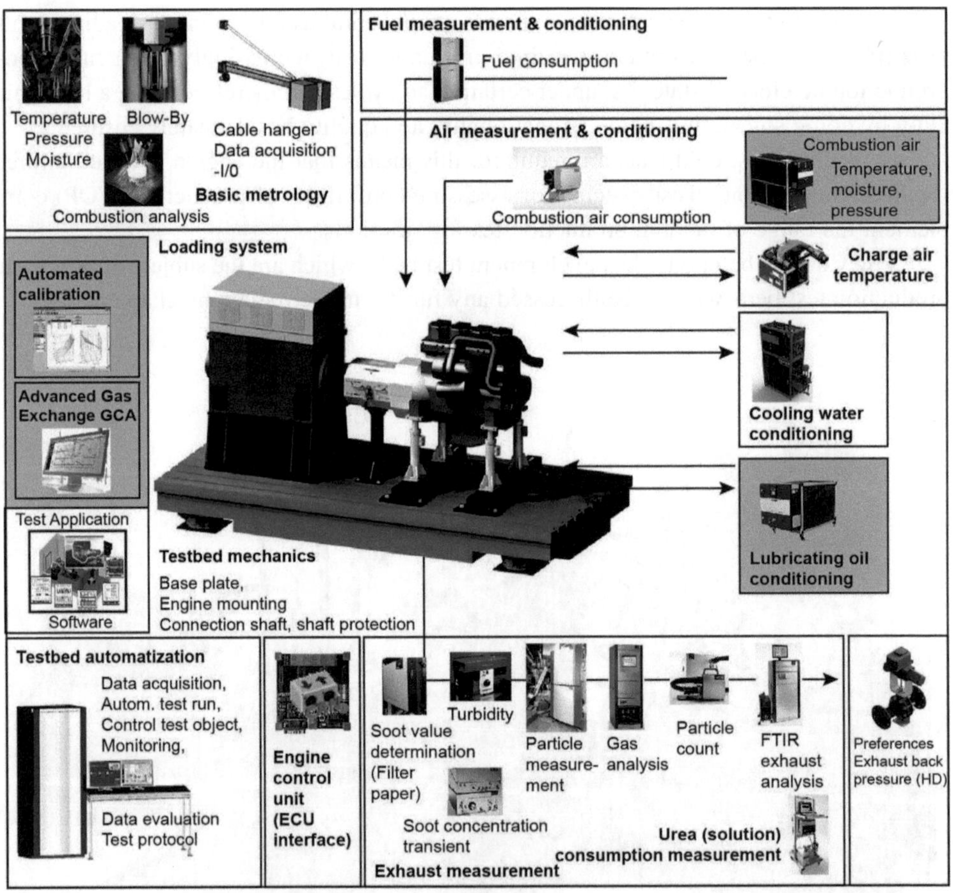

Fig. 2.2 Scope typically covered by an internal combustion engine development test bed

– Temperature and pressure measurement chains for measuring points in the engine periphery
– Instrumentation for combustion diagnostics
– Blow-by gas measuring device
– Emission measurement equipment
– Communication interface to the engine control unit (ECU)
– Test bed automation (control/simulation)
– Calibration tools to optimize control unit calibration

Dependent upon the employed test bed technologies, such as dynamometer and test bed automation system, we basically distinguish between the following categories of engine test beds:

– Steady-state test beds
– Non-steady-state test beds

2.1.3 Steady-State Engine Test Beds

A **steady-state test bed** is characterized by the following features (see Fig. 2.3):

– Load level adjustment by means of torque and speed value pair (T_d/ω) and by means of torque/alpha or speed/alpha value pair (alpha = throttle position)
– Control phase with defined tolerances
– Measurement following a stabilization phase and, if required, combined with additional criteria to be satisfied (e.g. the achievement of a certain oil temperature)
– Evaluation of the recorded measured values over a certain time interval (e.g. determined by the fuel consumption measurement)

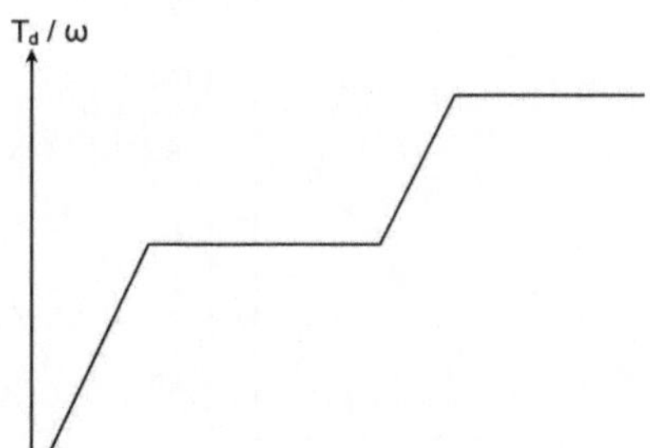

Fig. 2.3 Steady-state operating states

Steady-state test beds are used in research, development and in production. Typical applications for a steady-state operating point sequence include recording a fuel consumption map for an internal combustion engine (engine graph—see Fig. 2.4), recording a full-load curve (see Fig. 2.5) or use in the calibration of engine functions.

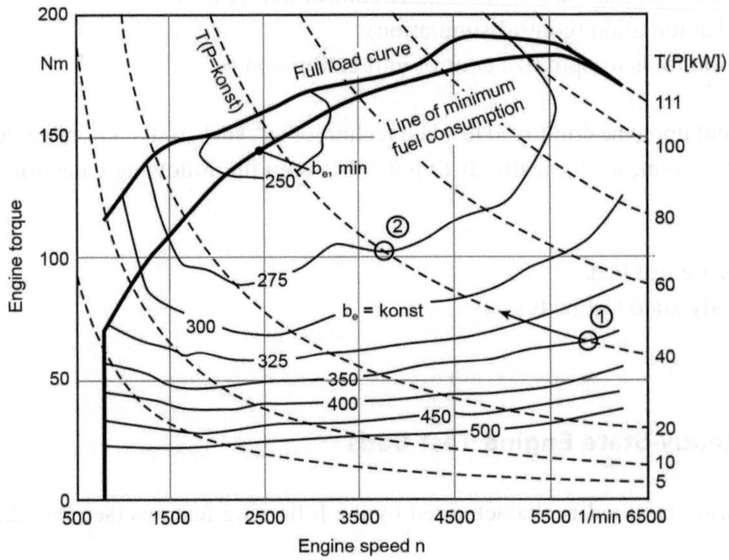

Fig. 2.4 Fuel consumption map of an engine

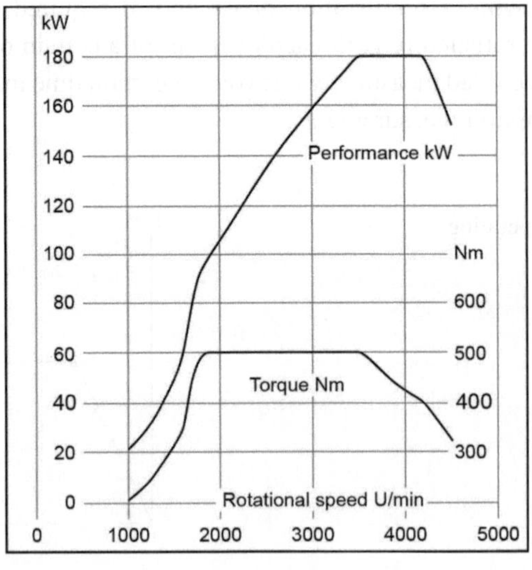

Fig. 2.5 Full-load curve of an engine

2.1.4 Non-Steady-State Test Beds

Test beds for non-steady-state tests on internal combustion engines are divided into the following types:

- Transient test bed
- Dynamic test bed
- High-dynamic test bed

Transient test beds (see Fig. 2.6) have the following features:

- Load-point definition by means of torque/speed value pair in time steps ≤ 1 s
- Continuous control of speed and torque within the required tolerances
- Continuous measurement and recording of measured and calculated values
- Time-resolved and/or interval-based representation of the recorded measured values

The application areas for transient test beds include the testing of engine response to sudden load variations, exhaust emission homologation tests on heavy-duty engines or off-road vehicles/machinery applications (construction machinery, tractors, etc.) or the examination of transient engine operation with regard to fuel consumption and emission production.

The features of a **dynamic test bed** (see Figs. 2.7 and 2.8) are:

- Engine testing in a simulated vehicle driving operation
- Load-point definition as a velocity/time profile
- Simulation of the vehicle powertrain (powertrain oscillations up to typically 8 Hz), the driver's behavior, the road load and the road gradient (see Fig. 2.8)
- Continuous measurement and recording, also during load point transition
- Time-resolved and/or interval-based representation of the recorded measured values

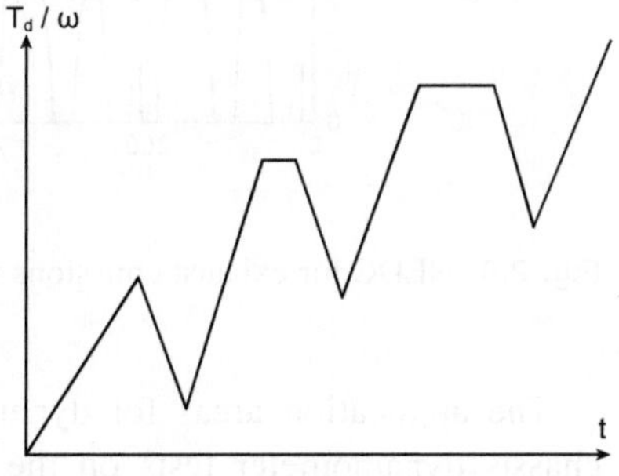

Fig. 2.6 Transient test cycle

Fig. 2.7 Transient, dynamic test bed with asynchronous machine

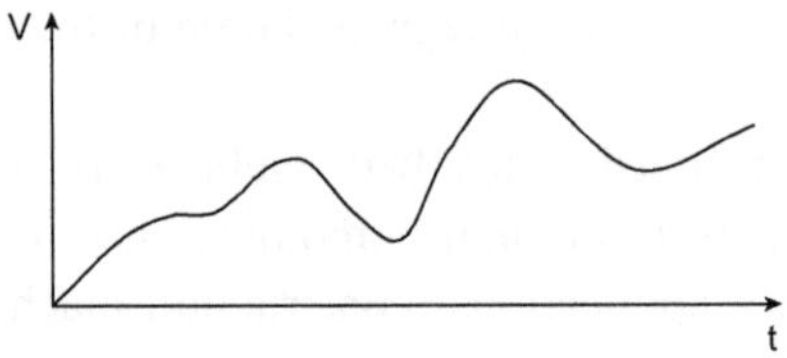

Fig. 2.8 Dynamic test cycle

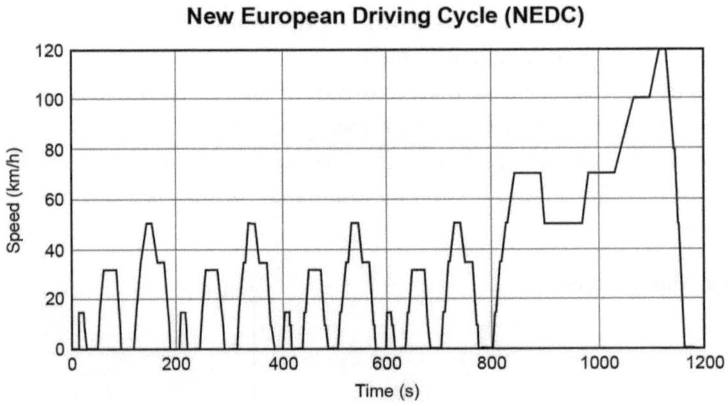

Fig. 2.9 NEDC for exhaust emissions testing in light-duty vehicles (passenger cars)

The application areas for dynamic test beds comprise the reproduction of vehicle chassis dynamometer tests on the engine test bed, the execution of legally required exhaust emission cycles for light-duty vehicles (passenger cars) (see Fig. 2.9) in compliance with international legislations (e.g. EU-NEDC, US EPA FTP-75, WLTP) to assess

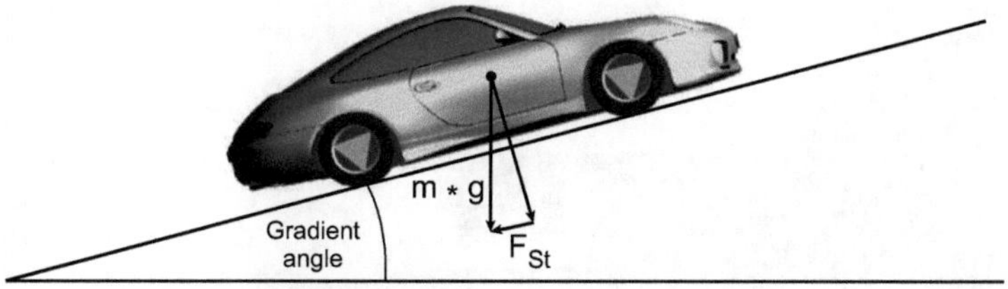

Fig. 2.10 Dynamic test bed simulation of road gradient

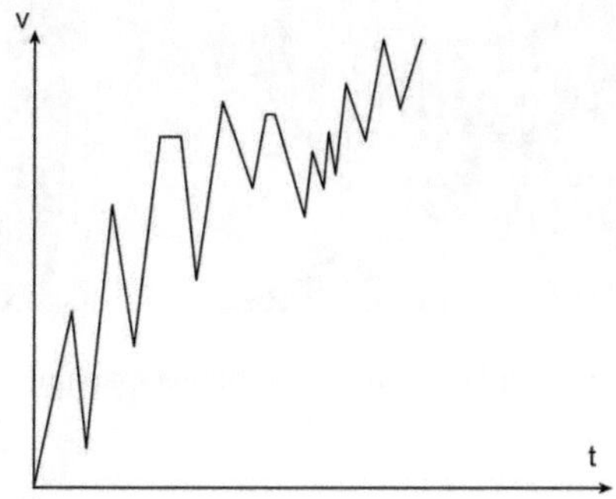

Fig. 2.11 High-dynamic test cycle

the engine exhaust emission behavior in the planned vehicle use at an early stage, and the optimization of fuel efficiency and exhaust gas emissions.

Features of the **high-dynamic test bed**:

- Testing of the internal combustion engine in the virtual vehicle, driver and road environment in a situation that is as close to reality as possible (real-life driving operation, see Fig. 2.10)
- Unit-under-test operation in the critical range between engine, starter and idle speed
- Powertrain simulation up to approx. 40 Hz
- Real zero-torque simulation during gear-shifting processes and idle operation phases

High-dynamic test beds (see Fig. 2.11 and Fig. 2.12) are used for calibrating the engine cold-start behavior without vehicle, tuning vehicle drivability without vehicle and without road, performing real fuel-consumption drives (real driving emissions—RDE) with a real engine and a virtual powertrain in a virtual environment (see Fig. 2.13).

The following figures illustrate the difference between a transient test bed and a high-dynamic test bed (see Figs. 2.14 and 2.15). A real-life engine start is compared with an engine start on the test bed with different dynamometers (shown by the example of an AVL APA-HI for transient transitions and an AVL DynoSpirit for dynamic transitions).

Fig. 2.12 High-dynamic engine test bed with synchronous machine

Fig. 2.13 Maneuver-based testing on the engine test bed

2.1.5 Research Test Beds

2.1.5.1 Single-Cylinder Engine Test Beds

Single-cylinder engine test beds are chiefly used for research purposes, e.g. for the combustion process itself. In special cases, a cylinder liner made out of glass and adapted to the cylinder head is used (see Fig. 2.16). This gives you visual access to the combustion chamber and allows you to observe the injection and combustion process by means of optical measuring methods. As a result, it is possible to examine the prospects of success

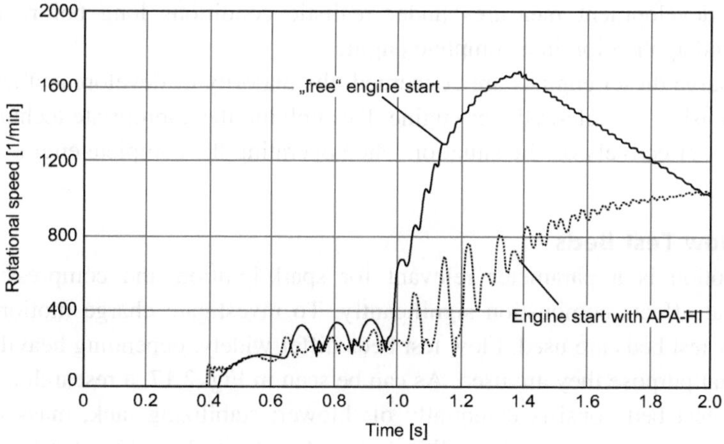

Fig. 2.14 Conventional transient test bed operation range (using the example of an AVL APA-HI)

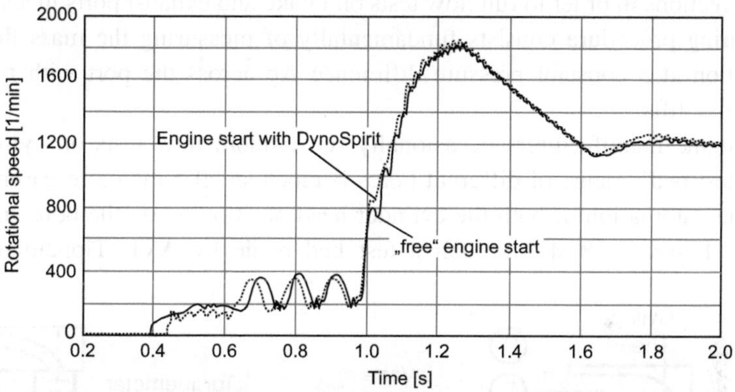

Fig. 2.15 High-dynamic test bed operation range (using the example of an AVL DynoSpirit)

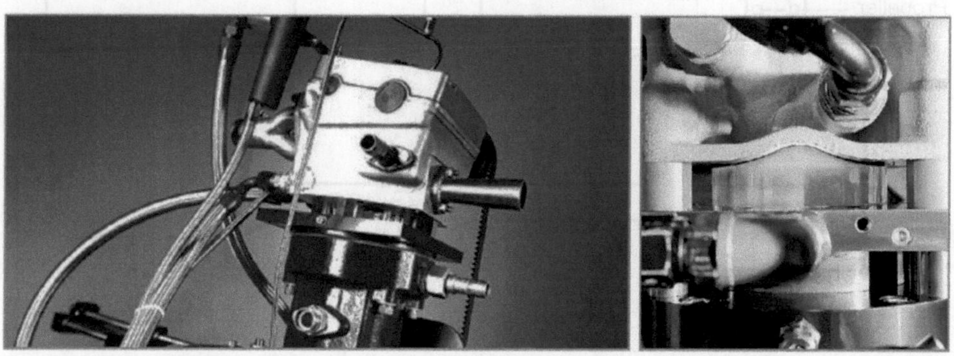

Fig. 2.16 Single-cylinder research engine

of certain development measures under realistic conditions long before any of the concepts find application in a complete engine.

Where large diesel engines are concerned, the majority of development activities are carried out using a single-cylinder engine. By applying the appropriate techniques, work focuses cost-effectively on the situation when operating the complete engine.

2.1.5.2 Flow Test Beds

Charge motion is a parameter relevant for spark-ignition and compression-ignition engines that affects combustion significantly. To investigate charge motion in further detail, flow test beds are used. Flow test beds differ widely, depending heavily on where and for what purpose they are used. As can be seen in Fig. 2.17, a research and development flow test bed consists essentially of: blower; stabilizing tank, mass flow meter; pressure, temperature and air humidity sensors (see Sect. 3.3). The stabilizing tank has built-in integrating measuring devices (paddle wheel and/or torque meter) to determine swirl and tumble (see also Sect. 3.3.15). The test bed should be capable of representing two flow directions in order to run flow tests on intake and exhaust ports in a single setup. The measuring procedure consists fundamentally of measuring the mass flow and the charge motion at a constant pressure difference Δp across the port with typically ten different valve lifts.

Some engine manufacturers occasionally run random flow tests on cylinder heads. As such a test bed's range of different bores is much smaller than a research test bed's, it allows the automation of both the cylinder head positioning on the bore and valve lift adjustment. Figure 2.18 shows such a test bed made by AVL Tippelmann. 50–100

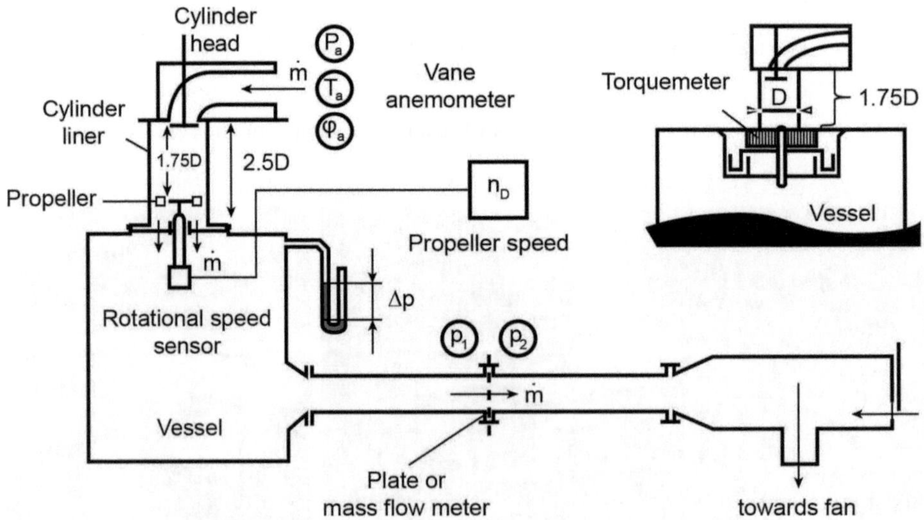

Fig. 2.17 Example of a steady-state flow test bed setup for measuring swirl and flow in intake and exhaust ports

Fig. 2.18 Advanced-development test bed for testing flow in single cylinder heads

measurements are possible per day. Other fully automatic flow test beds with a measuring capacity of approx. 1000 measurements per day can be found in production lines for cylinder heads.

2.1.6 Special-Purpose Engine Test Beds

To ensure the engine function and therefore proof of its functionality under specific use conditions, special-purpose test beds are used. This category of engine test beds includes the following types:

- Tilt test bed
- Climate test bed
- High-altitude/climate test bed

Tilt engine test beds are used to optimize oil-carrying components or oil foaming behavior (i.e. aeration), develop oil circuit or crankcase ventilation systems or for friction tests. For this purpose, the internal combustion engine is mounted on a special testing facility (see Figs. 2.19 and 2.20) and subjected to continuous dynamic tilting (e.g. two-axial, with adjustable tilting angles). Following certain driving profiles, the internal combustion engine can be tested both in fired and motored operation.

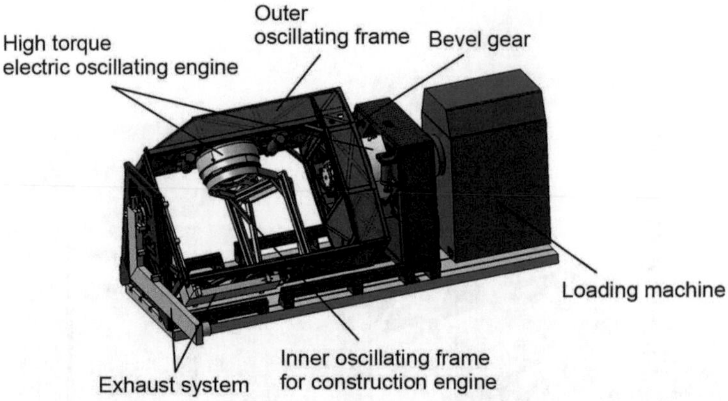

Fig. 2.19 Test bed setup for a tilt test rig

Fig. 2.20 Dynamic tilting frame

The **climate engine test bed** (see Fig. 2.21) serves to examine how the internal combustion engine behaves in different ambient-air conditions (e.g. ambient air corresponds to combustion air) in the form of temperature and humidity values in a laboratory environment, early on in the development process. For this purpose, temperatures approximately in the range of −40 °C to +50 °C are simulated (see also Sect. 3.2.4).

A subject to investigate would be, for example, the cold-start behavior of an internal combustion engine. In addition to testing the ignition behavior of the engine itself, such test systems are also used to examine and optimize the associated exhaust gas emissions.

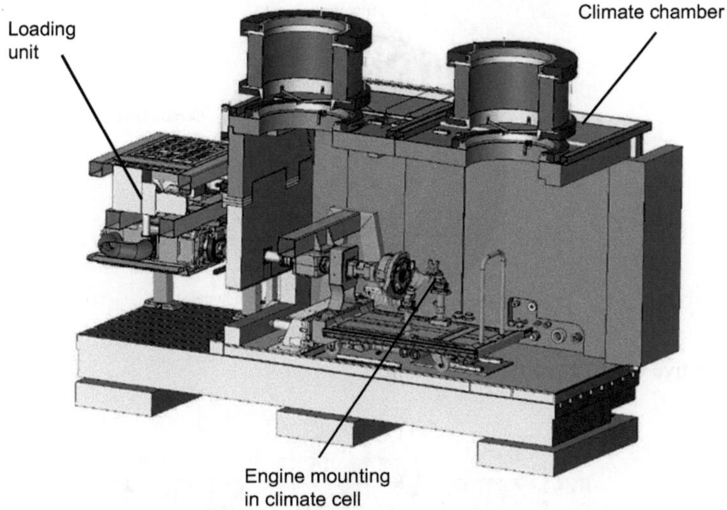

Fig. 2.21 Climatic engine test bed

Depending on requirements, the test bed's air-conditioning system is implemented either by placing the whole test bed into a climatic chamber or simply by using a climatic hood. With the latter only the internal combustion engine is given a climatic enclosure that is placed over the unit under test to allow the ambient climatic conditions to be set.

In special cases, **high-altitude/climate test beds** are used in the development of internal combustion engines to examine and optimize the influence of the ambient air pressure on the engine behavior, in addition to that of the temperature. The related optimization of maps in engine control units and the verification of exhaust emission behavior in such environmental conditions (e.g. not-to-exceed emissions in e.g. heavy-duty engines) are just a few examples worth mentioning.

High-altitude/climate test beds can be implemented as conventional complete chamber systems or as "optimized" solutions. In the former case, the entire engine test bed is installed inside a high-altitude/climate chamber. In the latter case, pressure-tight air supply and exhaust systems are fitted to all of the relevant engine connections and components (simulation of a high-altitude/climate chamber).

2.2 Component Test Beds

An advanced powertrain not only includes the conventional internal combustion engine but also electric power components. Such arrangements essentially consist of the following five elements (Fig. 2.22):

These elements are also referred to as components and, being part of the vehicle development process (Fig. 2.23), they have to be subjected to integration and validation

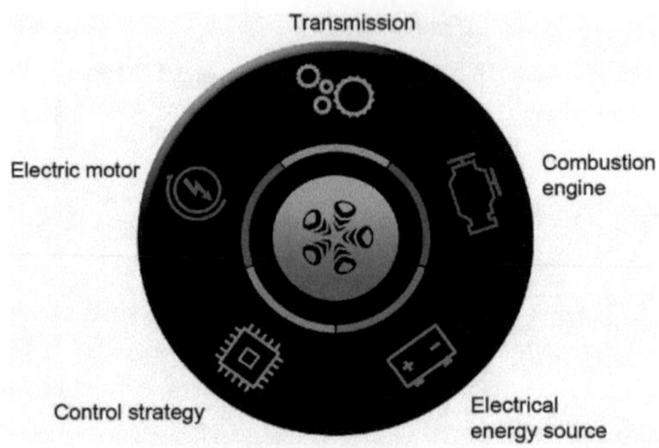

Fig. 2.22 The five elements of the powertrain

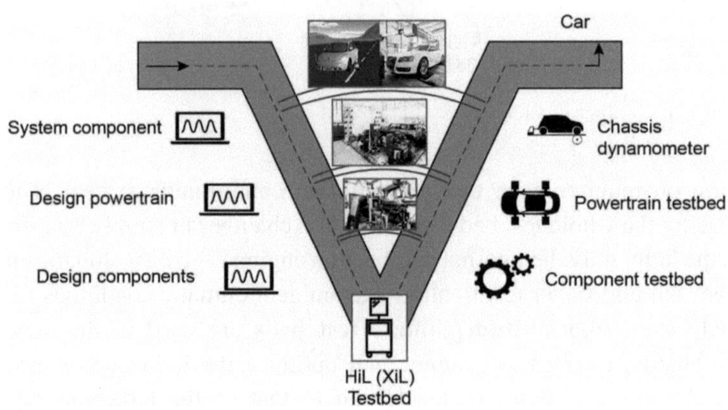

Fig. 2.23 Vehicle development process

tests on suitable test systems. An engine test bed is, for example, a suitable validation site for combustion engines.

2.2.1 Test Beds for Components of Internal Combustion Engines

Test beds for components of internal combustion engines are used to conduct selective function, performance and durability tests on such parts. The selective test permits the parallel development and optimization of such components before their integration into the engine.

As component test beds are generally used very flexibly, the demands on the measuring and automation equipment are very specific. Besides having sensors that are easy to configure, there must be a variable number of demand value channels with freely definable meanings and signal ranges. The automation system must allow the definition of freely definable limits and assigned reactions for any given measured/calculated value and, in the event of limit violations, execute freely definable reactions for any given demand value channel (see also Sect. 4.3.4). Several types of component test beds are discussed in further detail below.

2.2.1.1 Test Beds for Oil-Circuit Components

Setup

The key components in a motor vehicle's oil circuit are (see Fig. 2.24):

- Lube oil pump (pressure pump)
- Suction pump (in the case of dry sump lubrication)
- Oil filter

For testing such components, a test bed setup is needed that:

- provides the test medium (engine oil) with adjustable parameters (pressure, volume flow and temperature)
- provides an air separation system in the oil reservoir
- regulates back pressure (in the unit under test's oil return line)
- measures oil temperature, mass flow and density (also in the return line, wherever applicable)
- controls the speed of the drive system connected to the test components and measures at least speed and torque

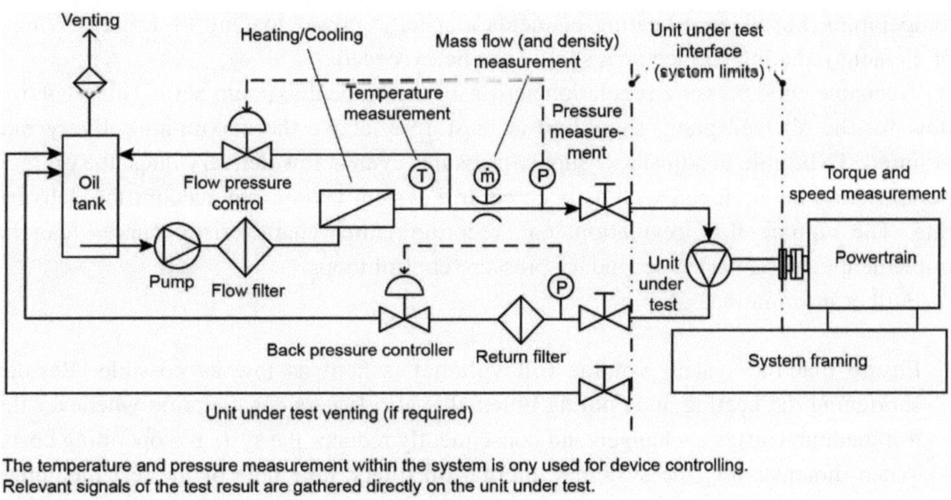

Fig. 2.24 Schematic diagram of an oil-circuit component test bed

- allows the adjustment of the admixture of air to oil (for tests with suction pumps; admixture ratio may be changed during the test run)

The following applications are typical examples for the use of such systems:

- Recording of pressure pump maps based on the parameters supply pressure, back pressure, temperature, speed. For cavitation tests (negative supply pressures) the system requires special vacuum regulators.
- Recording of suction pump maps based on the parameters supply pressure, back pressure, temperature, speed, oil density (degree of aeration), delivery volume. For supplying the test medium with the appropriate parameters (oil/air ratio) and the required consistency (oil foam), suitable additional systems must be provided.
- Filter characteristics based on the parameters temperature, flow rate (including oil density, if required)
- Flow tests on the engine block or cylinder head, or on various other components
- Durability tests

System Dimensioning

The system's cooling power must at least cover the power of the installed drive system connected to the unit-under-test component. However, it should also enable the system to cool down from the maximum temperature to 25 °C in no more than 15 min (at the end of the test run). Longer cooling times extend the waiting period until the unit under test can be disassembled, reducing the system's productivity. Similar times are recommended for the system's heating-up time (from ambient to test temperature). When dimensioning the heating elements, it is important to pay attention to the test medium's permissible temperature. For electric heating elements a specific power loading of 1.5–2.0 W/cm^2 of element radiating surface area should not be exceeded.

To enable good pressure regulation across all areas, the maximum permissible volume flow for the oil feed pump should be at least 15 % above the maximum delivery rate required. To be able to adjust low supply pressures even at low delivery rates, the oil feed pump should be equipped with a speed control system taking into account the delivery rate. The volume flow regulation, e.g. recording filter characteristic curves, can be implemented using nested secondary pressure control loops.

Further information:

- Ensure that the system volume (oil volume) is kept as low as possible. Besides shortening the heating and cooling times, this also lowers consumption whenever the test medium (oil) is exchanged and consequently reduces the system's operating costs.
- When dimensioning the system's internal oil filters, take note of the test medium's lowest temperature and lowest viscosity. Otherwise, low oil temperatures may lead to impermissibly high differential pressures at the oil filter.

- Ensure that it is possible to drain the system easily and completely; it must also be possible to change the oil filters in a fast and simple way.
- When choosing a system for the hydraulic interfaces to the unit under test, keep in mind that there may be pressure pulsations at the outlet of an oil pump (of the unit under test). For this reason, any quick couplings must be designed to resist such permanent pulsation.
- A malfunction in the internal oil pump must lead to an immediate standstill of the unit under test.

2.2.1.2 Test Beds for Water-Circuit Components

Setup

The key components in a motor vehicle's water circuit are (see Fig. 2.25):

- Water pump
- Radiator
- Engine block
- Thermostat

For testing such components, a test bed setup is needed that:

- provides the test medium (cooling medium) with controllable parameters (supply pressure, volume flow and temperature). In addition, the test bed must produce the static system pressure depending on the temperature.
- regulates back pressure (in the unit under test's water return line)

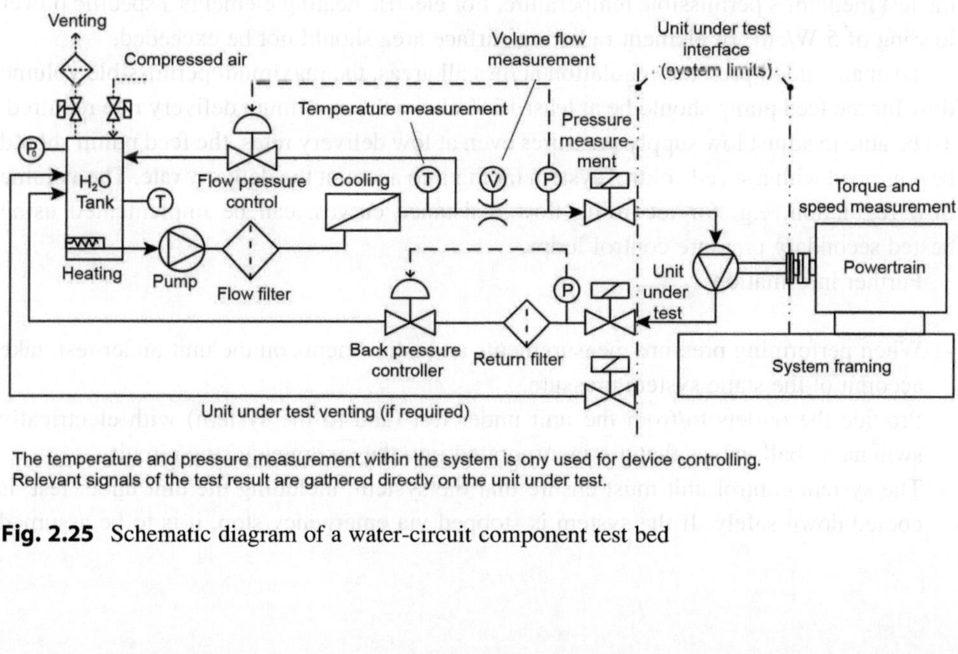

Fig. 2.25 Schematic diagram of a water-circuit component test bed

- measures water temperature and volume flow as well as supply and back pressure
- controls the speed of the drive system connected to the test components and measures at least speed and torque

The following applications are typical examples for the use of such systems:

- Recording of water pump maps based on the parameters supply pressure, back pressure, speed. For cavitation tests (negative supply pressures) the system requires special vacuum regulators.
- Flow tests (recording of characteristic curves) on the engine block, cylinder head, radiators or various other components
- Cooling power tests (these require an adjustable blower)
- Durability tests

System Dimensioning

The maximum test conditions determine the dimensioning of the system. For the upper temperature range, a margin of +20 °C to +30 °C is recommended. In low-temperature range systems, one can choose smaller margins to take account of the sharply rising costs (for the cold-temperature range). The system's cooling power must at least cover the power of the installed drive system connected to the unit-under-test component. However, it should also enable the system to cool down from the maximum temperature to 25 °C in no more than 15 min (at the end of the test run). Longer cooling times extend the waiting period until the unit under test can be disassembled, reducing the system's productivity. Similar times are recommended for the system's heating-up time (from ambient to test temperature). When dimensioning the heating elements, it is important to pay attention to the test medium's permissible temperature. For electric heating elements a specific power loading of 5 W/cm^2 of element radiating surface area should not be exceeded.

To enable good pressure regulation across all areas, the maximum permissible volume flow for the feed pump should be at least 15 % above the maximum delivery rate required. To be able to adjust low supply pressures even at low delivery rates, the feed pump should be equipped with a speed control system taking into account the delivery rate. The volume flow regulation, e.g. for recording flow resistance curves, can be implemented using nested secondary pressure control loops.

Further information:

- When performing pressure measurements and adjustments on the unit under test, take account of the static system pressure.
- Provide the outlets to/from the unit under test (and to the system) with electrically switchable ball valves that are incorporated into the emergency stop circuit.
- The system control unit must ensure that the system, including the unit under test, is cooled down safely. If the system is stopped via emergency stop, it is to be assumed

that the unit under test is overheated. In such cases, the appropriate safety measures have to be taken (there is a danger of excess steam and/or pressure build-up if the connection lines to the unit under test are opened too soon).

- It is essential to prevent vapor lock at any position of the system. Give particular attention to the following points: the position with the highest temperature (next to the heating) and the position with the lowest pressure (upstream of the pump).
- If the system is operated with overheated water, the setup can even pose a risk after the pumps have been stopped! Take adequate safety precautions.

2.2.2 Test Beds for Hot Gas Components

Due to the growing use of charging technologies in diesel and gasoline engines, precise thermodynamic testing on a test bed for hot-gas components is becoming a key tool for improving engine efficiency and reducing CO_2 emissions. To achieve optimum efficiency in the interaction between the charging process and the engine, it is necessary to determine charger maps, which are then used as input parameters to adapt the chargers to the engines and simulate gas exchange processes.

These maps are measured on a hot-gas test bed specifically designed for such a component. Accordingly, the purpose of this type of test bed is to operate and test the unit under test independently of the internal combustion engine in a normalized test setup in as wide an operating range as possible.

2.2.2.1 Physical Principles

The effective power output of an internal combustion engine is proportionally dependent on mean pressure, speed and piston displacement, as is shown in the formula below:

$$P_e = p_m \cdot \omega \cdot V_h \cdot \frac{1}{Z} \tag{2.2}$$

P_e Effective power [kW]
p_m Mean pressure [Pa]
ω Speed $\left[\frac{1}{s}\right]$
V_h Piston displacement [m^3]

Equation 2.2 Effective power

In a four-stroke engine, $Z=2$; in a two-stroke engine, $Z=1$. If the piston displacement is increased, the engine power increases too. However, the engine efficiency drops, e.g. due to increased frictional power. In addition, engine mass and size increase, which goes against lightweight design efforts.

An increase in power can be achieved by raising the speed, which, however, produces a disproportionate increase in frictional power and gas exchange loss. To ensure that no compromise is made on the power output, automotive engineers are focusing on increasing mean pressure, which, among other things, is implemented by employing boosting (downsizing and down-speeding) strategies. Adopting such an approach shifts the engine load point to ranges with improved specific fuel consumption.

Mean pressure and its dependencies can be expressed as follows:

$$p_m = \rho \cdot H_u \cdot \eta_e \cdot \frac{\lambda_L}{\lambda \cdot L_{min}} \tag{2.3}$$

Equation 2.3 Mean pressure calculation

The heating value H_u and the minimum air/fuel ratio L_{min} are characteristic values of the fuel and can be assumed as given. This means that the mean pressure is proportional to the air density ρ downstream of the compressor, the effective efficiency η_e and the volumetric efficiency λ_L and inversely proportional to the air/fuel ratio λ.

$$p_m \sim \rho \cdot \eta_e \cdot \frac{\lambda_L}{\lambda} \tag{2.4}$$

Equation 2.4 Mean pressure and its dependency on process quantities

The air density can be represented by the general gas equation:

$$\rho = \frac{p}{Rg \cdot T} \tag{2.5}$$

Rg Gas constant $\left[\frac{J}{kg \cdot K}\right]$
p Boost pressure [Pa]
T Temperature [K]

Equation 2.5 Density of air

To increase the density—and thus the engine power—it is possible to either raise the boost pressure or lower the air temperature. This is achieved by means of compressors or charging-air coolers.

2.2.2.2 Charging Principles

Basically, several types of charging principles are distinguished:

- **Mechanical supercharging**: When applying this charging principle, the compressor is mechanically connected to the crankshaft. Examples for the utilized compressors are Roots blowers, screw-type compressors and spiral-type superchargers. It is also possible to allow for variable gear ratios (e.g. the compressors are connected via CVT or similar).

- **Exhaust turbocharging**: By applying this charging principle, the compressor is connected to the engine both thermodynamically and mechanically. A turbine is driven by the engine's exhaust gas. Compressor and turbine are connected by a shaft. Consequently, the turbine covers the power requirements of the compressor which performs the charging.
- **Hybrid forms of charging**: With this charging principle, electric generators are used to generate electric power; the electric generators are powered by exhaust gas-driven turbines. The electric power is then used for electromotive supercharging (turbogenerator and e-booster). This means that there is no direct mechanical coupling between the turbine and the compressor, which enables more flexibility during turbocharger operation.

2.2.2.3 Turbocharger Test Bed

Setup

Turbochargers are usually operated under steady-state conditions, but some application cases in endurance testing also require transient operation.

In either type of operation, the following is required:

- Production of hot gas that is supplied to the turbine
- Oil supply for the (usually) friction bearings
- Control of the compressor throughput via a wastegate valve

It is additionally advisable to ensure:

- Intake-air conditioning to stabilize the operating conditions
- Cooling water supply (only in the case of water-cooled turbochargers)

Apart from the unit under test itself, the key components of a turbocharger test bed comprise the test bed frame, the tubing for coolants, fuel, process air and compressor air and the related control devices and measuring points. Process-air supply, fuel supply, oil and coolant conditioning, sensors, control units and an automation system are additionally required (Fig. 2.26).

These components can be divided into the categories turbine circuit, compressor circuit, conditioning systems, electrical systems, as well as automation and control systems. Figure 2.27 shows the relevant components of the compressor circuit and the turbine circuit, as well as the conditioning systems.

Fig. 2.26 Turbocharger test bed

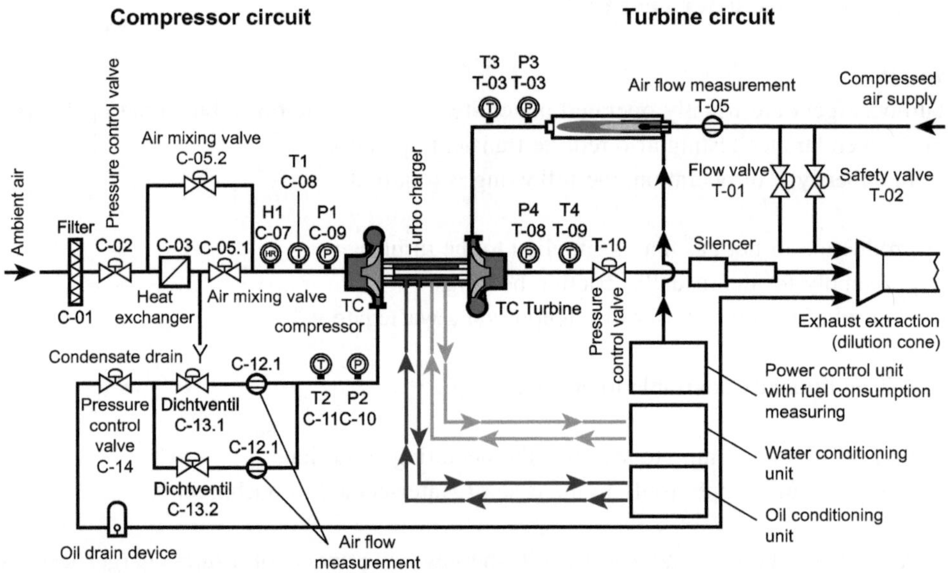

Fig. 2.27 Schematic diagram of a turbocharger test bed

Turbine Circuit

The hot combustion gas is produced by the combustion of dried compressed air and diesel or fuel oil, respectively, or compressed natural gas (CNG) inside a combustion chamber. The process results in the generation of a steady-state flow of high-temperature gas. Typically an electrically powered compressor supplies the compressed process air—optionally with regulation of volume flow or in-tank pressure. Sensors in the test bed setup measure online values such as pressure, temperature, turbocharger speed or mass flow. The controller for the compressed-air mass flow and mapping procedures relating to the turbine's input power level use these values.

Liquid fuels like oil, diesel or bio diesel have the advantage that they are less demanding on building services than natural gas. Aside from that, liquid fuel burners achieve a more homogeneous temperature distribution across the diameter than natural gas burners due to the distinctly different way these burners are designed—a fact which leads to an improvement in the turbine's energy balance quality. Conversely, the use of compressed natural gas as a fuel has the advantage of having a slightly higher hot-gas peak temperature and also being more economical in operation. There are only a few test beds that are operated with liquid pressurized gas (LPG).

The combustion chambers' typical power ratings range between 200 kW and 800 kW. In the partial-load ranges of smaller combustion chambers, even power outputs below 5 kW are sometimes implemented. The exhaust gas power output to the turbine can be expressed by the following equation:

$$P = \dot{m} \cdot \Delta h = \dot{m} \cdot c_p \cdot (T_E - T_A) \tag{2.6}$$

P Power [kW]

$\dot{m}\cdot$ Mass flow rate $\left[\frac{kg}{s}\right]$

Δh Specific enthalpy difference $\left[\frac{J}{kg}\right]$

c_p Specific heat capacity $\left[\frac{J}{K}\right]$

T_E Inlet temperature [K]

T_A Outlet temperature [K]

Equation 2.6 Exhaust gas power of turbine

The hot gas supplied to the turbine has a temperature level between 200 °C and 1200 °C. The power maps are measured at a temperature stabilized to 600 °C or 620 °C, respectively, which corresponds to the measurement conventions currently used. To be able to measure the entire application range of currently available automotive turbochargers, mass flow rates between 0.005 kg/s and approx. 1 kg/s are needed. This is barely possible with a single test setup (see Figs. 2.28 and 2.29).

The unit under test (turbocharger) is mounted at the outlet of the combustion chamber. The controlling system integrated in the housing of the exhaust gas turbine stage accelerates the hot-gas flow and induces swirl which is equal to kinetic energy. This swirl is then used by the subsequent turbine wheel to cause the shaft (and thus the compressor) to rotate. The turbine extracts some of the hot gas's internal energy—consisting of thermal, kinetic and pressure energy. The exhaust gas is then discharged from the test bed without having any retroactive effects on the unit under test.

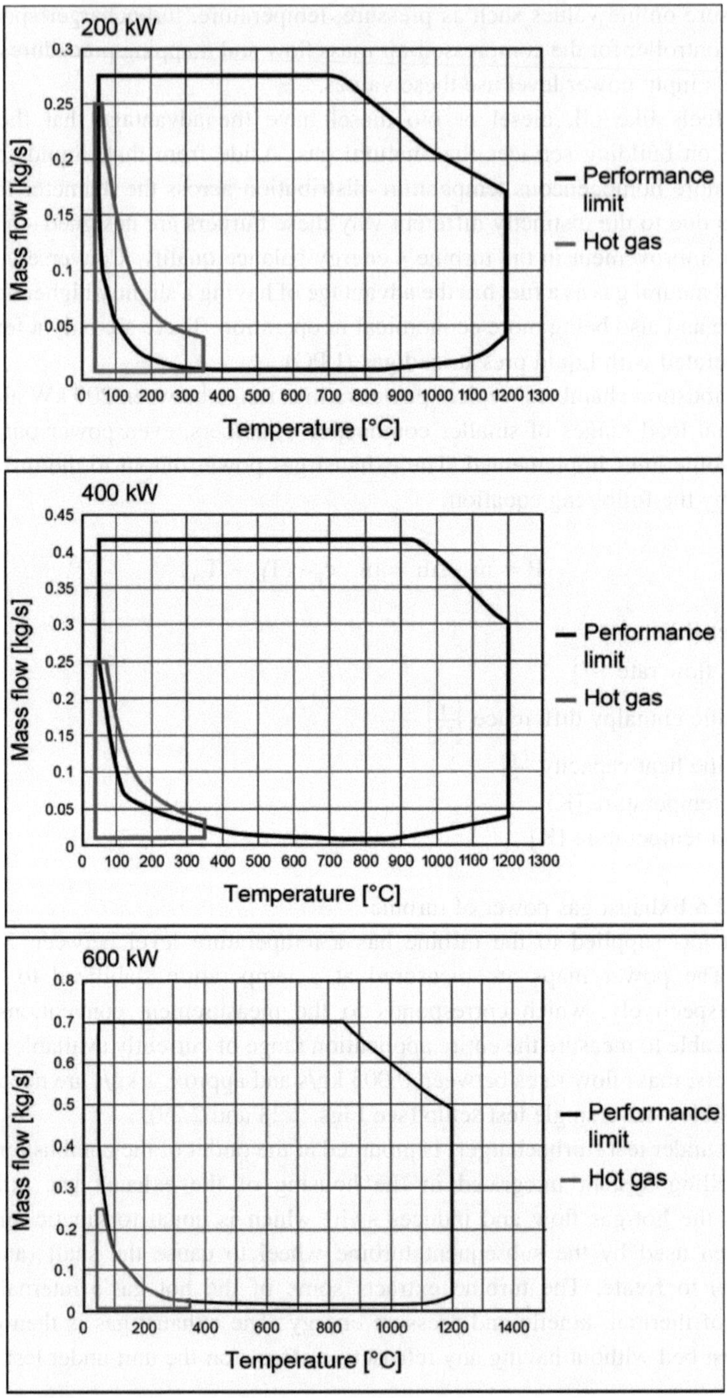

Fig. 2.28 Examples—mass flow rate versus temperature in CNG combustion chambers 200 kW, 400 kW and 600 kW

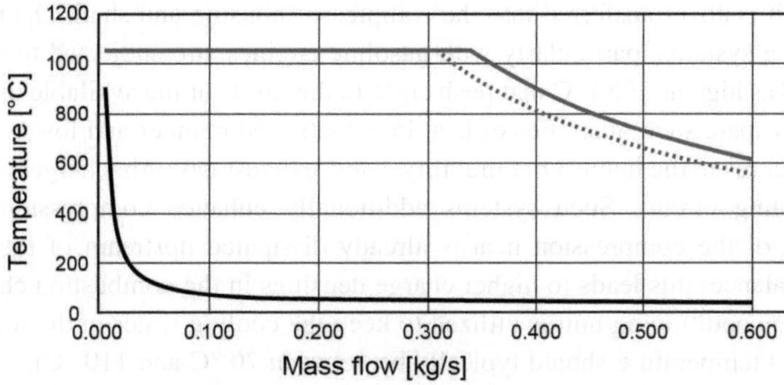

Fig. 2.29 Temperature versus mass flow in a 380-kW diesel burner

Compressor Circuit

The turbine drives the compressor via the shaft connecting compressor and turbine. The air taken in this way is pre-filtered with minor loss of pressure, automatically conditioned by means of a conditioning unit and maintained in the desired state. This helps to avoid external influences (and the corrective calculations that may be necessary as a result) and perform a non-trivial stabilization at the surge limit. To record a turbocharger map it is necessary to measure parameters such as humidity, temperature and pressure of the intake air. This is done using the relevant sensors in the intake system. Next, the air flow reaches the compressor stage. The compressor impeller then adds kinetic energy to the medium (in our case air), i.e. it is accelerated. In the subsequent diffuser, this added kinetic energy is converted into pressure energy by deceleration. As this is not achieved without loss, additional heat loss is generated. To accomplish compressor mapping, it is again necessary to measure compressed-air properties such as temperature, pressure or mass flow rate. To this end, the compressed air is subjected to a mass flow measurement before being passed through a throttle valve that regulates the mass flow.

Conditioning Systems

Usually, friction bearings support the shaft on which the turbine and compressor wheels are mounted. Lubricating and cooling oil is supplied to the bearings. A conditioning unit is used to adjust the oil to required operating conditions. The oil conditioning system regulates the oil's temperature, pressure and, if necessary, its mass flow rate. The achievable temperatures typically range between 40 °C and 130 °C, in particular cases between 20 °C and 150 °C. The adjustable range is between approx. 1.2 bar to 6 bar of absolute pressure. Some test beds are also used to perform a simulation of crankcase pressure, which involves changing the pressure difference via the shaft bearing. All of these values have a direct impact on the turbocharger's efficiency.

The turbine wheel and the housing reach very high temperatures due to the exhaust gas inside the combustion chamber. While a part of this heat is dissipated via the lubricating oil, some of it is also transferred into the compressor housing and shaft. Operated at full load, charging systems, particularly with gasoline engines, are subjected to exhaust gas temperatures as high as 1050 °C that push close to the limits of the available materials. To be able to dissipate such quantities of heat in a controlled manner and lower the component temperatures in the interest of durability, some exhaust gas turbochargers incorporate a water cooling circuit. Such systems additionally enhance compression efficiency because part of the compression heat is already dissipated upstream of the charge-air cooler. On balance, this leads to higher charge densities in the combustion chamber.

A separate conditioning unit is utilized to keep the cooling water in the required state (flow rate and temperature should typically be between 70 °C and 110 °C).

Electrical Systems

All of the electrical equipment, such as sensors and actuators, are connected to the automation system using I/O units. The test bed's electrical switch cabinets are usually located near the unit under test to ensure that the cables from the measuring point to the measuring amplifier are as short as possible. These switch cabinets connect the automation system (typically positioned in the control room) to the test bed.

Automation

The automation system (see also Sect. 4.6) acts as an interface between the physical test setup and the test bed operator (see Fig. 2.30). It ensures the stable control of the unit-under-test in different states and performs measured data acquisition, calculation and the storage of the measured results. In post-processing (see also Sect. 4.7), maps are calculated and displayed graphically. This is preferably done online during test execution. The automation system executes predefined test runs. Unmanned operation helps to achieve cost efficiency. Endurance tests may take up to 500 h. In such cases, unmanned and continuous operation is a fundamental requirement.

2.2.2.4 Test Operation

Test operation comprises two core areas:

- Function and performance testing
- Endurance and durability tests

Function Testing

Function tests are run to acquire maps in order to better understand the behavior of the turbocharger and its components. The operational requirements for such tests have already been described (see also page 34 and following). They need constant stabilized mass flows, constant temperatures and speeds as well as an unchanging pressure level.

Turbocharger compressor maps characterize different quantities, yet always with speed as a reference quantity. For every map, between 6 and 12 speed lines are defined, which are distributed across the turbocharger's operating range. These speed points are run with

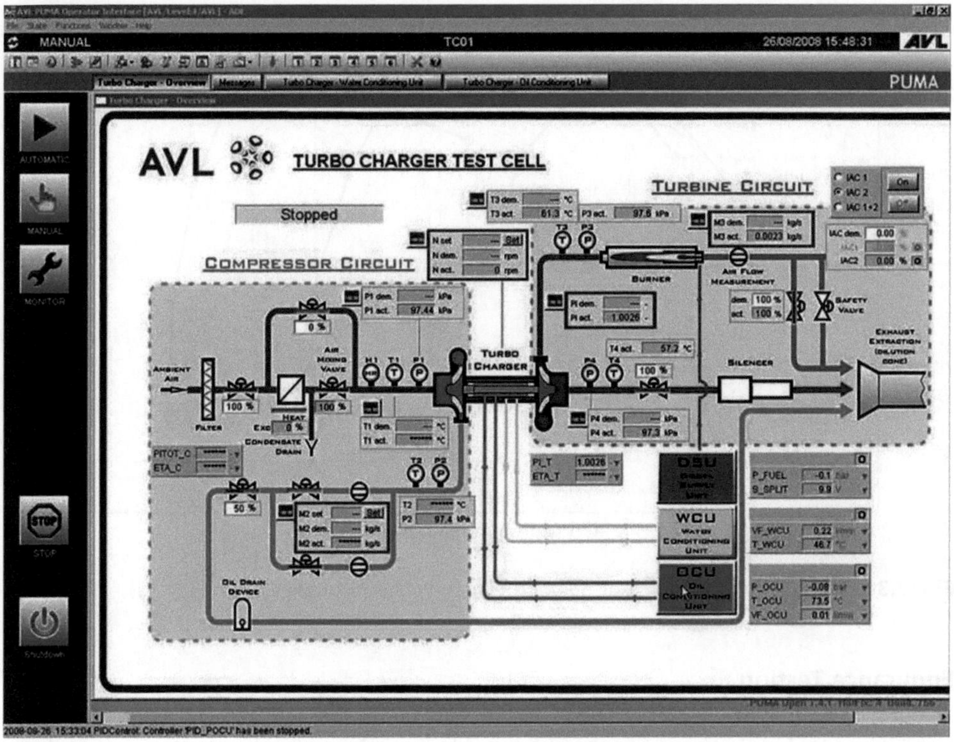

Fig. 2.30 Screenshot of an automation application for turbocharger testing

varying mass flow rates, resulting in the surge line at lower mass flow rates (i.e. an instable, surging flow through the compressor) and the choke line at maximum mass flow rates (i.e. the maximum permissible mass of air flowing through the compressor). Along these lines the testing points are spread evenly across the length of the curve.

Typically all measurement points are run automatically after steady-state conditions are achieved: after defined waiting times have elapsed or after specifically defined stability limits have been reached. In the stabilized state, several consecutive measurements are made and the arithmetic mean values are stored at the map points. By this method, up to 12 measuring points are set per speed line (see Fig. 2.31). The same procedure is applied to determine the turbine map points. It is common practice that the automation system combines closely spaced compressor and turbine points and measures them together.

To gain additional insight in turbine maps, compressor performance points can be varied by using a closed-loop compressor system. This method increases the compressor working medium's density (pressure boost upstream of the compressor inlet), which raises the power input of the compressor at a given speed. This allows measuring extended turbine maps.

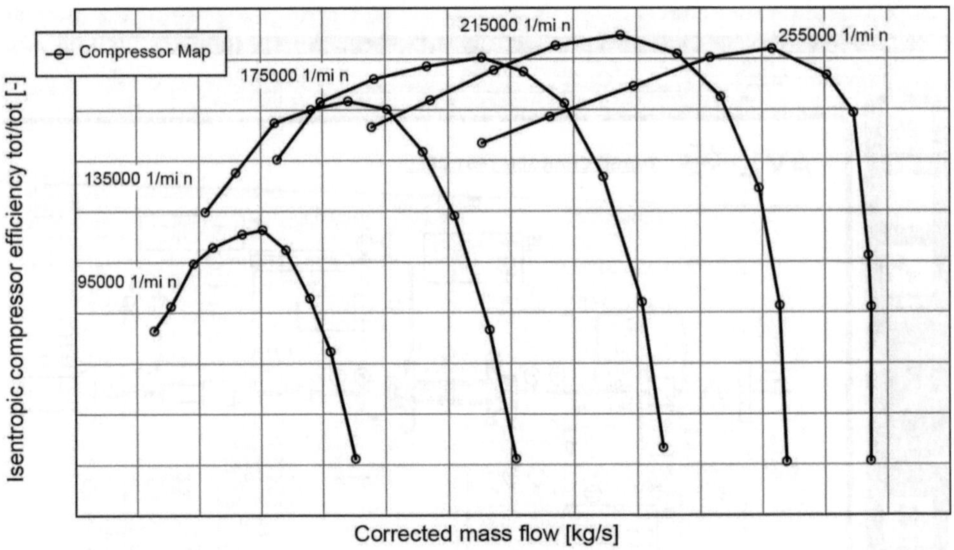

Fig. 2.31 Example of a compressor map, efficiency versus throughput

Endurance Testing

Endurance tests place high demands on the unit under test and the test bed. Some tests require hundreds of hours of test operation to run mass flow, speed and temperature ramps.

Examples of such test runs include (but are not limited to):

- Burst containment testing: conducted to test the robustness of the compressor housing in the event of a high-speed rotating impeller fragmentation.
- High-temperature creep test: used to measure turbine temperature stability.
- Thermal shock testing (hot and cold cycling with component temperature changes up to 250 K/s): to determine the thermo-mechanical fatigue of turbine parts. It is also possible to expose paired turbocharger assemblies alternately to hot and cold temperatures (via switching facilities), a method which not only raises the test bed service life (steady-state burner operating point) but also subjects the unit under test to maximum temperature gradients (smallest thermal inertias used because the burner is not involved in the temperature change). In addition, this increases the statistical lot size per test (2–8 turbochargers in the same test).
- Fatigue cycles: conducted to test component stability in transient load changes (speed variations). These make it possible to check critical resonant ranges, e.g. in bearings or similar components.
- Component tests: e.g. closing cycles for wastegate turbines or variable-geometry turbines under cyclically changing pressure and temperature conditions.

2.2.3 Test Beds for Transmission Components

During their development, transmission components are subjected to endurance strength tests and functional trials. Validating individual component assemblies can therefore be done individually, as explained in Sect. 1.2.1, and does not depend on the presence of other components, let alone the entire powertrain. We generally distinguish between structural-component test beds and powertrain-component test beds.

Individual structural components (e.g. toothed wheels) are tested at an early stage on special test beds. A structural-component test is primarily conducted to validate the choice of material for transmission components and lubricants. Structural-component test beds are specifically designed for a certain unit under test or development task. Due to the specific and, in part, very detailed nature of the tests on the respective structural components, particular demands are also placed on the test bed equipment (e.g. oil conditioning systems, actuators, etc.).

Examples for structural-component test beds are:

- Gasket test beds
- Tilt test beds for testing lubricants
- CVT thrust-belt test beds (Fig. 2.32)

Powertrain-component test beds allow the testing of entire groups of components. The requirements on the conditioning systems and actuators depend on the desired development tasks and the involved unit under test. These test beds are usually not merely endurance test beds but are also used for function tests (e.g. efficiency measurement). Examples for powertrain component test beds are:

- Shift test bed for testing the gear shift and synchronization elements
- Clutch test bed

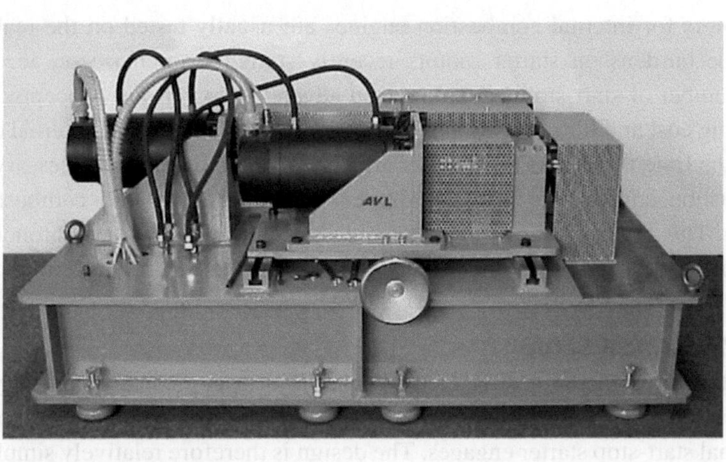

Fig. 2.32 CVT thrust-belt test bed

Fig. 2.33 Converter test bed

- Dual-mass flywheel test bed
- Converter test bed (see Fig. 2.33)

Component tests are usually kept as simple as possible, so that it is possible to achieve basic results on the characteristics of a component or its endurance strength within a short period of time.

2.2.4 Starter Motor Test Bed

Starter motors for internal combustion engines are usually tested on the real (physical) engine. The burdens on starter motors recently started to increase on account of the growing number of start-stop technologies in advanced powertrain concepts. The OEMs face growing cost and time constraints. The lack of availability of the internal combustion engine at the time the starter motor tests are conducted is making it necessary to replace the combustion engine as a test environment with a model-based combustion engine simulation. This is done using a high-dynamic permanent-magnet synchronous machine as load unit for the starter motor combined with advanced control algorithms.

2.2.4.1 Mechanical Setup

The essential setup of the testing facility consists of a permanent-magnet synchronous machine (see Fig. 2.34), a torque measuring flange and the starter ring into which the conventional start-stop starter engages. The design is therefore relatively simple and has a

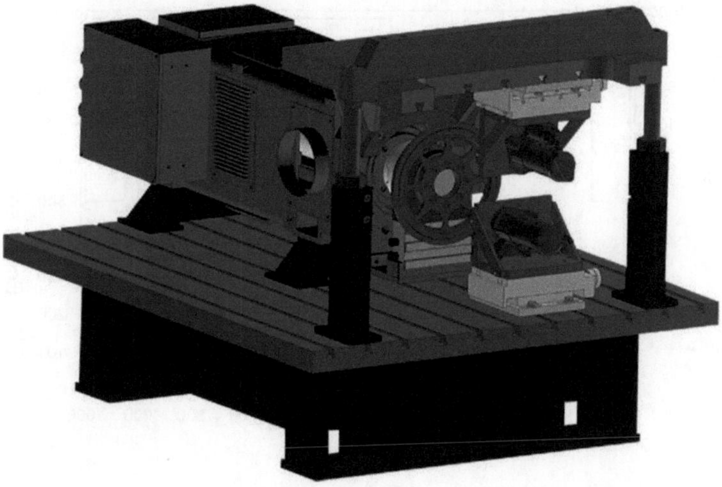

Fig. 2.34 Mechanical setup of a starter test bed

very low rotating mass, which subsequently has a positive effect on the system dynamics; in other words, the maximum required power of the electric motor used can be reduced significantly.

Since only the start-up procedure is relevant for testing (and not the engine rev-up), the test can be terminated as soon as the starter has disengaged from the starter ring gear. Based on this fact, a starter could theoretically be tested up to 20 times per minute. However, the internal thermal behavior of a starter motor limits testing to a frequency of three starts per minute. The resulting downtime of the test bed can be minimized by mounting four different starter motors on one component test bed and testing the individual starters alternately.

2.2.4.2 Control of the Permanent-Magnet Synchronous Machine

The realistic loading of the starter motor in the first milliseconds of the starting procedure is a major challenge for controller performance. With a real control loop made up of industrial components the delay times tend to add up to a significant order of magnitude. If the battery voltage and loading torque from the internal combustion engine are considered as input variables for the starter motor, and the starter motor current as output value, it is easy to understand that a delay in the loading torque will consequently have an equivalent influence on the starter motor current. This then no longer corresponds to the loading that occurs in the real internal combustion engine.

This means that both mechanical components, such as the planetary gear set in the starter motor and electrical components like brushes or the commutator, would be incorrectly loaded, resulting in false endurance test results.

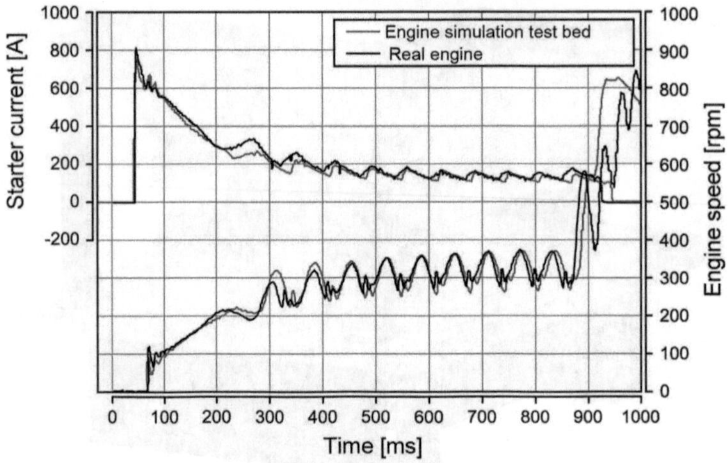

Fig. 2.35 Comparison of test bed and real internal combustion engine

The objective is therefore to build up the load unit as free from delay time as possible or to implement suitable control algorithms to reduce the effects caused by delay time. The delay time in this system is in the order of 2–3 ms, which may not seem much, but relative to the electrical components of the starter motor it falls within a significant order of magnitude.

2.2.4.3 Results

Figure 2.35 shows a comparison of a starting process executed on a starter motor test bed with a simulated IC engine and one with a real IC engine. Both the current drawn from the starter motor and the speed of the internal combustion engine/starter motor show a good agreement. All of this requires sophisticated control technology, which is described in Sects. 4.5.1 and 4.5.2.

2.2.5 Electric Motor Test Bed

The electric motor test beds addressed in this section are designed to test and characterize electric motors and their inverters, which are used in powertrains of hybrid and electric vehicles. These are belt-driven starter generators, integrated starter generators and traction motors.

Electric motor test beds are used for testing such components under realistic and easily reproducible conditions. We distinguish between function, performance and durability tests. Selective and realistic tests allow engineers to develop and optimize these electrical components while they are developing the other powertrain components, ensuring the proper product function before the components are integrated in the complete powertrain.

The two key electrical components required in hybrid and electric-vehicle powertrains are:

- Liquid-cooled electric motors; there exist a few models that are air-cooled.
- Liquid-cooled inverters; there exist a few models that are air-cooled.

2.2.5.1 Setup

The test bed setup (see Fig. 2.36) must satisfy the following requirements:

- Coolant conditioning (water, water/glycol, oil) with adjustable and controllable parameters (temperature, volume flow and pressure)
- High-voltage DC power supply for inverters, energy recovery capacity, with adjustable and controllable parameters (voltage, current and output)
- Electric load unit, asynchronous or synchronous, four-quadrant operation, with adjustable and controllable parameters (speed, torque and rotor angle)
- Measurement of temperature, volume flow and pressure of the cooling medium (if necessary also in the return line)
- High-frequency measurement of DC voltage and DC current at the inverter's DC terminals (DC stands for direct current)
- High-frequency measurement of AC phase voltages and AC phase currents between the inverter and the electric motor (AC stands for alternating current)
- Measurement of speed, torque and rotor angle at the electric motor shaft
- Speed control for the electric load unit with accurate and dynamic speed and torque measurement
- Mounting system for the electric motor and the electric load unit, which is mechanically separated from the test bed building to protect against vibration
- Shaft connection between the electric motor and the electric load unit, torsion-proof
- Shaft locking device as an additional option for short-circuit current testing
- Temperature chamber or climate chamber as an additional option for high and low temperature or climate testing, specifically required for durability tests

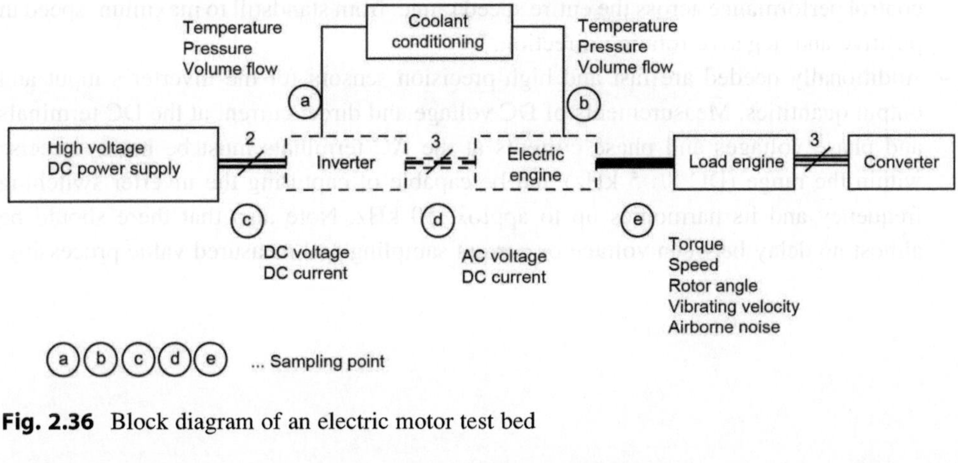

Fig. 2.36 Block diagram of an electric motor test bed

2.2.5.2 Applications

These types of test beds are designed for determining all mechanical characteristics (e.g. torque curves, vibrations, noises, torque quality), electrical characteristics (e.g. DC operating voltage range, short-circuit current, rated load points, overload points, open-circuit voltage) and thermal properties (thermal rotor and stator time constant). They are also useful for obtaining detailed technical data of the component on cooling or overload capacity (e.g. power factor, electrical and mechanical power) or the characteristic torque curve, noise emissions as well as axial and radial vibration.

2.2.5.3 Test Bed Requirements

The unit-under-test portfolio defines the selected performance and function range of the test bed. Dynamic testing tasks, i.e. executing rapidly changing demand values, require a corresponding short-term overload capacity or an oversizing of these test bed components to compensate for mass inertia and heat storage capacities. In case of an extensive unit-under-test portfolio it is advisable to check whether a single test bed is sufficient to achieve the accuracy and dynamic required for all units under test or whether several differently sized test beds would be more efficient.

Easy-to-make adapter systems allow a fast and fault-free connection of different-type test units to the test bed.

Instrumentation and functionality are adapted to the intended use. Test beds used for research have to be easily adaptable to a variety of testing tasks. Endurance test beds require a climate chamber capable of simulating ambient temperatures between −40 °C and +150 °C and air humidity levels between 10 % and 95 % of relative humidity. Test beds used in end-of-line production testing must be capable of fully automatically and reproducibly classifying units under test as "OK" or "not OK" based on measurement results.

2.2.5.4 Notes on Dimensioning

- An electric load unit is needed with preferably low torque ripple as well as high speed control performance across the entire speed range, from standstill to maximum speed in positive and negative rotating direction.
- Additionally needed are fast and high-precision sensors for the inverter's input and output quantities. Measurements of DC voltage and direct current at the DC terminals and phase voltages and phase currents at the AC terminals must be highly precise within the range (DC ... 5 kHz) and be capable of capturing the inverter switching frequency and its harmonics up to approx. 50 kHz. Note also that there should be almost no delay between voltage or current sampling and measured value processing.

2.2.6 Inverter Test Bed

2.2.6.1 Introduction

In automobiles, electric drive motors are controlled by drive inverters that combine the required power electronics with signal electronics and software in a single device. Depending on the particular vehicle design, this group of components is not only responsible for engine control but also for other capabilities such as battery charging or balancing between the high- and low-voltage system. To be able to test such drive inverters, it is necessary to connect the appropriate system components (or equivalent loads) to the electric device interfaces. Particular requirements result from the inverter's interface to the electric motor, which is typically formed by the phase outputs and the rotor position sensor signal.

An electric motor can be considered as a suitable load for inverter drives provided that the electric motor is coupled with an adequate mechanical load system (dynamometers—see Sect. 3.2.1 in order to generate the relevant load. With such an arrangement, an electric motor test bed (see Sect. 2.2.5) is used to test the inverter. Acting as an inverter test bed, such a setup, however, is subject to certain restrictions resulting from the mechanics. A load unit adds mass inertia which has to be compensated for. In addition, there are system limits to the control dynamics. At critical operating points, e.g. zero speed or maximum load, the test setup soon reaches its limits. Other issues are the limited reproducibility of tests (adjustment inaccuracies) and the missing possibility to enforce simple faults in the motor for testing purposes.

If the electrical aspect of an electric machine is perfectly mimicked and the electric machine's mechanics are simulated, it is possible to circumvent such limitations. This is the basic concept of the e-motor emulator. The e-motor emulator is a fully electronic model of an electric machine without any mechanical components, opening up new possibilities in inverter testing. At the electric connection points of the tested inverter the e-motor emulator replaces the real electric motor and reproduces the relevant current and voltage signals accurately and in real time, across all four quadrants while the electric machine's mechanics remain simulated. Likewise, a rotor sensor is emulated that transmits the machine's mechanical position to the inverter (Fig. 2.37).

2.2.6.2 High Flexibility and High Dynamics

The most remarkable feature of an e-motor emulator is its compact setup thanks to the missing mechanics, and its resulting versatility. This already becomes evident when choosing which motor is to be used to test the converter. As this motor only exists virtually as a software model, it can be changed in terms of its characteristics or replaced entirely at the click of a mouse, with no need of physical modification. As the test bed has no rotating shafts, the e-motor emulator allows tests with no mechanical constraints such as speed dynamics or wear. The tested inverter (unit under test) can be safely subjected to a drive unit's extensive range of fault modes while a current safety limitation in the emulator protects the unit under test.

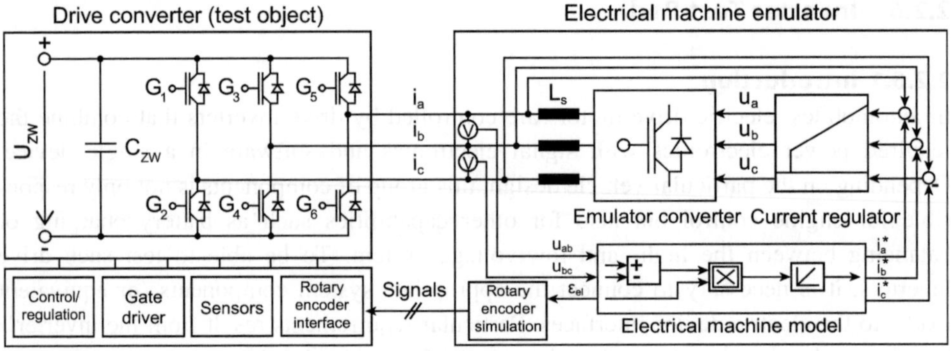

Fig. 2.37 Block diagram—test setup of emulator with inverter

The e-motor emulator technology is a bridging technology and ranks somewhere between the established hardware-in-the-loop technology (see also Sect. 2.3) and the dynamometer technology (see also Sect. 3.2.1). Today several in-the-loop methods are employed in inverter development: e.g. model-in-the-loop, software-in-the-loop or hardware-in-the-loop. With the latter method, it is of significance to consider "how much" hardware of the component to be tested is physically used in the test setup (see also Fig. 1.6 in Sect. 1.2.1). Up until recently, transferring this technology to power inverters required the integration of "bypass hooks" in the component. This means that in conventional HiL test beds the real inverters are replaced ("bypassed") by models with identical interfaces and properties very similar to those of the real inverters. For component testing based on ISO 26262, an approach like this is of no use because the unit under test is not present in its original state, but "bypassed."

When using an e-motor emulator, no bypass hooks are required anymore. This means that the inverter, in its original state, is operated in exactly the same way it would be in a testing procedure with an e-motor connected to a load unit, allowing tests in accordance with ISO 26262, including the execution of any fault modes. The advantages of both testing methods—HiL and load unit tests—are combined. Although the electric machine only exists as a software model, its full electrical power flow is ensured. The drawback, however, is the test bed's higher cost.

2.2.6.3 Test Setup Validation

When using an e-motor emulator, it is crucial that the electrical characteristics at the electrical connections are as similar as possible to those of the real e-motor. Such reproduction quality is of crucial importance because it provides the basis for how the emulator can be used for the various test applications.

Typical for such a "power" HiL system with e-motor emulator is the convenient way it works. Fault modes in the drive unit can be tested with little effort. It does not require a manipulation of the unit under test, which is typical of hardware-in-the-loop arrangements. The converter can already have series-production status and the corresponding e-motor is not even developed. This procedure allows complete converter tests in accordance with ISO 26262 (Functional Safety).

2.2.6.4 Typical applications

Possible test scenarios for inverters include:

- Rotor positioning sensor adjustment
- $^{d}/_{q}$ current feed testing
- Torque control testing
- Speed control testing
- Power control testing
- Recuperation testing
- Testing of critical operating points such as hitting the curb
- Controller design for active vibration absorption
- Monitoring testing (stimulation of electric machine faults)
- Active short-circuit testing
- Use of virtual electric machines in the test loop
- Round-robin testing of inverters

2.2.6.5 Complete Inverter Testing System

In a complete inverter testing system, further components, apart from the e-motor emulator, are needed to operate a drive inverter and conduct meaningful tests. A safety control system ensures safe operation (see Sect. 4.8). Aside from personal safety, device protection is a critical aspect in high-voltage systems. An entire system setup for a high-voltage inverter testing system consists of the following components:

- E-motor emulator with a power supply
- Battery emulator (DC supply for unit under test) (see also Sect. 3.2.3)
- Mounting system for the unit under test
- Cooling water supply (unit under test) (see also Sect. 3.2.4)
- Safety system (see also Sect. 4.8)
- Test automation system (see also Sect. 4.6)

Depending on requirements and the specific unit under test, there may be further components necessary, such as climatic chamber, power meter, low-voltage supply, residual bus simulation, or DC load sink. The following illustration shows a typical inverter test bed (Fig. 2.38).

Fig. 2.38 Typical inverter test bed for the high-voltage powertrain

2.2.7 Battery Test Bed

Batteries are the heart of the electrified powertrain. Accordingly, their testing and validation is of major significance. Basically, batteries are electrochemical energy storage devices. Battery test beds are therefore also suited for testing similar-typed units under test such as ultra- or super-capacitors. Various types of electrochemical energy storage systems must be supported. Lead, nickel-metal-hydride and, most importantly, lithium-ion battery technologies currently dominate the automotive market.

As an example, the following section addresses detailed requirements for testing on lithium-ion batteries.

We basically distinguish between tests of lithium-ion batteries on individual cells, modules and full battery systems. The demands on the required voltage level and the power supply at the test bed, as well as the complexity of the testing task, will differ widely depending on whether tests are run on individual cells or on entire battery systems. Table 2.1 gives an overview. Most tests are conducted in special temperature or climatic cabinets. Temperature, current and voltage monitoring is crucial to battery testing. To monitor temperature, particularly during the batteries' development stage, it can be useful to employ infrared (IR) cameras.

When conducting such tests, safety is of paramount importance because in the event of an accident cells may even explode. This can be caused either by defective cells, overcharging or excess voltage. For this reason, it is imperative to prevent the accumulation of explosive mixtures in the test chamber (e.g. by conducting the test in an inert gas atmosphere). We additionally recommend monitoring the gaseous atmosphere for flammable gases, and installing a fire suppression system, e.g. extinguishing systems based on water mist or carbon dioxide.

Table 2.1 Overview of the typical demands on cell, module and battery system tests

Unit under test	Typical power range	Special boundary conditions
Cell	Up to 6 V 50–400 A	– Temperature tests only – Additional temperature sensors
Module	Up to 60 V 50–600 A	– Temperature and climatic tests – If required, integration and monitoring of simple module monitoring electronics – If required, external cooling – Additional temperature and voltage sensors
Battery system	Up to 400 V, Partially up to 1000 V Up to 250 kW	– Temperature and climatic tests – Integration and monitoring of the battery management system – External cooling – If required, residual bus simulation for the battery control unit – Additional temperature and voltage sensors

The tests to be conducted depend on the defined validation procedures. Normally, application-related power/voltage profiles are executed under different temperatures or varying air humidity levels. To facilitate comparison, standardized test procedures are conducted too, as suggested by the ISO 12405 and ISO 62660 standards or other associations such as SAE, FreedomCAR, UNECE or VDA (German Association of the Automotive Industry). Also, accelerated aging tests are frequently conducted at elevated temperatures, but without electrical load. Cells or batteries subjected to such aging tests are regularly checked for capacity fade and an increase in the internal resistance.

To carry out any of these tests, it is beneficial to use an automation system. Such systems not only enable faster programming of test sequences and the automated evaluation of the test results; they also improve safety, as the warning and shutdown thresholds (such as individual cell voltages or max./min. temperatures) can be integrated with little effort. Figure 2.39 shows the typical test room setup, consisting of a container accommodating the climatic chamber, power electronics, control room and the cooling and safety equipment.

Besides the electrical and thermal performance and validation tests already addressed, cells and batteries additionally have to be tested under various ambient conditions, to validate mechanical stability and safety. The exact test conditions can also be found in the standards indicated above.

2.2.8 Fuel Cell Test Bed

Fuel cell technology (see also [1] enables the generation of electrical power (and heat) highly efficiently and without pollution, when based on hydrogen fuel cells (e.g. polymer electrolyte membrane fuel cells also known as proton exchange membrane fuel cells—PEMFC), or with minimum pollution, if high-temperature solid-oxide fuel cells are being

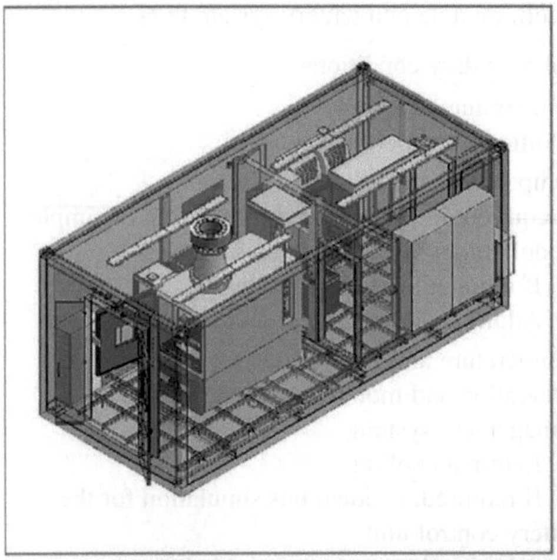

Top left: schematic view of the test container with climate chamber, control electronics, control room as well as cooling and safety system

Top right: photograph of a real container with view of the climate chamber

Bottom left: cabinet with control electronics

Fig. 2.39 Typical setup for conducting battery tests

used (solid oxide fuel cells—SOFC). The SOFC can be fueled with natural gas or reformate gases (also referred to as synthetic gases) and is suited for operation with diesel, gas, natural gas or methanol.

Fuel cell test beds essentially consist of two independent gas paths. The purpose of the cathode path is to make the oxidants (air or oxygen) and nitrogen or mixtures of these gases available. By means of the anode path, gaseous fuels, such as hydrogen, methane or CO, are supplied to the anode side of the fuel cell. The gases along both flow paths can be conditioned independently of each other; the required temperatures and pressures depend on the type of fuel cell and may reach up to 150 °C (with PEMFC) or 800 °C (with SOFC) or up to 5 bar (rel.). It is also possible to add water vapor or liquid water to either of the two gas streams, independently of each other. Liquid fuels, such as diesel or methanol, can be added by using a dosing pump (Fig. 2.40).

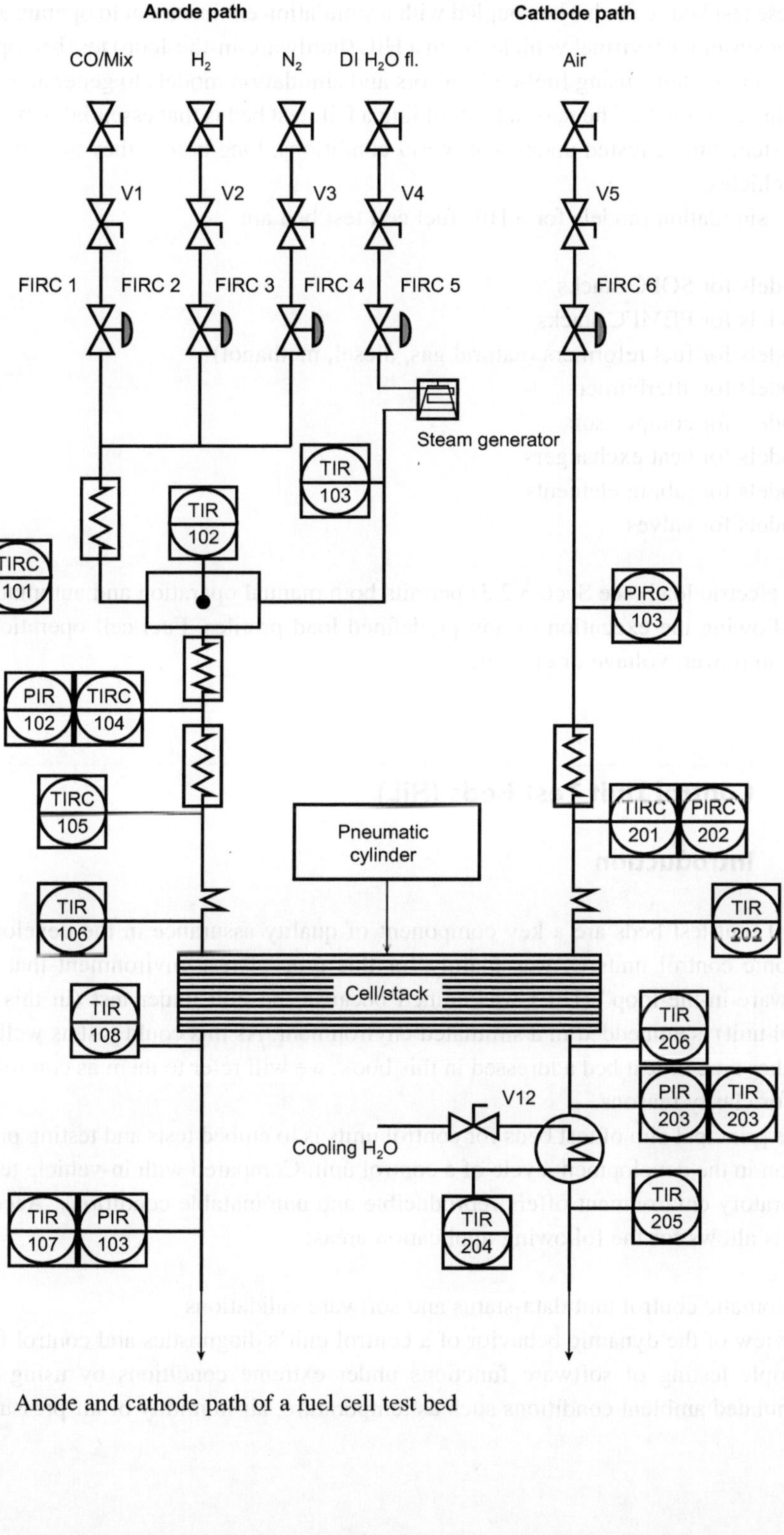

Fig. 2.40 Anode and cathode path of a fuel cell test bed

These test beds can also be coupled with a simulation environment to operate a fuel cell stack or system in a virtual vehicle. Such a HiL (hardware-in-the-loop) test bed operates in closed loop systems using fuel-cell sensors and simulation models to generate real-world operating conditions. The reason for building a HiL test bed is that essential parts of a fuel-cell system can be tested under real-world conditions, long before they are integrated in final vehicles.

Key simulation models for a HiL fuel cell test bed are:

- Models for SOFC stacks
- Models for PEMFC stacks
- Models for fuel reformers (natural gas, diesel, methanol)
- Models for afterburners
- Models for compressors
- Models for heat exchangers
- Models for tubing elements
- Models for valves

An electric load (see Sect. 3.2.3) permits both manual operation and automatic operation, allowing the execution of any predefined load profiles. Fuel cell operation can be based on power, voltage or current.

2.3 Control Unit Test Beds (HiL)

2.3.1 Introduction

Control unit test beds are a key component of quality assurance in the development of electronic control units. It was mainly for this type of test environment that the term "hardware-in-the-loop" (HiL) was coined because the unit under test (in this case the control unit) is embedded in a simulated environment. As this could just as well apply to any other type of test bed addressed in this book, we will refer to them as control unit test beds for clarity reasons.

The principal aim of test beds for control units is to embed tests and testing procedures early on in the development cycle of a control unit. Compared with in-vehicle test drives, a laboratory environment offers reproducible and automatable conditions. According to [2], this allows for the following application areas:

- Automatic control unit data-status and software validations
- Review of the dynamic behavior of a control unit's diagnostics and control functions
- Simple testing of software functions under extreme conditions by using specified simulated ambient conditions such as temperature, air humidity or air pressure

- Reproduction of extreme driving situations without putting test drivers or prototype vehicles at risk
- Specification of aging and failure situations in setpoint devices, sensors, actuators or cable connections
- Specification of erroneous or implausible states for testing the monitoring and diagnostic functions

Control unit test beds are used both for testing individual control units and for integration tests run on an entire network of control units. Using control unit test beds in combination with engine or powertrain test beds is also becoming increasingly significant, for example, for control unit calibration (see Sect. 2.3.6). After outlining the general setup of a control unit test bed, the various applications will be addressed.

2.3.2 Setup

The principle setup of a control unit test bed is shown in Fig. 2.41. The unit under test's cable harness is usually connected to the test bed hardware via a breakout box. The breakout box is where all signals can be tapped by additional hardware such as oscilloscopes.

The test bed's IO hardware components (see also Sect. 3.3) allow you to produce and measure standard signals such as analog, digital or pulse-width modulated (PWM) signals. In addition, special I/O cards with hardware signal generators are employed for generating crankshaft-synchronous signals for engine control units or speed signals for driving-dynamics or transmission control units, for example. If electric loads are to be reproduced and not physically installed, solutions are available for emulating electric loads, for example to emulate injection nozzles electrically.

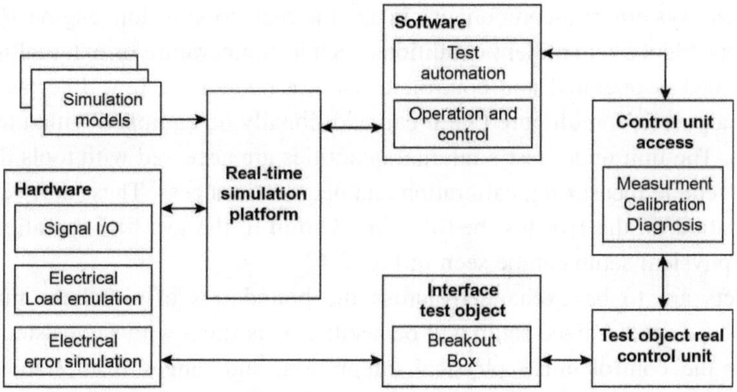

Fig. 2.41 Basic setup of a control unit test bed according to [3, 4]

To satisfy the diagnostics functions of modern control units, special signal conditioning systems are sometimes necessary that have real actuators, so-called real loads. An optionally available electric fault simulation tool can be used to produce short circuits based on battery voltage, mass or other channels, as well as line interruptions or increased inner resistances.

The simulation models needed to reproduce the control unit environment are executed on a real-time simulation platform. This concerns not only the physical vehicle component models or the models for driver and environment but also the components for residual bus simulation in order to supply the unit under test with all the necessary signals. Due to the growing complexity in models, more and more engineers have started to deploy multi-processor systems or FPGA solutions for executing models in real-time or for generating special signals.

There are different ways to create the required simulation models. Generally, we distinguish between physical and empirical modeling. A combination of the two procedures is considered a useful complement rather than a contradiction.

- **Physical modeling** refers to methods in which the laws of physics are applied in order to derive mathematical simulation equations. This technique usually leads to so-called white-box models because both the internal structures and the parameters are transparent and have a physical meaning.
- On the other hand, **empirical** modeling approaches use measured data from the inputs and outputs of the systems to be modeled to describe the input and output behavior on the basis of mathematical approximation procedures. The resulting mathematical equations and parameters have no direct physical interpretation. These models are also known as black-box models.

Empirical simulation models are the logical alternative when dealing with complex physical phenomena, which currently disallow real-time calculation with physical models or unknown correlations. Examples for this would be the simulation of exhaust aftertreatment system sub-components in an internal combustion engine or a battery system's dependence on ambient conditions such as temperature or external load.

The test bed is operated and controlled using a human-machine interface, typically running on a powerful multicore PC. It can additionally be equipped with a test automation system. The unit under test's internal quantities are accessed with tools that are also used for in-vehicle measuring, calibration and diagnostic access. These software tools are usually installed on the HiL test bed PC, in addition to the test bed operation software. A possible physical setup can be seen in Fig. 2.42.

If projects are to be exchanged across the boundaries of simulator manufacturer organizations, a standardized setup will be needed. This starts with a consistent connector concept for the control unit's physical integration, and ranges from standardized pin assignment for logical control unit integration to the standardized connection of breakout boxes and uniform load conditioning. These considerations may lead to a load drawer concept as shown in Fig. 2.43, which can be integrated into a variety of simulator types

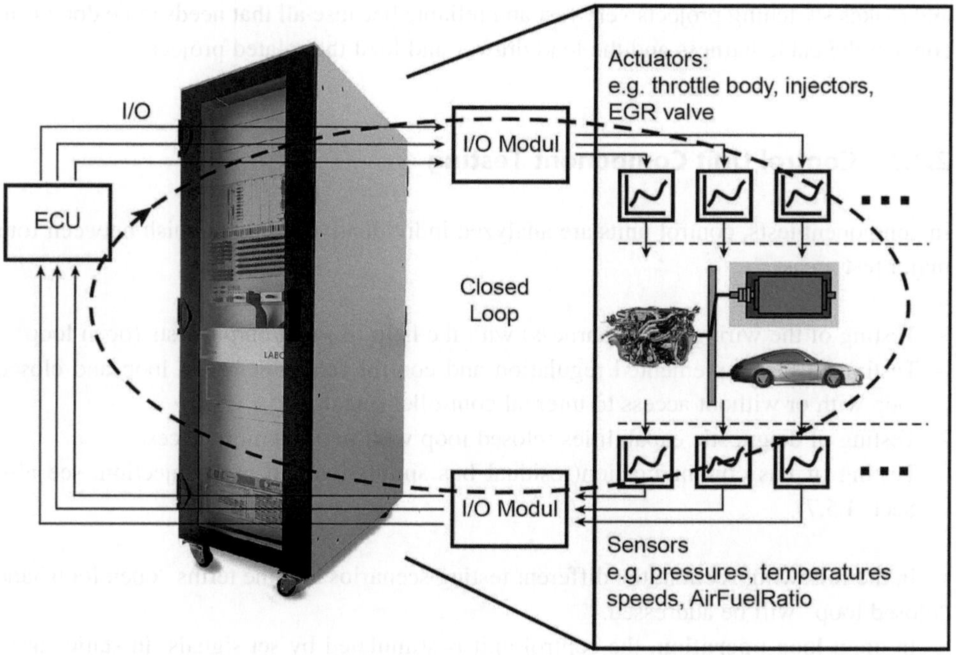

Fig. 2.42 Physical setup of a control unit test bed

Fig. 2.43 Load drawer concept for standardizing a HiL setup

and makes switching projects very fast and reliable because all that needs to be done is to connect the cable harness and the load drawer and load the related project.

2.3.3 Control Unit Component Testing

In component tests, control units are analyzed individually. We distinguish between four major test areas:

- Testing of the wiring (cable harness) with the help of input/output tests (open loop)
- Testing of the implemented regulation and control functions (open loop and closed loop with or without access to internal controller quantities)
- Testing of diagnostic capabilities (closed loop with fault memory access)
- Testing of bus communication(residual bus simulationwith error injection, see also Sect. 4.5.7)

In the following section, the different testing scenarios and the terms "open loop" and "closed loop" will be addressed.

In **open-loop operation**, the control unit is stimulated by set signals. In static cases, these might be pre-set values for checking the input/output behavior of the control unit and thus the cable harness. Where measured data are available from a test drive, these can be input to the control unit to test its response. Signal generators can be used to generate synthetic input signals. Usually, these signals are set as physical quantities, i.e. temperatures in °C or K, or torques in Nm (Fig. 2.44).

At the data conversion stage, these quantities then have to be adapted to suit the required electrical quantities (voltage values, pulse-width modulated signals, etc.) or

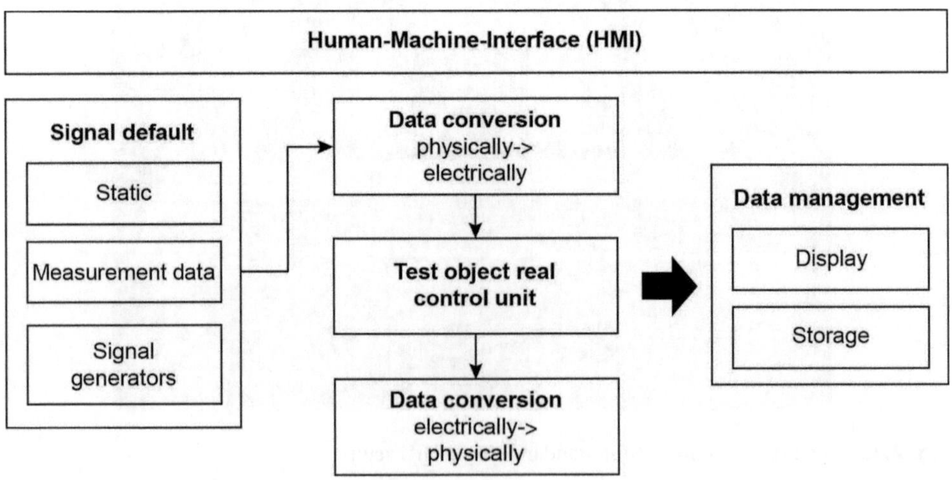

Fig. 2.44 Control unit test bed in open-loop operation

bus signals. If the data are to be recorded physically as well, the electrical control unit outputs need to be reconverted into physical quantities. When data are converted, it is possible to emulate further sensor faults, for example by superimposing an offset or a drift on a signal.

A key advantage of using open-loop operation is the low model requirements for test bed operation. Even so it is still possible to test the implemented functions for a large number of potential errors. Examples include diagnostic functions in the I/O components, I/O behavior of physical and electrical quantities or the testing of diverse open-loop control functions in the control unit. An example for an application is a program status and data status validation (see Fig. 2.45).

By comparing the functionality of old and a new xCU software and their input data, errors can be detected before the control unit is actually taken into operation in the real test carrier.

Unlike open-loop operation, closed-loop control functions require feedback from the system being controlled. Several diagnostic functions also rely on feedback signals.

In this **closed-loop operation** the entire control unit environment is represented (see Fig. 2.46). Aside from the actual vehicle model, it also includes the environment and the driver. If the control unit is networked with bus systems, the missing signals can be made

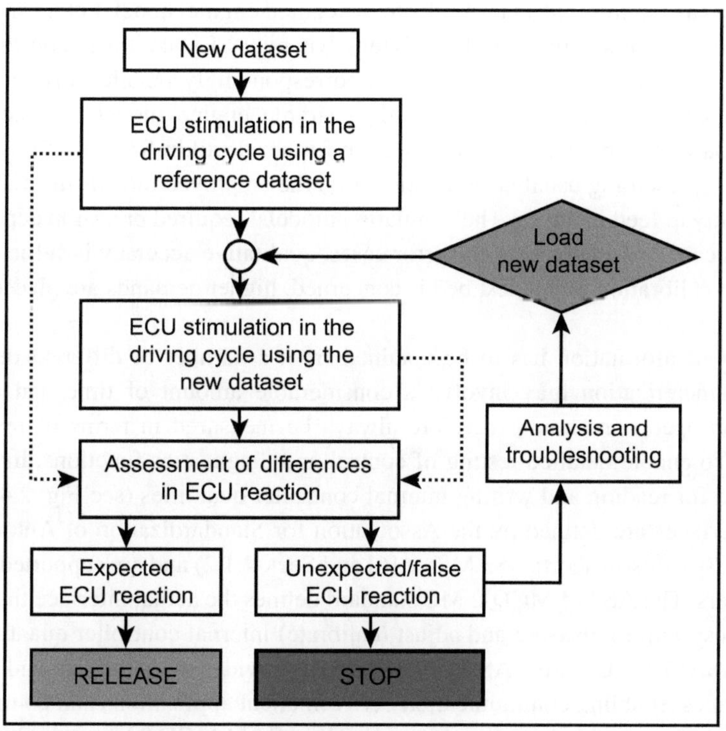

Fig. 2.45 Data-status validation (see also [5])

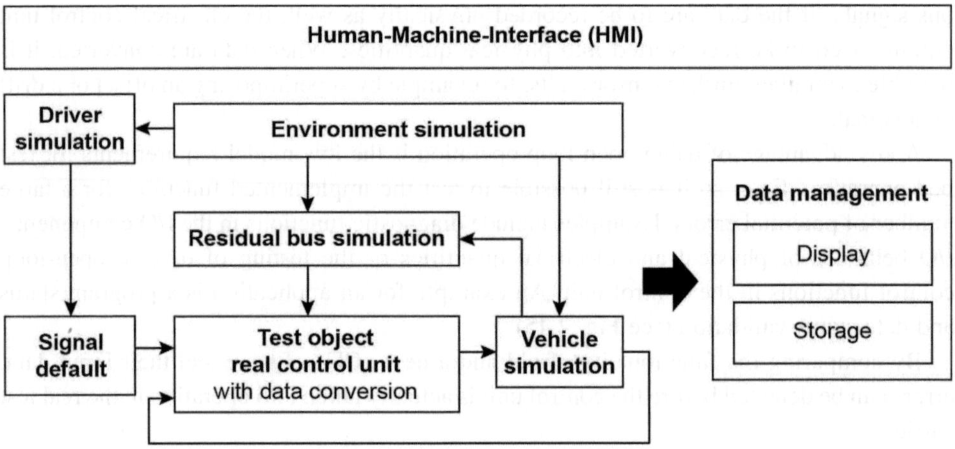

Fig. 2.46 Control unit test bed in closed-loop operation (see also [2])

available via a residual bus simulation. When reproducing the control unit environment, keep in mind to use a model that is adapted to the specific control unit; the model has to deliver the precise signals which the unit under test requires for operation and testing. Accordingly, an engine control unit calls for a detailed engine model, but it can do without a sophisticated reproduction of the driving dynamics. By contrast, control units for driving-dynamics control functions need a correspondingly detailed driving dynamics model, but can usually be operated with a simple engine model (a special case is maneuver-based testing, which is addressed in further detail in Sect. 4.5.5).

Besides representing usual-case functionality, the employed models must also provide the possibility to feed in errors. The simulation models' required precision depends on the application case. For many tests and error cases, qualitative accuracy is sufficient. Where control unit calibration on the test bed is concerned, higher demands are placed on model quality.

Given that information has to be obtained from a number of different departments, model parameterization may involve a considerable amount of time and effort, too. Closed-loop operation should therefore always be measured in terms of revenues and expenses. To enable detailed testing of control and diagnostics functions, interfaces are also needed for reading and writing internal controller quantities (see Fig. 2.47).

The interfaces are defined by the Association for Standardization of Automation and Measuring Systems(in short: ASAM, see [6] and Sect. 4.1.2) and are supported by various tool suppliers. The ASAM MCD-2 MC standard defines the actual interface that allows an application system to measure and adjust (calibrate) internal controller quantities. Building upon ASAM MCD-2 MC, ASAM MCD 3 MC provides measurement and application server services, enabling communication between client applications and control devices, independent from specific buses or protocols. The ASAM MCD 2 D standard (also known as ODX) enables access to diagnostics data, without the need of programming the external

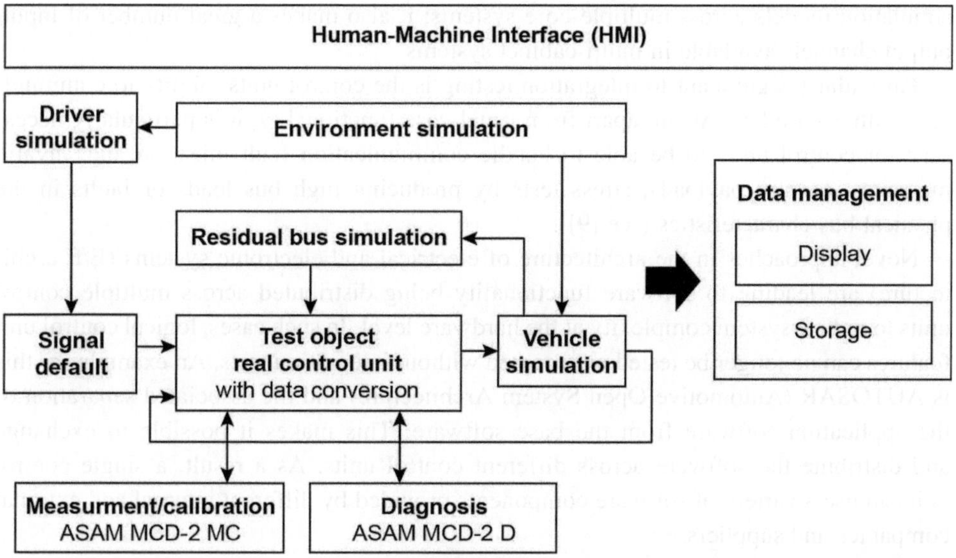

Fig. 2.47 Control unit test bed in closed-loop operation with access to internal controller quantities

diagnostics device in any particular way. ASAM MCD 3 D provides an interface for a diagnostics server.

By gaining access to internal controller quantities, particularly to the control unit's fault memory, it is possible to run test cases in closed-loop operation by specifying malfunctions and implausible states via the simulation model; these are then detected by the control unit and expected to produce an entry in its fault memory. The test checks whether the expected fault entries are in actual fact produced. This primarily applies to the legally required diagnostic capabilities, such as on-board diagnosis (OBD), or the European OBD (EOBD), by which all emissions-related systems have to be monitored. The required fault memory entries are defined in ISO standard 15031-6 [7].

The testing of in-vehicle bus system communication is a further testing priority, which is chiefly performed when testing several integrated control units in an integration test.

2.3.4 Control Unit Integration Testing

Once control units have been tested individually, they are integrated into a control unit network. Two different application cases require our attention. For the purpose of integration testing, it is possible either to set up a test system specifically for the integration test or combine existing control-unit component test beds. Dismantling should be possible without requiring long periods of time. According to [8], it is essential that such systems be scalable. This quality not only increases the computing power for simulation models across multiple-core systems; it also makes a great number of input/output channels available in multi-cabinet systems.

Particularly significant to integration testing is the control units' ability to communicate with each other. Again, apart from usual-case functionality, it is particularly necessary for control units to be able to handle communication fault injection, e.g. invalid messages, corrupt payloads, stress tests by producing high bus loads or faults in the physical bus characteristics (see [9]).

Novel approaches in the architecture of electrical and electronic systems (E/E architecture) are leading to software functionality being distributed across multiple control units to reduce system complexity at the hardware level. In such cases, logical control unit features can no longer be tested or accepted without integration tests. An example for this is AUTOSAR (Automotive Open System Architecture) and the associated separation of the application software from the base software. This makes it possible to exchange and distribute the software across different control units. As a result, a single control unit can use a variety of software components provided by different internal and external companies and suppliers.

Yet another trend is leading to more and more superordinate management systems of the kind that have already been employed in the heavy-duty vehicle industry for some time. An example for this is the torque-vectoring control unit. Driver assistance systems and energy management systems for hybrid and electric drives that are becoming increasingly complex are further examples. None of these new E/E architectures could be tested without having a control-unit network on the test bed.

Alongside internal control unit networks, communication to the outside is increasingly gaining significance too. Buzzwords in this respect are "vehicle-to-vehicle" (V2V) for connectivity between vehicles and "vehicle-to-infrastructure" (V2I) for connectivity between vehicle and infrastructure, as is required by modern traffic information systems or when accessing rental car services using smartphone apps. In future, the demands on the security of these interfaces will essentially shape the way vehicles are exposed to the outside world. For such applications, the significance of control unit integration tests is likely to increase even further.

Very often, the effort involved in setting up and operating a pure control unit network test bed within a virtual environment exceeds the economic benefit. One way to handle this issue is to combine real powertrain test beds with a control unit network. By doing so, certain jobs, such as the time-consuming assembly of cable harnesses or environment modeling, are significantly reduced or even avoided completely.

2.3.5 Test Automation

One of the major advantages of control unit test beds—both in component test operation and in integration test operation—is the possibility to automate tests and testing procedures. Considering the growing diversity of control unit variants, test automation is a key factor when it comes to quality assurance. Fundamental to test automation are standardized interfaces that can handle the remote control of all test bed components. These are defined in the former ASAM HiL standard (now ASAM-XiL) (see Sect. 4.1.2), for example, and are expected to allow the re-use of test cases on different types of test systems.

You will find a detailed description of the design and features of a test automation system, which are fundamentally similar in all test bed configurations, in Sect. 4.6.

2.3.6 Model-Based Calibration

In recent years, a new technology has evolved for control unit test beds: model-based calibration. The technology supports series-production vehicle applications by hardware-in-the-loop (HiL) simulation and enables calibration tasks to be performed faster and more cost-effectively on control unit test beds, without having to use internal combustion engine test beds or test vehicles. For this to be accomplished, technically mature models of the engine and exhaust aftertreatment components are needed that are capable of representing the combustion behavior with sufficient precision.

The method enables frontloading and even parallel development processes, as HiL systems allow program statuses of series vehicles to be validated and calibrated at an early stage, even before any engines or test vehicles are physically available.

To be able to use the HiL test environment efficiently, a "virtual" test facility is installed. The "virtual" test bed is defined by exactly the same automation and application system that is used on the real test bed (Fig. 2.48). This allows test engineers to efficiently

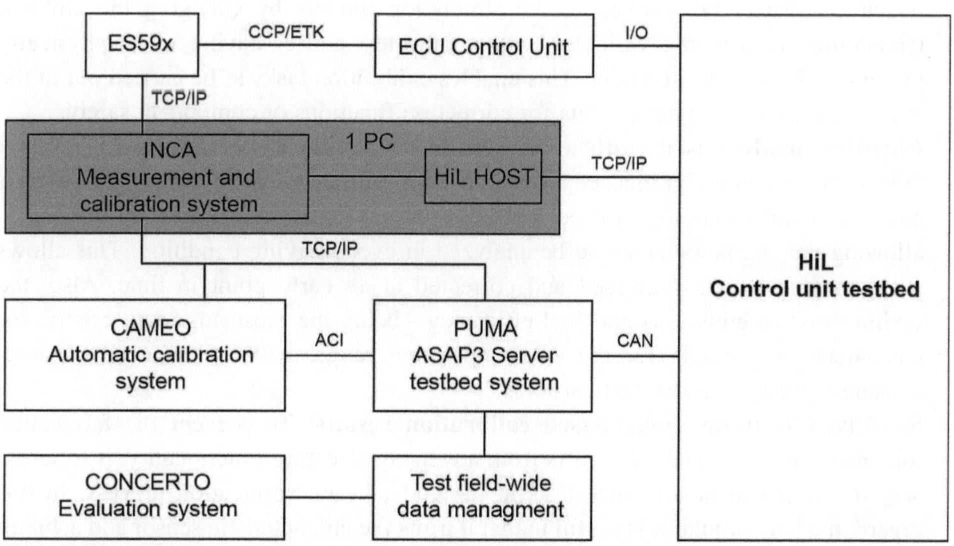

Fig. 2.48 Example of a virtual test bed for model-based calibration

carry out control unit calibration tasks both on the internal combustion engine test bed and the control unit test bed, without having to undergo extra training.

The test runs are programmed with the automation software and then executed (in the same way as they would on a real test bed). The automatic calibration system communicates with the test bed system and the control unit's measurement and calibration system via standard interfaces (see Sect. 4.1.2).

As would be done on a real test bed, the automation system sends controlling torques to the load unit, which is equally simulated by a model on the HiL system, from where the results of the engine model, such as torque, temperatures, pressures and emissions, as well as combustion characteristics (e.g. peak pressure, combustion center of gravity), are sent back to the test bed system. All measurement and simulation results are stored in the test facility's central data management system.

The application engineers can now analyze the measured data with the evaluation system they are accustomed to. Virtual test beds are consequently able to cover the entire tool chain, which ensures maximum re-use between development environments and leads to high acceptance among users.

2.3.6.1 Examples for Virtual Test Bed Applications

- **Model-based (pre-)calibration of control unit functions:** By applying high-precision simulation models—both of the engine and the entire powertrain—partial functions of the control units can already be calibrated and tested on the "virtual" test bed. Particularly where variant applications are concerned (i.e. when calibrating an engine that is used in various different vehicles), it is very useful to add a HiL system to the testing environment to test as many driving conditions as possible in a cost-saving manner. These mainly concern ambient factors, such as altitude or high and low temperatures, which can easily be adapted in the simulation models by changing the ambient temperature or pressure variables, without the test carrier having to be physically present in the test environment. This enables calibration tasks to be carried out in the control unit, such as entering data for correction functions or component safety.
- **Checking model-based calibration results in customer-specific driving cycles:** Given the growing significance of real-driving emissions (RDE), the "virtual" test bed is a useful technology for performing a variety of driving cycles automatically, allowing the engine behavior to be analyzed in every driving condition. This allows problem areas to be addressed and corrected at an early point in time. Also, the optimization of emissions and fuel efficiency—being the most important criteria for customers—is possible for the entire operating range, without the need of time-consuming tests with real test carriers.
- **Sensitivity tests on model-based calibration results:** To prevent the fabrication tolerances of sensors and actuators from adversely affecting functionality, it is necessary to run complex tolerance tests at the end of each application process. In this regard, the HiL simulator is useful in that it trims the characteristic sensor and actuator curves in the models, enabling a check of the xCU response.

In summary, it can be concluded that a HiL test bed is useful for supporting numerous calibration tasks in modern engine development processes. Its advantages are most importantly due its ability to simulate diverse environmental influences, component tolerances and possible malfunctions and wiring faults. Its high reproducibility and independence of component damage are further factors that make the virtual test bed an efficient test and calibration environment for application/calibration engineers.

Nevertheless, it should be mentioned at this point that there are indeed limitations to model-based calibration on the HiL test bed. These primarily concern hardware developments, cold-start optimization and endurance strength. Additionally, it must be kept in mind that the final validation and certification still have to be done on the real test carrier; this is because simulation models will always contain simplifications and never be able to represent reality as a whole. Likewise, the demands on the accuracy of simulation models are much greater than they would be with HiL-Systems for test applications.

In spite of this, HiL systems in a "virtual" test environment can be very useful for series application projects because they provide an inexpensive way to conduct multiple calibration tasks, either at an early stage or in parallel to standard test environments such as engine test bed, chassis dynamometer or road. Frontloading calibration into HiL simulation contributes toward quality enhancement and reduces development duration significantly. For more background on model-based calibration, please refer to Sect. 4.6.3.

2.4 Powertrain Test Beds

2.4.1 Powertrain Test Beds with Internal Combustion Engine

Growing pressure to innovate and increasingly short development times are the big challenges in powertrain development. To tackle these challenges, parallel development of the different groups of components is just as much required as the testing of the powertrain as a whole (see also Sect. 1.2.1).

The significance of powertrain development in vehicle development has risen sharply in recent years. Today, major advancements in consumption and emission development can only be achieved with an optimized powertrain—a fact that brings engine development and powertrain development much closer together.

This integral development has also had an impact on powertrain testing. In the past, the main focus of transmission testing used to be on endurance strength testing with regard to the mechanical system. Jobs, such as function tests, control unit calibration, driving-comfort and gear-shifting fine-tuning, as well as misuse tests, were exclusively performed in road tests, with few exceptions. Today, the increasing complexity of powertrain components and their management, along with the interaction between the different control units for the engine (ECU), transmission (TCU) and hybrid drive (HCU) with intelligent driver assistance systems (e.g. ESP stability control, etc.), require early validation of the complete system, even without a prototype vehicle having to available.

State-of-the-art powertrain test beds are conceived as multi-configuration test beds. On the one hand, the term "multi-configuration" means that different kinds of powertrain configurations (see Fig. 2.49) can be tested with a single test bed setup. It also refers to the fact that different types of powertrain and transmission concepts need to be tested, e.g.

- Manual transmission (MT)
- Automatic transmission (AT)
- Dual-clutch transmission (DCT)
- Continuously variable transmission (CVT)
- Hybrid powertrain concepts

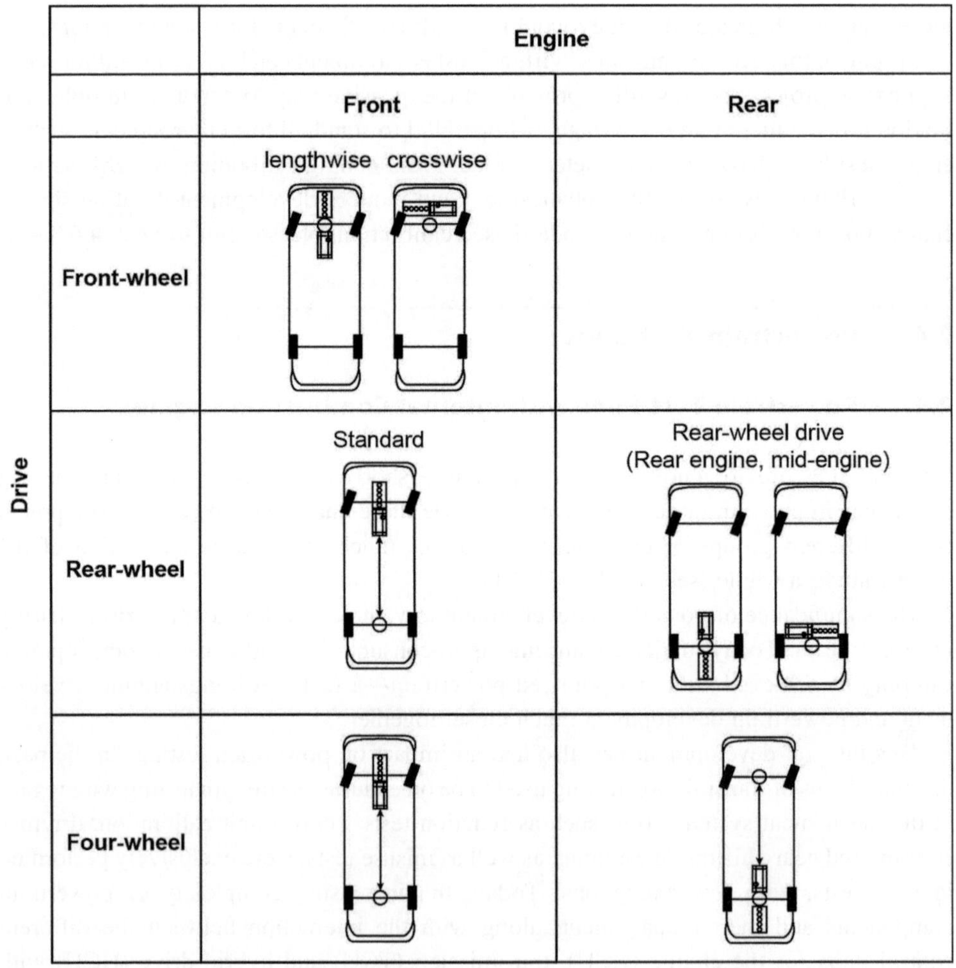

Fig. 2.49 Possible powertrain configurations

The powertrain configurations that require testing are the variant driver for test bed setup. In the area of light-duty vehicles (passenger cars), up to four machines are used to apply a load to the powertrain at the wheel shafts or cardan shafts. Typical test bed setups for a standard-configuration powertrain (front engine—rear-wheel drive) can be seen in Figs. 2.50, 2.51, 2.52, 2.53.

The test bed setup shown in Fig. 2.50 is intended for testing the internal combustion engine and transmission only. In this case, an appropriately suited electric machine is used to apply a load to the unit under test's cardan shaft. The setup outlined in Fig. 2.51 allows

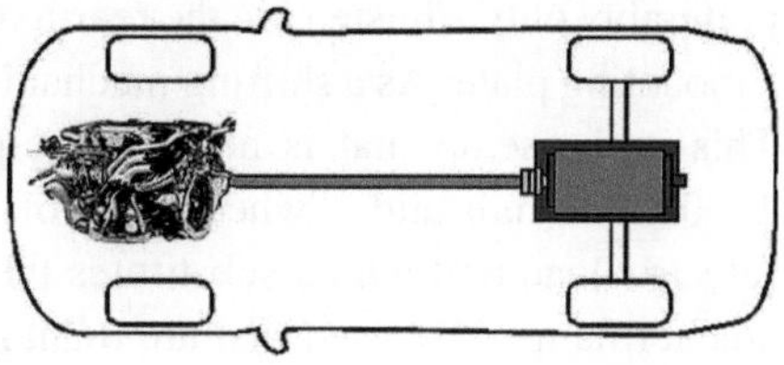

Fig. 2.50 Powerpack test bed for standard drives

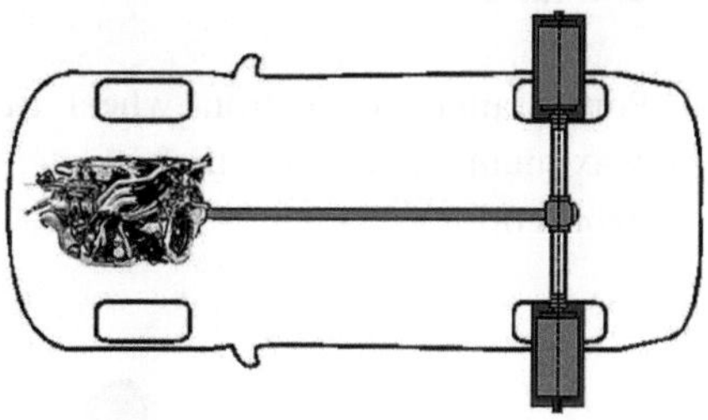

Fig. 2.51 Powerpack test bed for standard drives

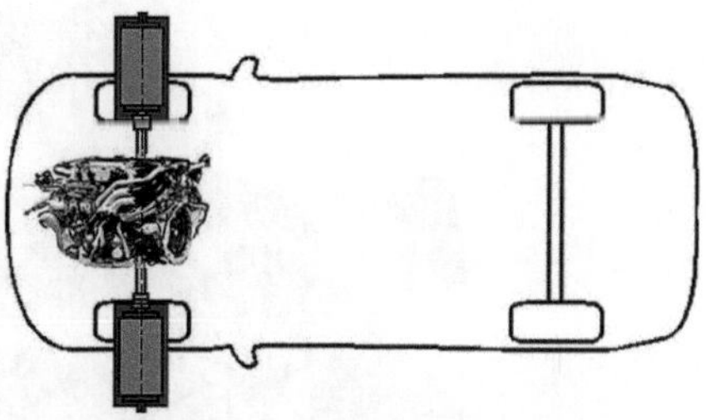

Fig. 2.52 Powertrain test bed for front-wheel drives

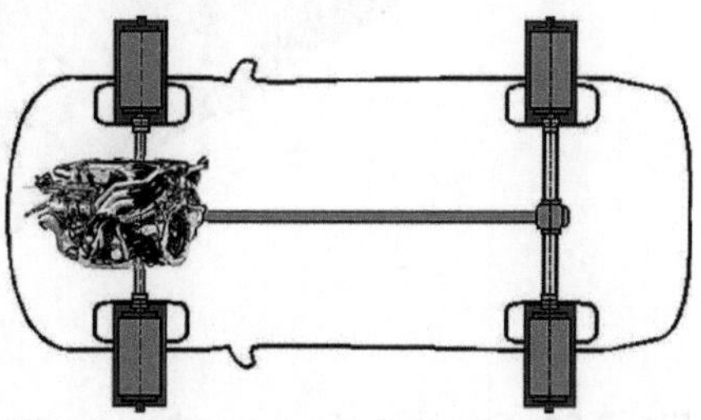

Fig. 2.53 Powertrain test bed for four-wheel drives

testing of the transmission and the rear-axle differential. Figure 2.52 presents the typical test bed setup for vehicles with front-wheel drive. The test bed setup with four dynamometers shown in Fig. 2.53 is suitable for testing four-wheel drives.

What all powertrain test beds (see Fig. 2.54) have in common is that the test bed setup must allow flexible adaptation to the varying track widths and wheel bases of the individual vehicles. For this purpose, two or four load units are mounted on an intermediate frame. Usually, the working height, i.e. the shaft center point, is fixed.

Axial displacement of the machines, i.e. the track width setting, is generally performed by means of a shifting mechanism integrated in the intermediate frame. The wheel base is preferably only adjusted via the rear-axle wheel machines. Usually, the frame is shifted on a mounting plate. As a shifting mechanism, a spindle drive is built into the mounting plate. This is the setup that is normally required for testing vehicles with a track width of 1300–2000 mm and a wheel base of 2200–3200 mm. To connect the wheel machine (electric load unit which substitutes the wheel) to the unit under test's axle shaft, special adapter flanges are used that are manufactured specifically for each unit under test.

The choice of load units and the right torque measuring flange should be made with several factors in mind:

- Powertrain concept (front-wheel drive, standard drive, four-wheel drive)
- Maximum speed of drive system (internal combustion engine, electric motor, hybrid drive)

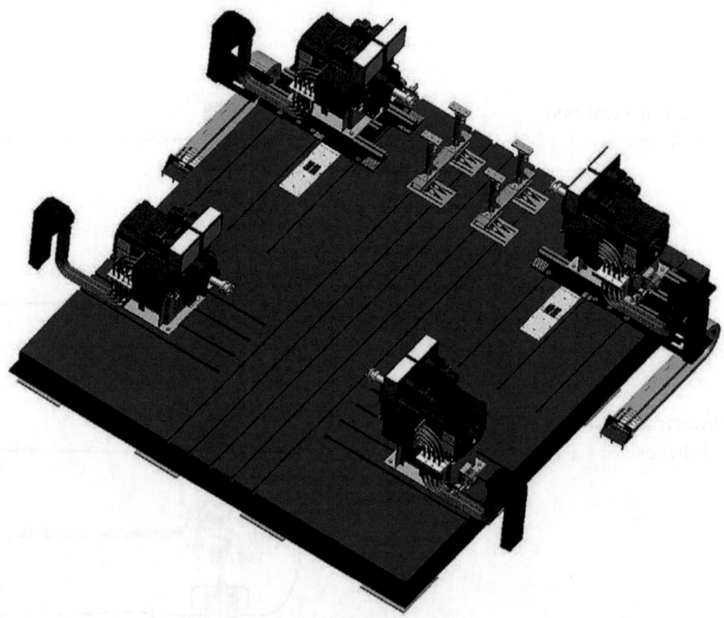

Fig. 2.54 Example of a four-wheel powertrain test bed

- Maximum torque of drive system (internal combustion engine, electric motor, hybrid drive)
- Total gear ratio spread
- Torque increase by clutches or torque multiplication by torque converters
- Maximum traction at the wheel
- Virtual test drive on different road surfaces (e.g. snow, tarmac, etc.)

Typical values for the maximum torque per wheel are 1300–3500 Nm. The maximum wheel speed is usually within the range 2400–3000 rpm. A virtual test drive on different road surfaces requires low-inertia high-dynamic wheel machines, so that the powertrain's resonance behavior on the test bed remains identical to what it would be in an on-road test.

For unmanned, automated operation, the equipment integrated in the test bed automation system includes the following devices:

- Gear-shifting system for mechanical actuation of the gear-shift mechanism (manual transmission, automatic transmission, CVT, dual-clutch transmission) and vehicle clutch or an interface to a control unit that can trigger a gear shifting process (e.g. via CAN message)
- Clutch actuator for mechanical actuation of the vehicle clutch or an interface to a control unit that can trigger clutch engagement/disengagement
- Mechanical accelerator pedal actuation by means of an actuator or an interface to a control unit that can set an accelerator pedal value (e.g. via CAN message)
- Conditioning equipment for various media (e.g. coolants, engine oil, transmission oil, combustion air, fuel, etc.)

In addition, depending on the testing task, a variety of measuring and monitoring devices have to be integrated, which may include:

- Measuring system for early damage detection to be able to shut down all systems and perform a damage analysis to prevent the unit under test from being destroyed in endurance tests
- Real-time classification to determine the load spectrum (see also Sect. 4.7)
- Consumption and exhaust emission measurement devices (see also Sect. 3.3)
- Tools for calibrating engine and transmission control units (see also Sect. 4.6.3)
- Tools for maneuver-based testing (see also Sect. 4.5.5)

Nowadays, maneuver-based testing, in particular, contributes significantly toward making the development process for advanced powertrains more efficient. A consistent simulation environment allows engineers to assess and optimize the vehicle (or powertrain) and the way in which it interacts with its environment throughout all stages of the development process (see also Sect. 1.2.1).

2.4.2 Powertrain Test Beds with a Prime Mover as Drive Unit

Shorter and shorter development times require the parallel development of internal combustion engine and transmission. When the first transmission test is run, the internal combustion engine itself is usually at prototype stage and still without the degree of maturity (in terms of durability and performance data) it would need for transmission testing. Substituting an electric machine (prime mover) for the internal combustion engine solves this conflict and additionally leads to reduced costs for prototypes.

The prime mover is above all required to reproduce damage-related loads on the powertrain. In this regard we must distinguish between loads produced by the engine's average torque output and the internal combustion engine's cyclic irregularity as a result of its discontinuous working cycle.

The engine's average torque output can be described in simplified terms as a function of speed and accelerator pedal position. In addition, the internal combustion engine's response behavior at load alteration must be taken into account by including a delay element discrete in time or depending on the crankshaft angle. The loads on the transmission, however, are not exclusively determined by the driver's activities and the corresponding road load, but also by the behavior of the various control units.

Control unit-related loads result from control units communicating with each other and lead to certain demands on the engine control unit, e.g. torque reduction during gear-shifting. Such signals are usually transmitted on the CAN bus (see Sect. 3.5.2). This means that it is necessary to represent any engine control unit components having an impact on the load (see Fig. 2.55). A further aspect that must be kept in mind is that missing control units may lead to plausibility errors in CAN messages, resulting in emergency operation mode with a limited scope of functions. This is why it is imperative to perform a CAN residual bus simulation (see also Sect. 4.5.7) which generates plausible CAN messages.

The cyclic irregularity is produced by forces acting on the crankshaft drive. These are generated by combustion (gas forces) and the non-uniform movement of masses (mass forces). The gas force is primarily determined by the burned mass fraction, whereas the

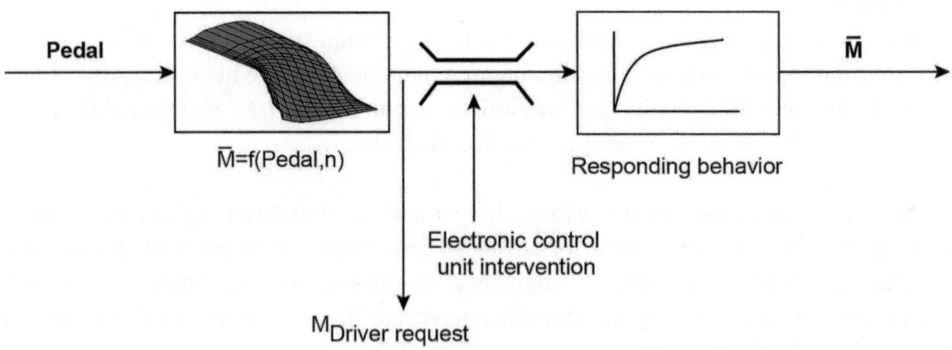

Fig. 2.55 Model of an internal combustion engine for the average torque output

mass force depends on speed. The instantaneous value of the resulting force is determined, among other things, by engine design, number of cylinders and combustion method. The combustion process is a crank angle synchronous process; the working cycle of a four-stroke engine covers 720 °CA.

Cyclic irregularity x(t) cannot be expressed by a simple mathematical function. But based on a working cycle, cyclic irregularity can be approximated by means of a Fourier analysis.

$$x(t) = a_0 + \sum_{i=0,5}^{\infty} \left(a_i \cdot \cos\left(i\omega_0 t + \varphi_{a,i}\right) + b_i \cdot \sin\left(i\omega_0 t + \varphi_{b,i}\right)\right) \tag{2.7}$$

Equation 2.7 Cyclic irregularity

Cyclic irregularity can consequently be described as a superposition of a fundamental oscillation (number of working cycles per revolution) and harmonics (orders of magnitude or multiples of the fundamental component). Cyclic irregularity leads to the excitation of torsional vibrations which affect the complete powertrain, e.g. jerking, transmission rattle, etc.

For a four-cylinder four-stroke engine, 6000 rpm leads to a frequency of 200 Hz for the fundamental oscillation (ignition frequency, 2nd order) and 400 Hz for the next harmonic (4th order) (see Fig. 2.56).

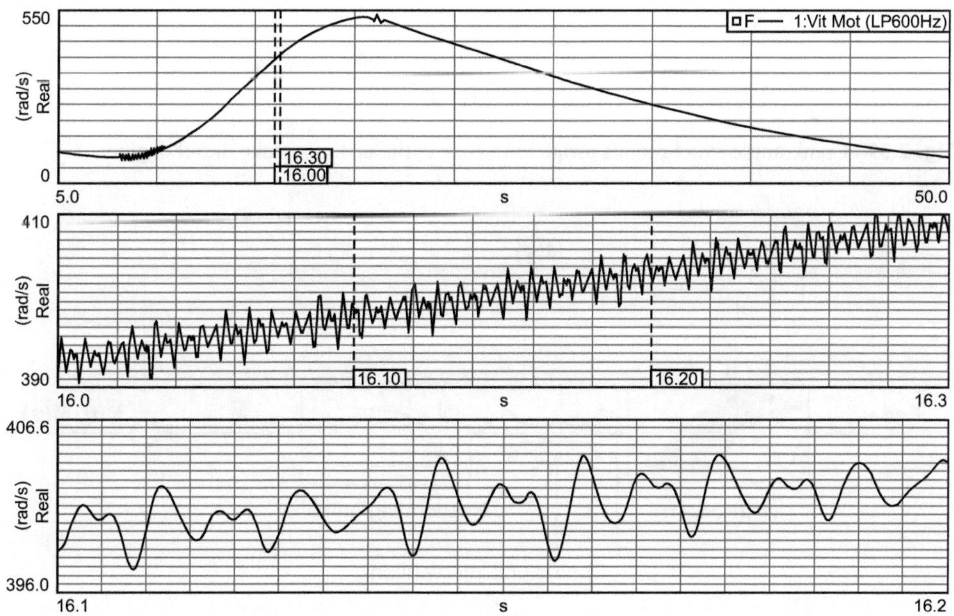

Fig. 2.56 Cyclic irregularity of a four-cylinder diesel engine

To reproduce cyclic irregularity, the mean engine torque is superimposed by the torque resulting from the zero-mean gas and mass forces (see Fig. 2.57). It is necessary to take into account that the cyclic irregularity is a crank angle synchronous signal (see Fig. 2.58).

For the electric drive unit to be able to represent the cyclic irregularity, it must satisfy several design criteria:

– The polar moment of inertia of the machine's rotor should be as close as possible to the internal combustion engine's (the crankshaft drive's) polar moment of inertia of (see Fig. 2.59).

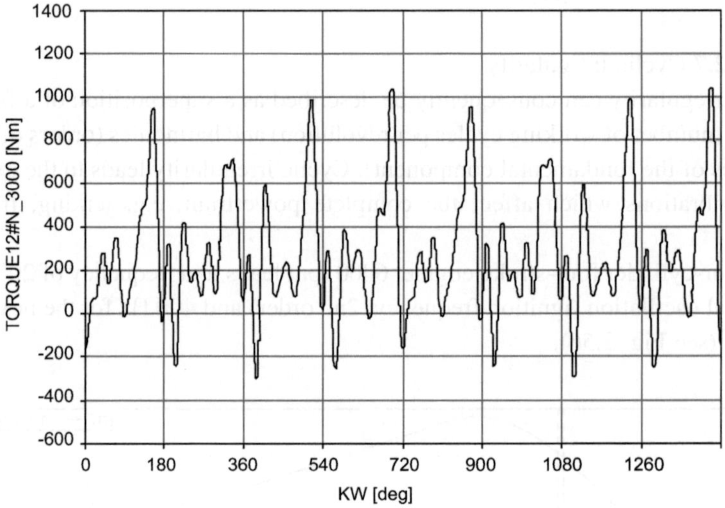

Fig. 2.57 Crank angle synchronous torque of a four-cylinder four-stroke diesel engine

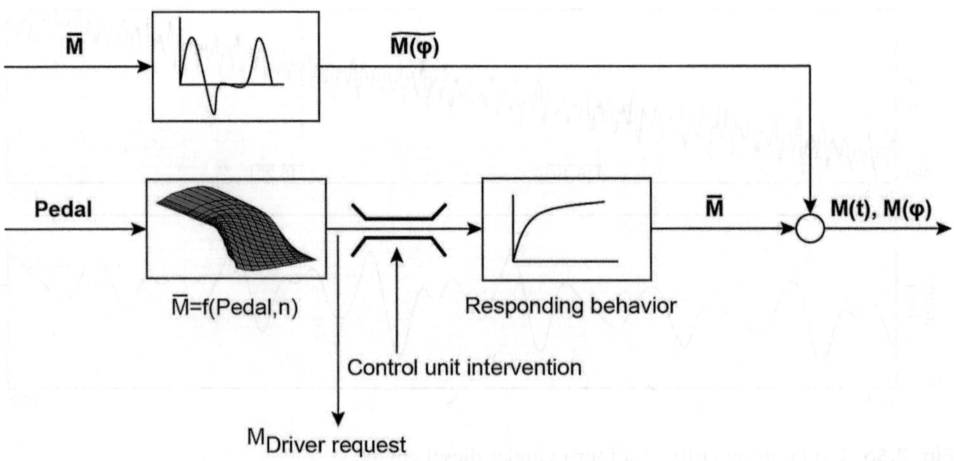

Fig. 2.58 Model of an internal combustion engine with cyclic irregularity

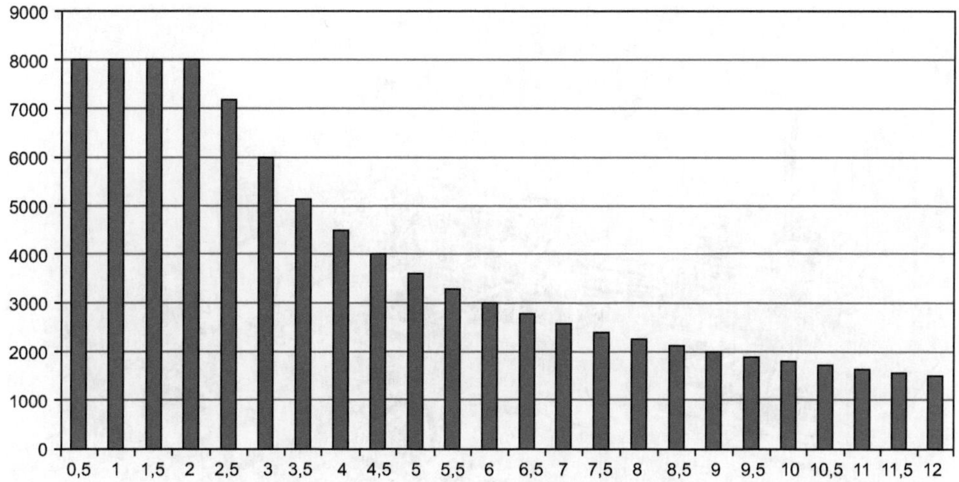

Fig. 2.59 Representable maximum speed as a function of the engine order for 300-Hz cut-off frequency

- The electric machine should be specifically designed so that it can be replaced by a transversely mounted front engine. Particularly where transversely installed engines are concerned, the installation space is limited in terms of the distance between transmission input and output shaft and the maximum length available. This means a prime mover is needed with maximum compactness, i.e. short overall length and small diameter (see Fig. 2.60).
- It is essential to use a prime mover with minimal torque control times and signal delay times, allowing it to reproduce the internal combustion engine's cyclic irregularity.

To reproduce cyclic irregularity on the test bed, a high-dynamic, permanent-magnet three-phase AC synchronous machine is used (Fig. 2.61). Typical performance data are:

- Rated power 370 kW
- Maximum speed 10,000 rpm
- Rated torque 650 Nm
- Max. overload torque 1300 Nm
- Mass moment of inertia 0.118 kgm^2
- Length 1150 mm

The prime mover's overload capability is of particular significance. Unlike simulations with electric drive units, those performed with a prime mover to simulate cyclic irregularity have torque peaks on a permanent basis. That is why this electric machine has to be dimensioned for uninterrupted, continuous operation with speed changes (Duty Type S8 in compliance with IEC 60034-1).

Fig. 2.60 Four-wheel test bed with prime mover

Fig. 2.61 FWD test bed with prime mover

It is also necessary to choose a suitable reference point for continuous loading. The machine's maximum constant load is generally based on the maximum permissible temperature increase. A constant torque produces the same temperature increase in the machine as the root mean square (RMS) value of a periodic torque signal would. This means that the prime mover will fall under Duty Type S8 if the RMS value of the torque signal M(t) is smaller than the machine's rated value.

Another factor that has to be considered is that impressing periodic signals can induce eddy currents in the rotor, which, as a result, produce a local rise in the rotor temperature. To prevent damage to the rotor, eddy currents have to be reduced, i.e. the maximum representable frequency must be limited to 300 Hz (see Fig. 2.59).

These simulation frequencies can only be transferred if the coupling between the prime mover and the transmission is sufficiently stiff.

2.4.3 Hybrid Powertrain Test Beds

Powertrain test beds for hybrid drives are no different from the powertrain test beds described in Sect. 2.4.1 because the drive units can always be regarded as an entity. Drive units always transfer torque and speed at the interface, regardless of whether an internal combustion engine alone is being used or an entire hybrid configuration (internal combustion engine with serial or parallel electric motors, range extenders with electric drives, fuel cell systems in electric vehicles, etc.).

2.5 Vehicle Test Beds

As integration test beds, chassis dynamometers (roller-type dynamometers) can be used very flexibly for research and development activities. To obtain type approval, vehicles must undergo certification to meet regionally specific standards with respect to a variety of aspects, the most important of which are fuel consumption, exhaust emissions, engine power, driving noise and electromagnetic compatibility. In some areas (e.g. exhaust emissions) it is mandatory to use chassis dynamometers, while for other areas (e.g. electromagnetic compatibility) chassis dynamometer testing is optional. Driving noise, for example, has to be determined in on-road tests, but it in this respect it is possible to use the chassis dynamometer as a development platform.

It must generally be noted that chassis dynamometers are increasingly moving away from being certification tools to becoming mechatronic development platforms for calibration, system optimization and system validation tasks [10].

When it comes to units to be tested, chassis dynamometers are extremely versatile. Their scope covers everything from motorcycles or subcompacts to passenger cars and light-duty commercial vehicles and even heavy-duty commercial vehicles and mobile industrial machinery (see Fig. 2.62).

Fig. 2.62 Scope of vehicles on chassis dynamometers

Depending on application case and structural conditions, a variety of roller and engine configuration concepts are put to use:

- 1 × 1 roller: One roller on one axle with one dynamometer (example: motorcycle)
- 2 × 1 roller: Two rollers on one axle with one dynamometer (example: vehicle with front-wheel drive, application: exhaust emissions certification)
- 2 × 2 rollers: Two rollers, each with a separate dynamometer; rollers are not mechanically connected (example: vehicle with front-wheel drive, application: NVH—noise, vibration, harshness—see Sect. 2.5.4)
- 4 × 2 rollers: Four rollers, two on each axle with one dynamometer (example: vehicle with four-wheel drive, application: exhaust emissions certification)

- 4 × 4 rollers: Four rollers, each with a separate dynamometer; rollers are not mechanically connected (example: vehicle with four-wheel drive, application: NVH)

For specialty applications, further configurations are possible that allow vehicles with more than two axles, for example, to be operated on chassis dynamometers. With regard to engine arrangement, there are also various configurations:

- In-line design: The motor is positioned next to the roller(s). The motor is either coupled directly or by means of a gear unit, for example to increase tractive force.
- Motor-in-the-middle design: The motor is positioned between the rollers. The rollers are mounted on the motor shaft itself.
- Motor-in-the-roller design: The motor is built into the roller as an external-rotor motor.

To allow safe and user-friendly operation, a variety of peripheral devices are available, examples of which are given below:

A central component is the **vehicle fastening system**, which prevents the vehicle from moving off the rollers during the test. Depending on the vehicle type, the kind of application and the measuring process, different fastening concepts are applied (Fig. 2.63).

To align the vehicle on the rollers, special **centering systems** are used, which allow the tire to be centered at the highest point of the roller. Using the centering system, the entire vehicle can even be raised so that the test bed can perform automated verification runs without the vehicle (see Fig. 2.64).

Moreover, various kinds of headwind blowers, conditioning modules, battery emulators, fueling systems, climate sensors and the like are put to use according to requirements.

The application areas for chassis dynamometers are diverse:

- Chassis dynamometers for emissions development and certification
- Chassis dynamometers for fuel consumption and performance testing
- Chassis dynamometers for endurance and durability testing
- Chassis dynamometers for NVH analysis
- Chassis dynamometers for EMC analysis
- Chassis dynamometers for advanced applications

Fig. 2.63 Vehicle fastening concepts on chassis dynamometers

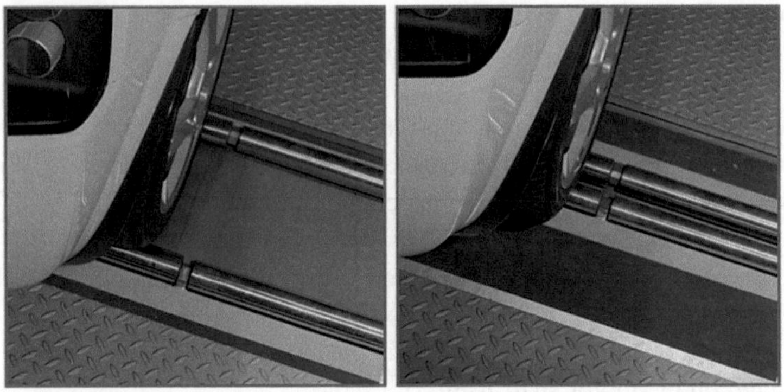

Fig. 2.64 Centering system open (*left*), closed (*right*)

2.5.1 Chassis Dynamometers for Emissions Development and Certification

One of the key areas in which tests are performed on chassis dynamometers concerns exhaust emissions certification. As this procedure is required by law, there are clear boundary conditions and regulations that define in detail what is required of such chassis dynamometers. The "Dynamometer Performance Evaluation and Quality Assurance Procedures for 48-Inch, Single-Roll, Electric, Light Duty Chassis Dynamometers" define the required criteria and related acceptance procedures, ranging from roller diameters, measuring and regulating accuracy levels to vehicle fastening system requirements. Further specifications are incorporated in the legislative texts. Below is a list of key criteria:

- Roller diameter and true-running accuracy
- Surface roughness
- Time, velocity and tractive force measurement accuracy
- System behavior relating to response time, overshoot and transient response
- Accuracy of road load simulation (comparison between theoretical demand values and simulated actual values)
- Stiffness of vehicle fastening system (the criterion here is vehicle movement under strong acceleration)
- Headwind blower (homogeneity and volume flow)

Additionally, the test bed has to be equipped with the measuring and automation systems needed to run tests according to the statutory requirements. Pursuant to the current legislation, standardized, region- and country-specific cycles are used to measure emissions and fuel economy. Figures 2.65 and 2.66 show the NEDC (New European Driving Cycle) for Europe and the FTP-75 (Federal Test Procedure) for the U.S. In the

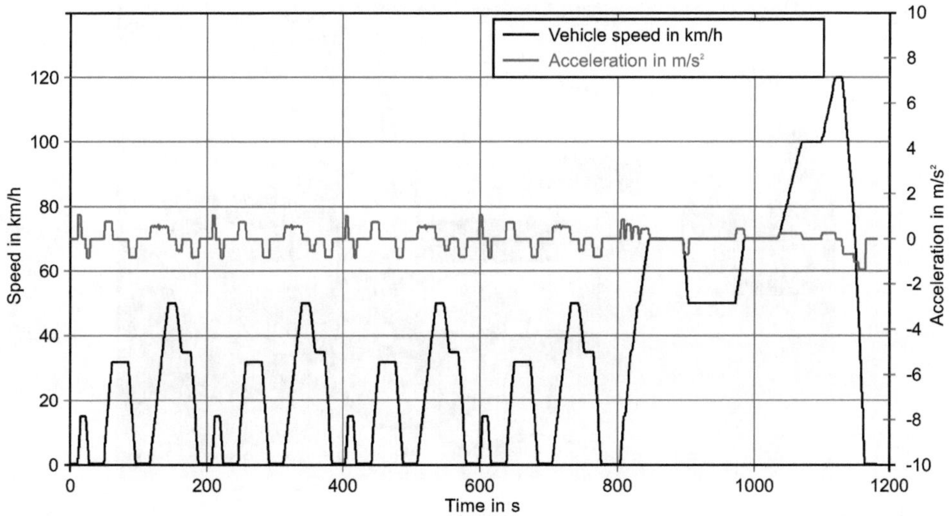

Fig. 2.65 European Driving Cycle

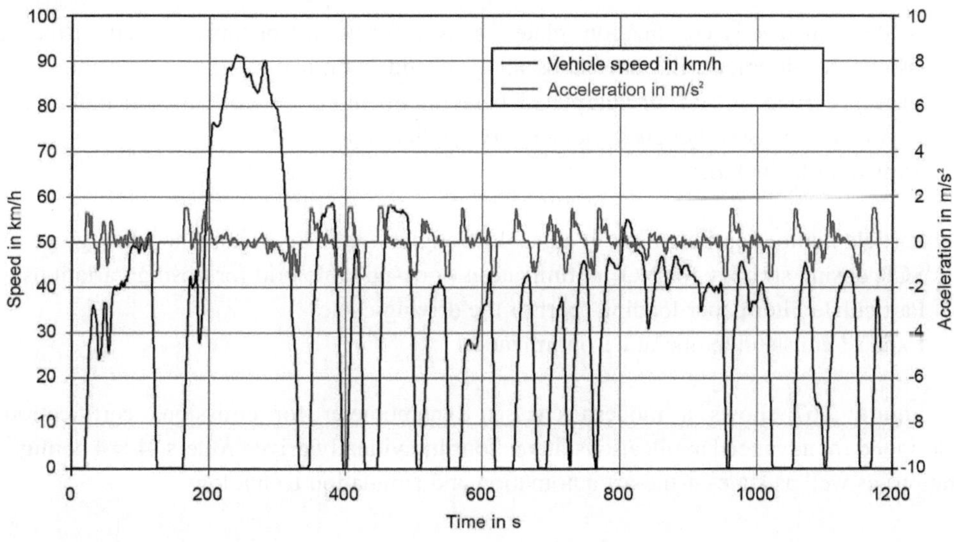

Fig. 2.66 Federal Test Procedure

near future, the NEDC is to be succeeded by the WLTP (World Harmonized Light Vehicles Test Procedure).

All three are basically velocity-time profiles which have to be driven by a human driver. The technology used includes driver guidance systems that display the profile to the operator and provide a preview and relevant information concerning the gears, etc. Then the velocity profile has to be reproduced within a specified tolerance band.

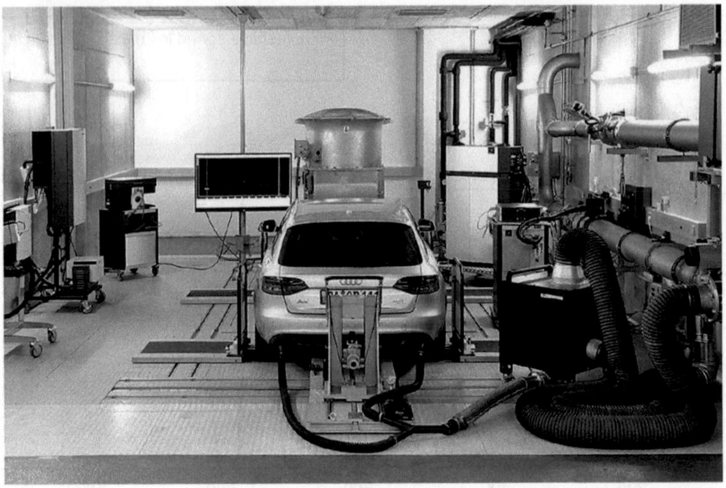

Fig. 2.67 Modern 4 × 4 chassis dynamometer with focus on emissions certification (*source*: TÜV Hessen Automotive)

Aside from strictly certification-related tasks, a further major topic concerns exhaust emissions development, i.e. emissions analysis and optimization. Tasks like these are frequently carried out on an automated basis in overnight sessions, which makes the deployment of chassis dynamometers a very efficient option.

Typical tasks include:

- Particle number optimization in a gasoline engine for EU6 (part of the driving cycle)
- SCR dosing strategy for NO_x optimization (look-up table grid for dosing variations)
- Particulate filter paper loading (part of the driving cycle)
- Exhaust emission temperature optimization

Figure 2.67 shows a modern chassis dynamometer for emissions certification. Designed for advanced applications, it has four individually driven rollers (4 × 4 configuration) as well as state-of-the-art automation and simulation technology.

2.5.2 Chassis Dynamometers for Fuel Consumption and Performance Testing

Chassis dynamometers for fuel consumption and performance testing have very much the same requirements as exhaust emission chassis dynamometers. As a vehicle's consumption values can be calculated from the emissions, it is not uncommon for consumption measurements to be performed on these test beds.

However, where fuel economy is to be investigated at a more detailed level, i.e. largely in research and development, there are a series of further requirements which the test bed must satisfy. These primarily concern the measurement of losses in the powertrain and tires, from which the experts can then draw conclusions about fuel consumption.

To this end, a test bed is needed that provides very high measuring accuracy and an adapted roller diameter. Basically, 48-in. rollers are suitable for an application of this kind. To settle for a compromise between running behavior and small wheel base, however, rollers are also available that have a diameter of 69 in. Figure 2.68 shows the example of a chassis dynamometer designed for efficiency analysis. The advantages are the rigid design, the high levels of accuracy and the 69-in. rollers mentioned previously.

The following formula describes, approximately, the relationship between rolling resistance on the road and on the roller:

$$F_{xR} = F_x \left(1 + \frac{r}{R_r}\right)^{1/2} \tag{2.8}$$

F_{xR} Rolling resistance on the roller
F_x Rolling resistance on the road
r Tire radius
R_r Roller radius

Equation 2.8 Rolling resistance

Figure 2.69 shows the same situation again diagrammatically. The radius of a tire is assumed to be 0.3 m. If this tire is run on a 25-in. roller, the formula shows that it has approximately 40 % more rolling resistance than it would have on a road.

There are many different ways of measuring performance on chassis dynamometers. The objective is to obtain the kind of engine map an engine test bed would provide,

Fig. 2.68 69-in. roller chassis dynamometer for analyzing efficiency

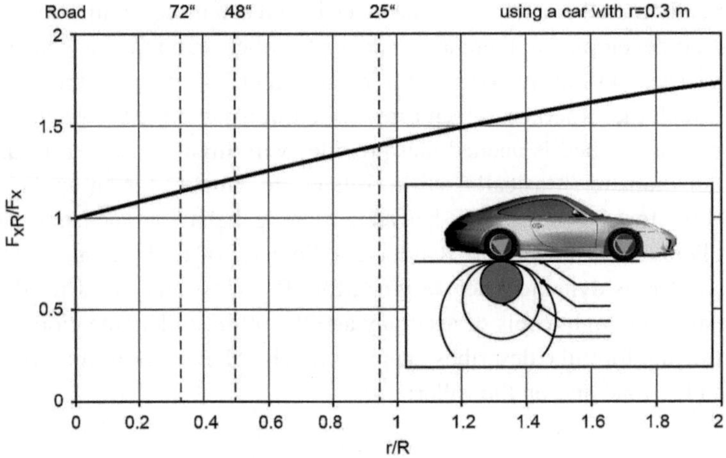

Fig. 2.69 Relationship between rolling resistance on the road and on the chassis dynamometer

Table 2.2 Methods of measuring performance on chassis dynamometers

	**Performance measurement in road load simulation (dynamic):** In a road load simulation, a vehicle performs full-load acceleration up to the required velocity and coasts to a stop to determine the powertrain losses.
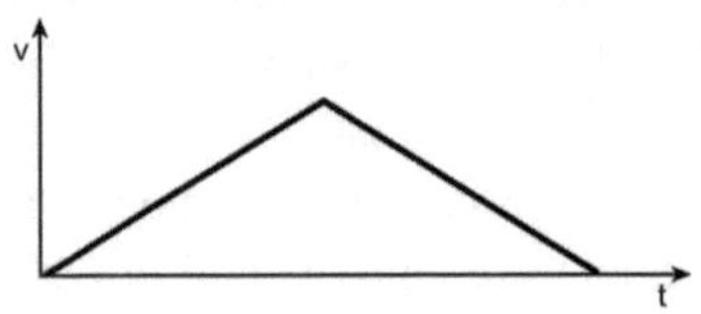	**Performance measurement in velocity control mode with ramp (quasi static):** The test bed is operated in velocity control mode while the vehicle is set to full throttle position. The velocity is ramped up very slowly (quasi statically). To capture the losses, this is followed by a ramp with negative sign.
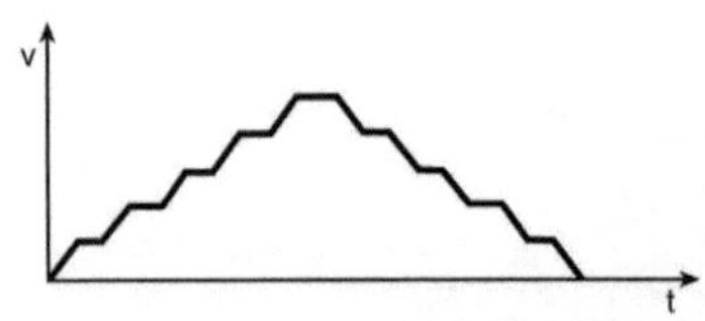	**Performance measurement in velocity control mode with steps (static):** The test bed is operated in velocity control mode while the vehicle is set to full throttle position. At defined speed points, steady-state speeds are set. Tractive force measurements and performance calculation also take place at these points in steady state.

showing the power output and torque versus velocity or speed, respectively. There are several different methods to measure the map (see Table 2.2).

Each of the methods has its own advantages and shortcomings in relation to feasibility and accuracy. For example, if the measurement is performed dynamically, there will be influences stemming from powertrain-component acceleration. Such effects can directly be seen in the power calculated from the measured tractive force and the velocity. An exemplary calculation for a powertrain and the impact of inertial acceleration on power is described in Eq. (2.9).

Inertia of engine:	$\theta_{ICE} = 0.6\ \text{kgm}^2$
Inertia of transmission:	$\theta_{transmission} = 0.2\ \text{kgm}^2$
Inertia of differential:	$\theta_{diff} = 0.05\ \text{kgm}^2$
Inertia of wheel/tire:	$\theta_{wheel} = 1.7\ \text{kgm}^2$
Acceleration:	$a = 2.7\,\frac{m}{s^2}$ (i.e. from 0 km/h to 100 km/h in 10 s)
Dynamic tire radius:	$r_{dyn} = 0.23\ \text{m}$

Inertia of powertrain:

$$\theta_{PT} = i^2_{diff} \cdot i^2_{transmission} \cdot (\theta_{ICE} + \theta_{transmission}) + i^2_{diff} \cdot \theta_{diff} + 2 \cdot \theta_{wheel} = 4.25\ \text{kgm}^2$$

To make an estimate, the gear ratio i is set to 1.
This yields the resulting force at the tire/roller surface:

$$F = \theta_{PT} \cdot \frac{a}{r^2_{dyn}} = 4.25\ \text{kgm}^2 \cdot \frac{2.7\,\frac{m}{s^2}}{(0.23\ \text{m})^2} = 217\ \text{N}. \tag{2.9}$$

Resulting power at 120 km/h:	$P = F \cdot v = 217\ \text{N} \cdot 33.3\,\frac{m}{s^2} = 7226\ \text{W}$
Resulting power at 60 km/h:	$P = F \cdot v = 217\ \text{N} \cdot 16.6\,\frac{m}{s^2} = 3602\ \text{W}$

Equation 2.9 Exemplary calculation of the inertia-related influences on performance

An essential aspect that requires careful consideration when measuring performance on chassis dynamometers is the specific tire slip involved. Simple corrections based on the speed difference between tire and roller are not necessarily effective because the slip condition dissipates considerable heat which the engine must produce. Slip depends on a variety of boundary conditions such as air pressure, roller diameter and the tire's position on the roller. Especially where light vehicles or motorcycles are being tested, the driver's influence (the driver's mass and the resulting normal force exerted on the roller) is clearly discernible in the performance values.

2.5.3 Chassis Dynamometers for Endurance and Durability Testing

Another important factor in view of vehicle certification is long-term and aging behavior. For this particular purpose, there are also especially designed test cycles. Particularly where tests of this kind are concerned—the vehicles are required to cover a minimum distance of 100,000 km—it is not advisable to resort to a real driver. In such cases, driving robots are used instead, which operate all interfaces from the vehicle to the roller.

As the test runs are fully automatic and executed over long periods of time, the chassis dynamometer test environment must be capable of fulfilling specific requirements. These include advanced monitoring sensors, e.g. to detect tire blowouts; fueling systems, etc. The available cycles again depend on the legislation. Examples for such cycles are the AMA Cycle (Approved Mileage Accumulation Cycle)—Fig. 2.70—or the SRC (Standard Road Cycle—Fig. 2.71). Emission measurements are made after defined distances (e.g. 5000 km) in order to analyze long-term behavior.

Another related application carried out in this test environment is the investigation of durability and component deterioration. This is primarily done in the context of research and development activities.

Several approaches are taken: On the one hand, real distances are reproduced on chassis dynamometers and, on the other, for increased efficiency, accelerated-deterioration procedures are applied. By eliminating less-damaging sections of the load spectrum, it is possible to save time and costs.

There are numerous accelerated-aging approaches available to examine different components of the vehicle powertrain. Two approaches applied in the area of clutch development are examples worth mentioning:

- Hill start test: This type of test involves starting the vehicle on hills with different gradients with or without a trailer.
- Standard program: In this case, several thousand gear-shifting processes are carried out continuously at a defined engaging frequency.

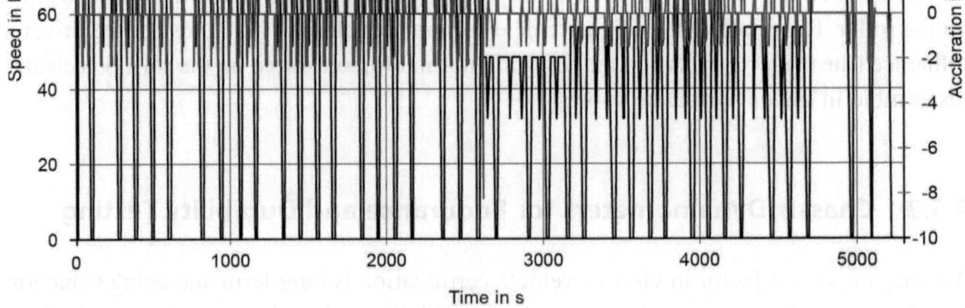

Fig. 2.70 AMA Cycle

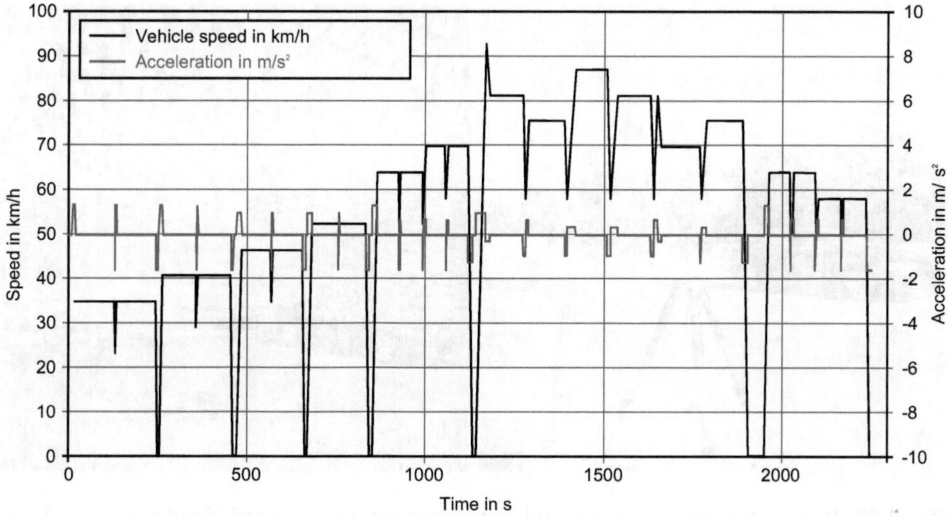

Fig. 2.71 Standard Road Cycle

Nowadays, these kinds of tests are usually conducted on component test beds (see Sect. 2.2.3). It is also possible to run validation tests on chassis dynamometers.

2.5.4 Chassis Dynamometers for NVH Analysis

The analysis and optimization of NVH behavior is a very wide-ranging field. The acronym NVH stands for noise, vibration and harshness (harshness: 20–100 Hz). Basically, we distinguish between exterior and interior vehicle acoustics. Depending on the application, there are specific requirements which NVH chassis dynameters and the related facilities have to satisfy. Several special features are listed below:

- Vibration-absorbing support system for the base plate
- Road surface simulation using road shells
- Noise level <50 dB(A) at 100 km/h
- Work platform with closable pit underneath the vehicle (e.g. for making modifications to the vehicle)

With regard to the exterior acoustics, the simulated pass-by noise test is a common application. On-road pass-by noise testing is a test run which is also relevant for obtaining type approval. For certification purposes, however, it is exclusively conducted in on-road operation (see Fig. 2.72 on the left). To this end, a microphone is positioned at a defined

Fig. 2.72 Pass-by testing on a real-world test course (on the *left*) and simulated on a chassis dynamometer (on the *right*)

distance from the vehicle. The test vehicle drives past this microphone at full-load acceleration, starting out from a defined velocity.

To optimize and analyze this behavior, the test can also be conducted on NVH chassis dynamometers. The setup is very similar. However, since the vehicle on the chassis dynamometer is unable to move past the microphone, several microphones are set up (see Fig. 2.72 on the right) and the test is evaluated as if the vehicle were moving past them. For this to be accomplished, a comparatively large room is needed that satisfies these requirements.

For component analyses and interior acoustics, the demands on the test cell are much lower. Nevertheless, there are other things that need to be taken into consideration. For instance, with acoustic investigations focusing on the exhaust gas system, a compromise has to be reached between a sufficiently effective exhaust gas extraction system and unobstructed discharge. The vehicle cooling system may have an adverse impact on measurements too. That is why cooling systems with pop-up flaps or noise-reduced fans are frequently utilized. For some measurements, it may even be necessary to temporarily turn off the cooling system completely.

In the field of acoustics, automation is a major factor. A great number of analyses are conducted in the form of state, steady-state or map measurements. Careful attention should be given to reproducibility and to the selective setting of such states, etc. Particularly where vehicles with complex drive systems are concerned, however, such adjustments have become very difficult for human drivers to accomplish.

NVH chassis dynamometers are frequently available in 4×4 or 2×2 configurations because in order to simulate road excitation, e.g. using cleats, the different rollers have to be activated individually and phase-controlled and because the acoustic encapsulation of the electric machines is easier to achieve.

2.5.5 Chassis Dynamometers for EMC Analysis

Chassis dynamometer EMC analysis (EMC: electromagnetic compatibility) is another field that falls in the category of homologation-related vehicle tests. EMC chassis dynamometers (see Fig. 2.73) have to fulfil particular requirements. The electromagnetic emissions have to be extremely low. Furthermore, the vehicle must not be affected by the electrical connections to the test bed.

This is why, for example, pneumatic robot drivers are used, or belts instead of chains for fastening the vehicle to the test bed. The vehicle is exposed to electromagnetic emissions by employing an elaborate technique. This is why many EMC chassis dynamometers are integrated into turntables. It is easier to rotate the test bed including the vehicle rather than the provided measuring and simulation equipment. The turntable rotation is additionally capable of stimulating the accelerometers inside the vehicle.

The main goals of EMC chassis dynamometers are to determine electromagnetic emissions and verify electromagnetic immunity. This area of application is chiefly interesting where electric and hybrid vehicles are concerned, as these are provided not only with numerous control units, but also with a great number of built-in high-voltage components (consumers, power lines, etc.).

It must be ensured that all of these electronic functionalities operate faultlessly, even when they are subjected to substantial levels of exposure. In the simplest of cases, specific functionalities are just switched on in different driving situations and maneuvers, whereas in more complex cases the functionalities are triggered by certain maneuvers. One example for this would be the excitation of the ABS system. For this particular purpose, there are a number of test runs that simulate slip or speed differences. This is why EMC chassis dynamometers are almost exclusively designed as 4 × 4 test beds.

Fig. 2.73 EMC chassis dynamometers

2.5.6 Chassis Dynamometers for Advanced Applications

Many functions and fine-tuning adjustments cannot be finalized or validated until they are integrated with the overall system. Beside the powertrain test bed, the chassis dynamometer is a test environment that allows the complete vehicle to be tested under reproducible laboratory conditions. New approaches enable development engineers not only to shift activities from the real test drive to these types of test beds, but also to verify their fine-tuning modifications in the overall vehicle.

For this to be accomplished, a consistent overall system is needed that is designed to execute complex tasks on chassis dynamometers. The roller mechanics and electrics and the chassis dynamometer's control and automation systems must all interact optimally and be based on a modular design to permit transferability and the integration of a wide range of tools. For this to be achieved, it is of paramount importance that test bed suppliers have in-depth expertise in vehicle and drive systems. To include some examples, two applications are described in further detail:

Nowadays, **drivability tuning** is largely performed on real roads. This is extremely time-consuming and results in correspondingly high costs. With a suitable test environment and tool chain, such tests can also be run on a chassis dynamometer. Figure 2.74 shows an **automatic transmission calibration process**. It can be seen that the real track is exclusively used for verification purposes. With the appropriate tools for drivability assessment, test scheduling and optimization, extensive portions of the test procedure can be carried out fully automatically on the chassis dynamometer or even in the office.

One critical aspect of drivability evaluations on chassis dynamometers is the way longitudinal acceleration is measured, as, obviously, the vehicle does not move. By utilizing a special fastening concept (see Fig. 2.75) involving a load cell, longitudinal acceleration can be derived from the measured longitudinal force.

Another example is the calibration and **development of driver assistance systems** that, among other things, use navigation chart data, or communicate with their environment using special sensors (e.g. radar, V2V, V2I, etc.) in order to achieve an energy-efficient operating state. Testing such capabilities on a real track is very time-consuming and scarcely reproducible because, apart from the basic conditions (temperature, road effects), there is also a level of uncertainty regarding conditions in the extended environment (traffic, etc.) which interact with each other).

Chassis dynamometers in conjunction with the relevant environment and road simulations are suitable for testing and fine-tuning such capabilities reproducibly under laboratory conditions. For this purpose, the simulation model reproduces the road including the environment conditions and the traffic situation (see also Sect. 4.5.5). The model calculates all the relevant information and is thus able to communicate directly with the vehicle. It is even possible to supply positioning data for the navigation system, for example, or the distance and relative velocity of the vehicle ahead. It is equally possible to run the vehicle to be tested on a chassis dynamometer in complex V2x scenarios (V2I, V2V) and have them interact with each other (see Fig. 2.76).

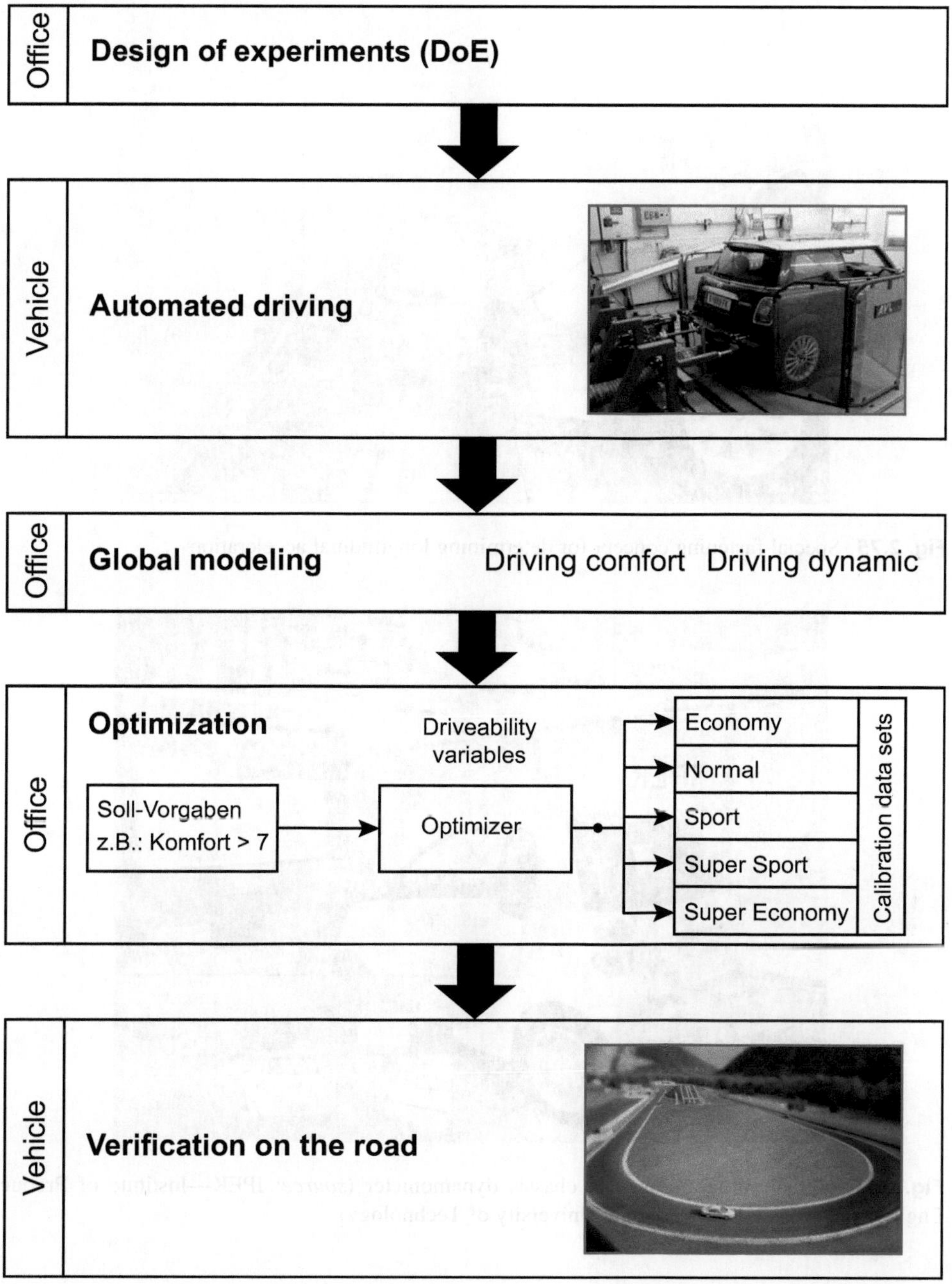

Fig. 2.74 Automatic transmission calibration process

Fig. 2.75 Special fastening concept for determining longitudinal acceleration

Fig. 2.76 Vehicle-in-the-loop on a chassis dynamometer (*source*: IPEK—Institute of Product Engineering at the KIT—Karlsruhe University of Technology)

2.6 Racing Test Beds

In their basic setup, racing test beds differ only marginally from the test beds the manufacturers use for series vehicles. Even the development tasks of racing organizations are the same as the ones carried out by manufacturers of series vehicles.

They design, develop, test and manufacture engines, powertrains and entire vehicles on a regular basis.

The most essential difference between race car manufacturing and series production lies in the fact that—apart from the different demands on the equipment—the product lifecycles of racing vehicles are much shorter. Vehicle components are used for one race or for one racing season at the most. The number of engineers and technicians available for development work is much smaller too.

But unlike series vehicles, the application case is limited and very precisely defined. This means that a large number of aspects featured in a broad series development, such as component reliability over years, or safety-related requirements, are not needed. These observations essentially result in two demands on the testing equipment:

It must meet the highest possible standards in terms of repeat accuracy, control performance and system dynamics, and it must provide maximum versatility. All of the test bed systems mentioned so far that are distinguished by their units tested (engine, component, transmission, hot gas, etc.) and different applications (R&D, endurance run or end-of-line) are also employed in motor sports, yet under different boundary conditions.

To give some examples, we have included a description of engine test beds and component test beds and the typical challenges encountered when using them in motor sports.

2.6.1 Engine Test Beds for Racing

Aside from basic investigations concerning development goals, steady-state and slightly transient racing engine operation on the test bed also serves quality assurance purposes (end-of-line, or, as a motor sports expert would put it, check-out or break-in), control unit calibration and performance testing.

A critical advantage of racing applications lies in the fact that the exact load spectrum of both vehicle and components is predictable and known down to the last detail, a fact that allows all vehicle parts to be optimized and tested for the loads occurring during the race. The limitations posed by practical on-road tests (racing track) make development activities difficult for the racing team. A crucial factor in this respect is having a test facility (test bed) that is capable of simulating the race. On the test bed any racing track can be simulated, allowing the unit under test to be optimized for any requirement and tested for durability. Furthermore, the test bed can also be used to prove theoretically achievable enhancements, e.g. by new engine components, and to verify their durability for their specific use before integrating them into the vehicle. The use of real-time co-simulation on a test bed eliminates all secondary influences, resulting in absolutely reproducible unit-under-test loads.

In lap simulation on the racing engine test bed, the remaining powertrain, the entire vehicle's driving dynamics, the track and the driver are simulated by a corresponding mathematical model. This means that the power transmitted to the crankshaft in the form of real torque has to be emulated mechanically by high-dynamic and high-speed load units.

This places the following demands on the complete mechatronic system consisting of engine, shaft, dynamometer, control unit and simulation model:

- It must be possible for the engine power to be sufficiently transmitted.
- The system dynamics must be sufficiently capable of not only simulating the maximum load and speed gradients occurring in the real vehicle, but also of emulating them on the drive shaft using high-dynamic load units.

The last requirement, in particular, is pushing the limits of the testing equipment currently available. By employing the latest converter technology and permanent-magnet synchronous machines, effects can be emulated that reach up to approx. 500 Hz in closed loop control and up to 1200 Hz in open loop control. A racing engine's damaging torsional and linear vibration characteristics are both above this critical frequency. That is the reason why phenomena not capable of being reproduced directly by the load unit have to be appropriately emulated. With regard to the linear vibrational behavior, for example, this is achieved by using special adaptive coupling elements in the engine mounting system. A suitable shaft connection allows the engine's torsional resonance behavior to be adjusted as required.

The racing engine in the vehicle is directly connected to the built-on transmission by torsionally stiff coupling. Consequently, on the test bed, the engine is coupled using a dummy gearbox and a suitably designed torsion-bar stabilizer, which emulates the mass and stiffness conditions in the real engine-transmission system, thereby reproducing the correct vibrational behavior. Figure 2.77 shows a setup with an engine mounting system, a dummy gearbox and a shaft.

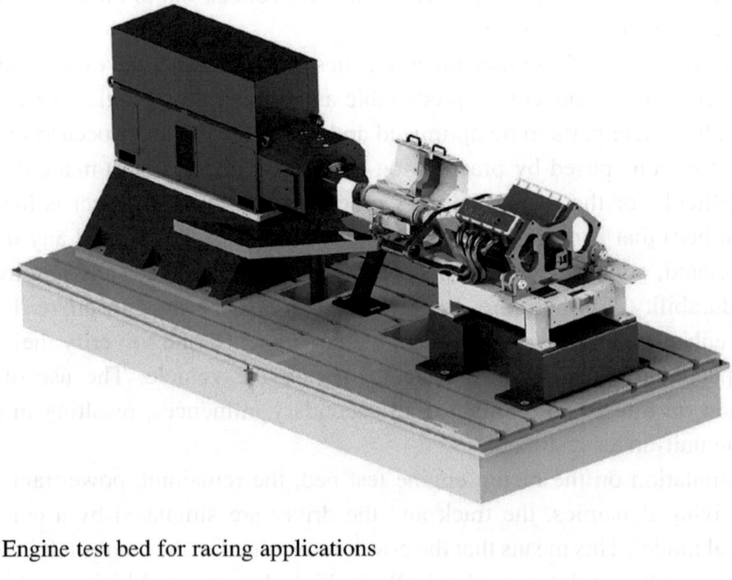

Fig. 2.77 Engine test bed for racing applications

In modern racing transmissions, gears can be changed without interruption of traction. In a Formula One car, gears are shifted even without reducing the internal combustion engine's torque. So while gear shift simulation places extraordinary demands on the testing facility, all of the other operating modes require little more performance of the electric load unit than what is provided by the internal combustion engine. As the gear-shifting process takes barely a fraction of a second, the test systems are designed with high overload capacity that is largely supplied electrically by an intermediate circuit in order to keep the required connected load on the power supply to a minimum. Characteristic values for designing load units for the Formula One are a rating of 800 kW power output for the dynamometer, but more than twice as much power output for short-term overload.

2.6.2 Component Test Beds for Racing

Component test beds for racing are of major significance primarily due to the regulations governing the individual classes in motor sports. Regulatory motor sport organizations, such as the FIA or the NASCAR, want to make races exciting events for the spectators. This is achieved when as many contestants as possible compete at more or less the same level.

The reasons for defining technical regulations are twofold:

- They are supposed to prevent technical advantages that might cause individual teams to dominate the racing track.
- The minimum budget required for a competitive vehicle should be kept within fixed limits to allow enough teams to compete in the relevant racing series for a certain period of time. The top teams, on the other hand, must be able to run a successful organization with their income from prizes, sponsoring arrangements and merchandizing.

It is for these reasons that individual components, materials or technologies are periodically excluded from the regulations. Teams, understandably, tend to shift much of their development focus to components where a certain degree of freedom is permitted and whose structural or functional improvement can lead to a competitive advantage in racing. Apart from that, the complete engine test bed is unsuitable for detecting the optimization potential of certain components because particular sensors are often not available and certain measuring points are very difficult to access. Ultimately, cost optimization is another major factor in component development. All of this results in an increasing demand for component test beds for racing applications.

The increased interest in electrifying the racing classes, in particular, is also driving the development in the area of electric and hybrid technologies. The demands for achieving maximum performance with minimum mass and the most compact design are pushing technological developments in a certain direction. The automotive industry derives no direct benefit because both durability and cost-efficient production are not in the foreground. Even so, technologies proven on the racetrack have frequently been passed on to the mainstream sector where they have been employed to benefit series development significantly. Prime examples for this are the active chassis technology or traction control systems. When it comes to electrification, it is assumed that, once properly adapted, fundamental insights will also find their way into tomorrow's series vehicles.

The Formula One's regulations for the 2014 season, for example, permitted not only the use of a motor generator unit to boost power for short periods or recuperate energy generated under braking but also a motor generator unit connected to the turbocharger.

This has several advantages:

- The turbocharger can be engineered as large as required, and dimensioned for optimal full load. In partial load or, most importantly, at load alteration the electric motor speeds up the turbocharger and prevents turbo lag by providing a short energy boost. This eliminates the need for a bypass valve (i.e. wastegate) or a pop-off valve to regulate boost pressure because the electric power unit connected to the turbocharger takes over this job. This measure improves the thermodynamic efficiency of the internal combustion engine while generating electric energy at the same time.
- The motor generator unit connected to the turbocharger is additionally capable of generating power from the exhaust mass flow and passing it on to the electric machine that is directly coupled to the crankshaft as additional source of mechanical energy, which is delivered to the powertrain. Despite the fact that the measure reduces exhaust gas back pressure (and thus combustion power) and that energy conversion will always come with losses, the procedure results in optimal total drive power transmitted to the powertrain.

In light of these considerations, the need for component testing becomes clear. Especially for providing a realistic test procedure for the electric motor connected to the turbocharger, highly complex simulation models are required because turbocharger speed is determined by the torque from the electric motor and the exhaust mass flow. The exhaust mass flow, on the other hand, depends on boost pressure (= turbocharger speed) and combustion. This immediately results in an implicitly closed and very complex system that contains at least the thermodynamic and gas dynamic components of the turbocharger, the intake path, the cylinder and the exhaust gas path.

Figure 2.78 outlines the topology of the current Formula-1 hybrid technology while Fig. 2.79 shows a picture of a test facility for testing the two electric drives based on this integrated engine simulation concept

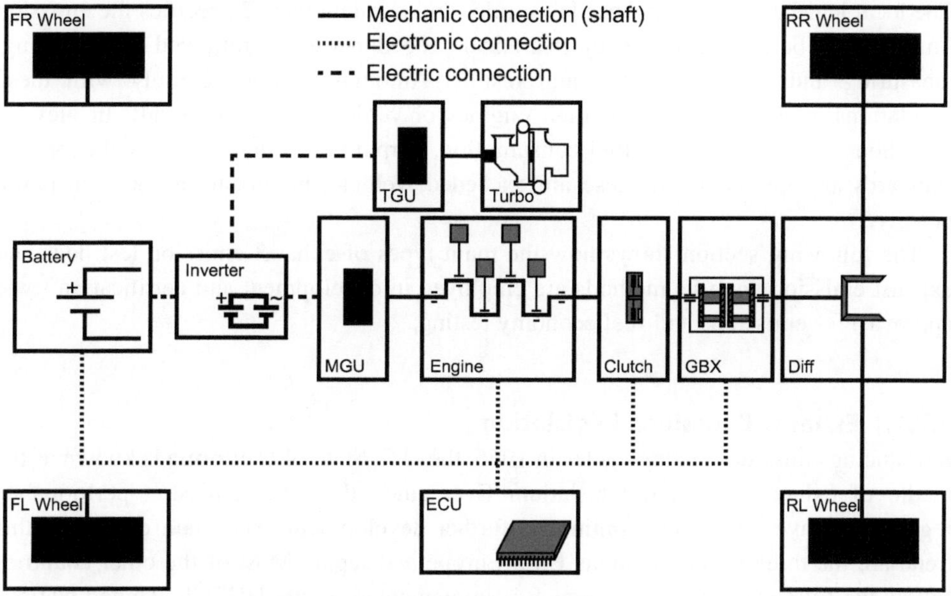

Fig. 2.78 Formula 1 hybrid technology

Fig. 2.79 Formula-1 hybrid test bed

2.7 Emission Test Beds

2.7.1 Overview

Huge efforts of auto manufacturers are required to ensure environmental compatibility in the new engines and vehicles they develop. Such requirements include compliance with the relevant country's exhaust emission, greenhouse gas and fuel efficiency regulations.

The local legislator enforces these laws, rules and regulations and specifies the emission limits not to be exceeded and the technical procedures to be followed when testing, measuring and calculating the final results. Automakers must comply with these regulations, if they wish to sell their engines or vehicles (type approval). In view of greenhouse gas emissions or fuel consumption, corporate average (fleet) fuel economy standards are applicable. If these are exceeded, vehicle manufacturers have to pay a penalty.

The following section shows how the main types of exhaust emission test beds and exhaust emission analysis methods are employed in development and certification (type approval) for emissions and fuel economy testing.

2.7.1.1 Exhaust Emissions Legislation

Introducing emissions testing cycles in 1966, the U.S. State of California is known as the cradle of exhaust emission legislation. Ever since then, the emission performance regulations have undergone continuous further development. The main drivers of this trend are the three economic areas U.S., Europe and Japan. Most of the other countries follow the European legislation, which is implemented by the UNECE (United Nations Economic Commission for Europe). Certain regions combine these statutory regulations, for example in South Korea, Brazil and, in all likelihood, also in China in the near future.

Most of the statutory regulations cover the following emission sources:

- Exhaust emissions from the tailpipe (internal combustion engine and exhaust aftertreatment system)
- Crankcase emissions
- Evaporative emissions from the fuel tank

Depending on the local legislation applicable, manufacturers have to achieve compliance with the specified limits for pollutants under certain conditions:

- Low temperatures (−7 °C), ambient temperatures (20–30 °C) and high temperatures (+35 °C)
- Various velocity profiles representative of different regional traffic conditions
- Different levels of altitude

The core elements of effective exhaust emissions legislation are:

- Definition of maximum emissions in a prototype vehicle (type approval)
- Testing of compliance with limits in durability runs (e.g. in Europe the durability requirement is 160,000 km)
- Verification of compliance with exhaust emissions limits in production by statistical methods (Conformity of Production, COP)

- Self-diagnostic capability of detecting deterioration or malfunctions occurring in individual emission-related systems and sensors in the vehicle (on-board diagnostic system, OBD)
- Regular checks of the state of each vehicle in use
- Modern legislations also check compliance with the defined limits in real-world driving (real driving emissions or in-service requirements)
- Quality standard requirements for the sold fuel

Limits are specified for the following pollutants due to their adverse effects on health and the environment:

- Unburned hydrocarbons are regulated as either THC (Total Hydrocarbon Emissions), NMHC (Non Methane Hydrocarbons) or CH_4 (Methane).
- Nitrogen oxides (NO_x) play a major role in the formation of smog and acid rain; nitrogen dioxide (NO_2) can have adverse effects on human health.
- Considered a powerful respiratory poison, carbon monoxide (CO) interferes with the oxygen transport system in the human body.
- Known to be the cause of respiratory diseases and cancer, particulates are limited either by mass (PM) or, in addition, by number (PN).
- Greenhouse gases are controlled because of their negative impact on the climate. The exhaust gas component in this respect is carbon dioxide (CO_2) which is produced whenever fossil fuel is burned. Some standards also limit nitrous oxide (N_2O or laughing gas) and methane (CH_4).
- Several exhaust emission components are additionally subjected to limits depending on the fuel or exhaust aftertreatment system being used. These include, for example, formaldehyde or alcohol if bio-fuels or mixtures are being used in the U.S. or ammonia (NH_3) if SCR exhaust emission aftertreatment systems (Selective Catalytic Reduction) are being used.

2.7.1.2 Units Under Test

Meanwhile exhaust emission standards have been issued for almost all internal combustion engine applications, starting with miniature engines as they are found in chain saws, for example, to any kind of motor vehicle, ranging from mopeds to heavy duty commercial vehicles, agricultural and construction machinery, rail vehicles, ships and large industrial engines. Basically, one can divide this broad variety of applications into three categories.

- Light-duty (LD) vehicles: They are mainly used for transporting people and range from two- and three-wheelers to passenger cars and light commercial vehicles (light trucks). In this category, it is always the complete vehicle that is examined and tested. Testing is performed on a chassis dynamometer (see Sect. 2.5) and the emission limits are defined in mass of emitted gas per unit of distance (g/km or g/mi).

– Heavy-duty (HD) commercial vehicles: This group comprises trucks and buses. Only the engine and the exhaust aftertreatment system are examined and tested in this category. The testing is performed on the engine test bed and the exhaust emission limits are defined in mass of emitted gas per unit of energy (g/kWh).
– Non-road machinery: Non-road engines and equipment comprise the broad range of all engine applications that are not intended for road transportation such as chainsaws, lawnmowers, emergency power generators, excavators, snowmobiles, construction machinery, agricultural vehicles and machinery, locomotives, marine engines and even large industrial engines for producing energy. In these vehicles and machinery, the only components examined and tested are the engine and the exhaust aftertreatment system as in the category above. The testing is performed on the engine test bed and the exhaust emission limits are defined in mass of emitted gas per unit of energy (g/kWh).

2.7.2 Exhaust Emissions Testing for Passenger Cars on the Chassis Dynamometer

Motorcycles (two- and three-wheelers), passenger cars and light-duty commercial vehicles are usually certified on chassis dynamometers. The amounts of emitted pollutants are measured during predefined test cycles. Normally a cold start is simulated followed by a driving schedule. Chassis dynamometer systems simulate the driving resistance, as it is encountered on the road, which enables the simulation of realistic driving (Fig. 2.80).

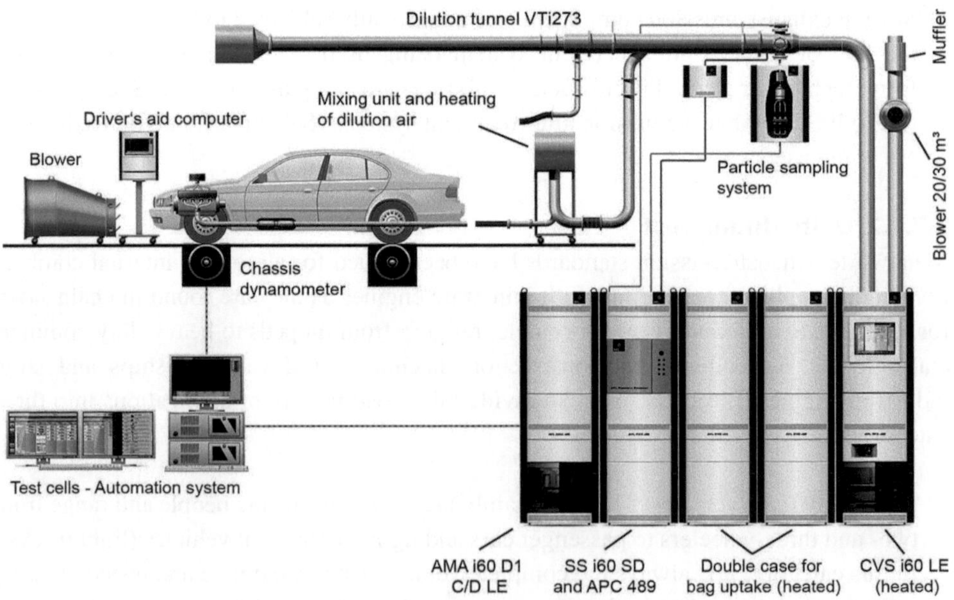

Fig. 2.80 Illustration of a chassis dynamometer with CVS exhaust dilution system for diesel and gasoline vehicles

2.7.2.1 Challenges

An advanced vehicle with exhaust aftertreatment system produces the highest emissions during the first 10–30 s after starting. This is because the exhaust aftertreatment system does not become effective until the exhaust temperature has reached several hundred degrees Centigrade. From this point on the exhaust aftertreatment system reduces emissions produced by the engine by up to 98 %. This is the reason why the test also includes the vehicle start-up. Prior to testing, the vehicle has to be conditioned at an ambient temperature between 20 °C and 30 °C for at least 6 h without being started.

Once the exhaust aftertreatment system has reached its minimum operating temperature, the emitted harmful substances remain at very low levels, whereas greenhouse gases are produced continuously. In heavily polluted ambient conditions, the exhaust emitted by the vehicle might even contain lower amounts of harmful substances than are contained in the air taken in by the engine. These are either burned in the combustion process or removed by the exhaust aftertreatment system. Such low emission levels are, understandably, a challenging task for the measuring equipment. In addition to high-precision analysis systems, a highly sophisticated system capable of integrating the involved test bed components is needed, along with well-coordinated test bed operation and maintenance capabilities.

2.7.2.2 Workflow

A chassis dynamometer (Fig. 2.81) must be designed in such a manner that it is possible to run about one vehicle test per hour. As the unit under test is the complete vehicle, and the adaptation to the test bed can be achieved fairly quickly (i.e. connect exhaust manifold to the tailpipe, push the vehicle onto the test bed, align and fasten), numerous different vehicles can be tested consecutively. This means that all vehicle- and test run-related information, such as test parameters, used units, predefined driving cycles or calibration data, has to be instantly available to the test bed operator. A high degree of automation throughout the testing center ensures highly efficient test runs and maximum investment productivity.

2.7.2.3 Simulation Features of Test Beds

The test bed must fulfill the following requirements:

- The driver follows a selected predefined driving cycle which defines the required velocity of the vehicle over time. In doing so, the driver must ensure that the vehicle speed remains within a certain corridor of the set value trace.
- A chassis dynamometer simulates the engine load the engine would be exposed to on the road: rolling resistances, air resistances and the load needed to accelerate the vehicle are components of the simulated load.
- A blower simulates the wind in front of the vehicle to realistically reproduce the road conditions with regard to engine cooling. The vehicle's air resistance is simulated as part of the simulated load from the chassis dynamometer.

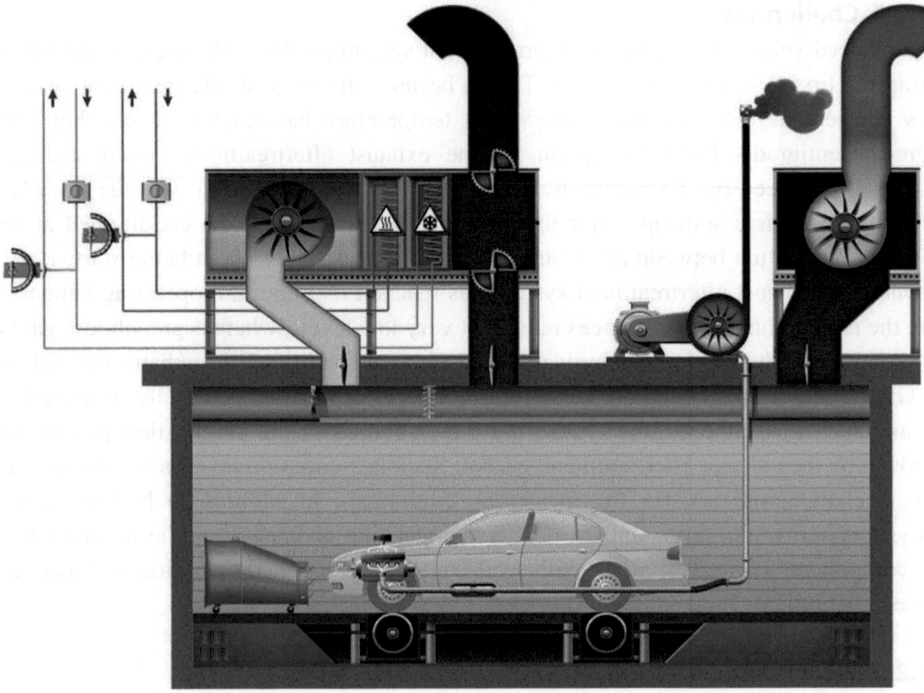

Fig. 2.81 Chassis dynamometer with air conditioning system

- The test bed's air conditioning system provides for controlled temperature and humidity conditions in accordance with the statutory regulations. Normally the temperatures for emissions testing need to be within a range of 20–30 °C, which can frequently be achieved by a simple air conditioning system. The new European legislation requires a demand temperature of 23 °C, though, and any deviations are only accepted as fluctuations around that demand value. As a result, all test beds now require full air conditioning. Apart from that, some national laws even require that exhaust emissions measurement be conducted at low temperatures (e.g. 10 °C or −7 °C) or at higher temperatures, which might even include solar radiation simulation (e.g. for the U.S.), to test the exhaust emission behavior of vehicles with active air conditioning systems.

2.7.2.4 Vehicle Preparation

The actual exhaust emission measurement is conducted after the vehicle has been brought to a defined initial condition:

- The vehicle's fuel tank is filled with a defined quantity of an exactly specified fuel. This step is also needed when preparing measurements of evaporative emissions.
- The evaporative canister for fuel vapors is filled.

- The vehicle drives several kilometers to ensure that everything is properly run in. Prior to the measurement, precisely defined test cycles are executed on the chassis dynamometer to pre-condition the vehicle.
- Immediately before the test, the vehicle is conditioned at a minimum temperature of 20–30 °C (future European regulations will require a temperature of 23 °C). During this period, it is not allowed to make any changes to the vehicle or switch on the ignition. The ignition must not be actuated until the actual start of the test. By switching on the ignition, the fuel pump would go on and this could result in start emissions different from those produced by a fuel line that was not completely full, as might be the case in real life if you started a vehicle early in the morning.

2.7.2.5 Performing an Emission Test

Starting the vehicle is already part of the emission test. The driver (or driving robot) drives the vehicle in accordance with the driving cycle required by the selected regulation. The cycle is designed to reflect an average drive and is represented by a velocity profile over time. In accordance with the European emission standards, the NEDC (New European Driving Cycle), which is to be replaced by the WLTP (World Harmonized Light Vehicles Test Procedure) in 2017, is applied in Europe. The U.S. emission regulations require that the FTP-75 (EPA Federal Test Procedure of 1975) be applied (see Figs. 2.82, 2.83, 2.84). How precisely and realistically the driving cycles must be reproduced is largely defined in the respective regulations (e.g. how the gas pedal must be actuated), but some minor deviations are allowed.

Throughout the test procedure (with the exception of the 10-min soak break during the FTP-75) the exhaust emissions are sampled continuously to determine the total mass of the individual pollutants emitted during the test. These total masses of emissions are then represented as units of mass per distance covered and output as g/km or g/mile (or as

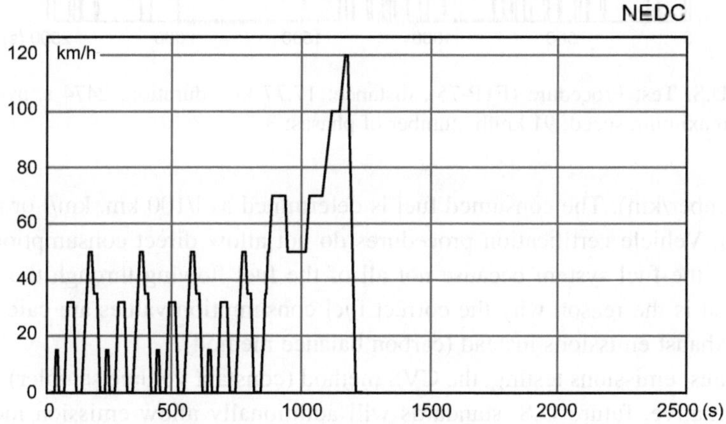

Fig. 2.82 New European Driving Cycle (NEDC), distance: 11.007 km, duration: 1180 s, average speed: 34 km/h, maximum speed: 120 km/h, number of phases: 2

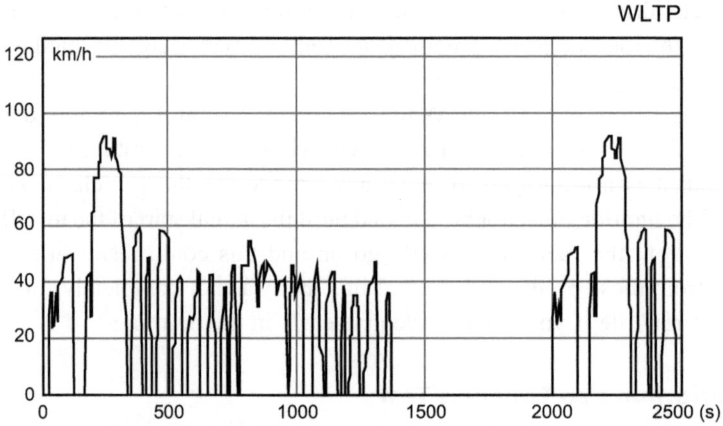

Fig. 2.83 World Harmonized Light Vehicles Test Procedure (WLTP, Class 3a), distance: 23.2 km, duration: 1800 s, average speed: 46 km/h, maximum speed: 125 km/h, number of phases: 4

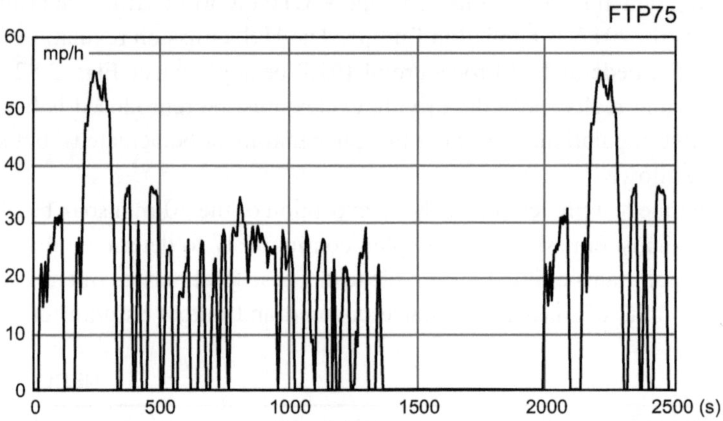

Fig. 2.84 U.S. Test Procedure (FTP-75), distance: 17.77 km, duration: 2474 s, average speed: 34.1 km/h, maximum speed: 91 km/h, number of phases: 3

particle number/km). The consumed fuel is determined as l/100 km, km/l or mpg (miles per gallon). Vehicle certification procedures do not allow direct consumption measurement inside the fuel system because not all of the fuel flowing through the fuel line is burned. That is the reason why the correct fuel consumption values are calculated back from the exhaust emissions instead (carbon balance method).

For exhaust emissions testing, the CVS method (constant volume sampler) is required. As an alternative, future U.S. standards will additionally allow emission measurement based on partial flow dilution or undiluted (raw) emissions (CFR-1066). Such alternatives are expected to remain merely alternatives for a long time and be accepted in the

U.S. only. Further details on exhaust emissions testing can be found in Section Exhaust Emission Measurement.

2.7.2.6 Development Work

A chassis dynamometer is usually not only used for certification testing but also for development and vehicle calibration tasks. Chassis dynamometers are therefore equipped with additional measuring instruments:

- Multiple-line exhaust emission analyzing systems: These systems are used for measuring raw exhaust gas concentrations upstream and downstream of the exhaust aftertreatment systems. Such measurements always require the determination of modal emissions too. Modal emissions are the individual emission masses calculated throughout the test in g/s, which directly relate to a vehicle operating mode (e.g. acceleration, deceleration, cruising).
- FTIR exhaust emission analyzing systems (Fourier Transform Infrared Spectrometer): These instruments can, for example, measure the reduction components from the urea solution upstream of the SCR emissions control system (Selective Catalytic Reduction) or allow to determine a variety of (currently) non-limited exhaust gas components. This might be done for environmental reasons or also to gain a better understanding of the processes occurring inside the engine or the exhaust aftertreatment system.
- Additional particulate measuring devices: These include soot sensors (e.g. Micro Soot Sensors MSS) and opacimeters and help to gain a better understanding of particulate emissions.

A relatively simple way to measure modal emissions is to determine the exhaust emission concentrations from the diluted exhaust gas. The process to determine modal emissions from raw exhaust gas is somewhat more complex. But this is necessary when measuring upstream of an exhaust aftertreatment system as this requires the continuous sampling of the engine exhaust gas volume or mass flow determination. The emitted masses are multiplied by the time-aligned engine exhaust gas flow and the respective exhaust gas component density. To determine the engine exhaust gas volume or mass flow, there are several possibilities:

- CO_2 tracer technique: In conjunction with a CVS system, the dilution ratio of the CVS is determined by the CO_2 tracer technique. By adopting this approach, the dilution ratio is determined by comparing the CO_2 concentrations in the raw exhaust with the CO_2 concentrations in the diluted exhaust. Based on the CVS flow rate (diluted exhaust gas) and the dilution ratio, the exhaust gas volumetric flow is calculated. During phases in which the engine is not running, such as overrun fuel cutoff, automatic start-stop operation or in hybrid vehicles, the engine emits no CO_2, so the tracer method is not applicable during these intervals.

- Together with a CVS system, an additional measurement of the dilution air flow into to the CVS can be performed in order to determine the exhaust flow from the engine, which results from the difference between the CVS total flow rate and the dilution air flow rate. However, calculating the difference between two relatively high measurement values to obtain a relatively small figure can rapidly lead to large errors due to propagation of error.

A better method than the indirect measuring techniques addressed above would be to measure the engine's exhaust flow directly. Certain methods are common practice on the engine test bed, but less frequently used on the chassis dynamometer. With some methods, the instruments used are too imprecise or insufficiently robust. Further measuring methods are:

- Intake air and fuel consumption: These masses supplied to the engine are equally emitted by the engine and not affected by combustion (principle of mass conservation). Although standard on the engine test bed, the method is not used on the chassis dynamometer because of its somewhat complex application for the overall vehicle.
- Intake air and air/fuel ratio (lambda) of combustion: This method is relatively uncommon and is also rather difficult to apply on the vehicle.
- Fuel consumption and air/fuel ratio (lambda) of combustion: This method, too, is rather uncommon, as it is just as difficult to apply on the vehicle as the method mentioned previously.
- Exhaust flow signal from the engine control unit: This method is used in mobile measuring equipment in on-road tests, but not on test beds. If it is used on the road, it additionally requires a check using a test-bed method to verify accuracy.
- Direct exhaust flow measurement at the tailpipe: This method is suited well for the chassis dynamometer and is occasionally employed. With view to aspects such as accuracy, robustness and costs, however, the systems currently available are not yet sufficiently mature for widespread usage.

2.7.2.7 Hybrid Electric Vehicles (HEVs)

In view of exhaust emissions testing, hybrid electric vehicles can be divided into two categories. The first category covers vehicles that do not rely on any external battery charging facility. These are tested in the same way as conventional vehicles would be. The only additional requirement is the determination of the battery's state of charge (SOC). If, after a test run, the state of charge is identical to what it was before the test run, the test results are used as they were actually measured. If the state of charge is different from what it was before the measurement, this difference is incorporated into the results by employing a mathematical method. If, for example, the battery contains less energy after the test than it did before, energy consumption is extrapolated with the aid of mathematical models because the battery will have to be recharged sometime in the future. Recharging is accomplished by driving and this consumes fuel.

The second category comprises hybrid electric vehicles that can also be charged from an external facility. These are frequently referred to as plug-in hybrid electric vehicles (PHEVs). Being considerably more complex in their technology, testing such vehicles requires a significant amount of extra work. The regulatory requirements define additional measurements:

- Determination of the range that can be achieved in all-electric operation
- Determination of the amount of electric energy that is needed to recharge the battery after the entire all-electric distance has been covered
- Determination of fuel consumption with fully charged and depleted battery
- Determination of emissions; the emission limits have to be met in all tests, regardless of whether the battery is fully charged or depleted

For hybrid electric vehicle testing, extra equipment is needed to determine the battery's energy balance and to detect the on/off phases of the internal combustion engine. Integrating two different drive concepts requires attention to further rules. It is necessary to ensure that the measured values are always correct, independent of the operating conditions that involve e-motors and combustion engine.

2.7.2.8 Evaporative Emissions

Aside from the well-known exhaust emissions produced by the combustion process and emitted by the tailpipe, a vehicle also emits evaporative emissions from the fuel system. Especially when using gasoline, which may even evaporate at ambient temperatures, fuel vapors (THC—total hydrocarbons) are emitted to the environment from unpressurized fuel tank systems whenever the fuel tank is heated up, e.g. when a car is parked in the hot sun. Without suitable countermeasures (THC control systems or evaporative emission control systems) these emissions would be much higher than the THC emitted from the tailpipe of modern vehicles.

To reduce these emissions, vehicles have built-in THC control systems, for example carbon canisters. These keep the fuel vapors from escaping the tank together with the air-fuel vapor mixture and are occasionally purged against flow direction when the engine is running. In addition, the fuel vapors purged back into the engine are burned. This means, they are prevented from escaping into the environment and are returned to the engine as fuel. According to what has been calculated by the German Federal Environmental Agency, this contribution to fuel savings roughly translates to 6–7 l of fuel per year and vehicle.

In vehicle exhaust emissions testing, evaporative emissions testing is an equally significant part of the exhaust emissions legislation. The test also involves measuring THC emissions resulting from other sources too, e.g. from paints, underbody coating, lubricants, tires, plastics or rubber parts.

Similar to exhaust emission tests, evaporative emission tests are very much dependent upon the entire vehicle conditioning process (conditioning duration, ambient conditions, fuel quality, etc.) and are essential for meeting the limits defined in g/test.

There are currently four types of tests in use, all of which are conducted in conjunction with test runs on emission chassis dynamometers. In terms of pre-conditioning, these tests are essential for the subsequent evaporative emissions testing.

- Parking/day simulation (diurnal emission test): The vehicle is parked for 24 h. During this period, the fuel is exposed to certain temperature cycles.
- Hot-soak simulation: Following a test run (cold start cycle), the vehicle is stopped on the chassis dynamometer; the hot soak phase takes 1 h.
- Fuel dispensing spitback test (ORVR [Onboard Refueling Vapor Recovery] + spitback): Fueling simulation at the fuel station with (active) or without (passive) vapor recovery
- Driving simulation (running loss test): The evaporative emissions are determined on the chassis dynamometer while an exhaust emission test is being conducted. For this purpose, the chassis dynamometer is integrated in a SHED chamber (Sealed Housing for Evaporative Determination) (see Fig. 2.85). As an alternative, it is also possible to use the point-source method, according to which the emissions are only measured at vehicle points at which evaporative emissions typically occur.

Particular test systems are needed to perform evaporative emission tests:

- SHED testing chamber in which the vehicle is tested. During testing in this gastight chamber, the temperature is controlled according to specified profiles. The SHED system requires a volume compensation system to compensate for any increase in pressure (VV/VT SHED [variable volume, variable temperature]). An unrealistic

Fig. 2.85 SHED testing chamber (Sealed Housing for Evaporative Determination) in which a vehicle's evaporative emissions are determined

increase in pressure inside the enclosure would otherwise cause the evaporation process to appear different from the way it would in reality.

- Gas analysis system for THC measurement
- Canister loading system required to condition the vehicle's built-in carbon canister. For reasons of efficiency, the conditioning is performed outside of the vehicle and not inside.
- Fuel conditioning system: It is required to ensure that the vehicle is supplied with fuel conditioned to the required temperature.
- RL SHED chamber for driving simulation test (running-loss SHED). This is accomplished by combining a SHED with a chassis dynamometer system.
- Point-source system as an alternative to a RL SHED testing chamber when using a normal chassis dynamometer
- Automation system

2.7.3 Exhaust Emissions Testing for Commercial Vehicles

Unlike passenger car testing (light-duty vehicles), which is done on complete vehicles, emissions testing and certification testing on medium-duty and heavy-duty commercial vehicle (trucks and buses) is done on the engine and the exhaust aftertreatment system only. The testing is performed on the engine test bed and the exhaust emission limits are defined in mass of emitted gas per unit of energy (g/kWh).

2.7.3.1 Test Bed Equipment and Exhaust Emissions Testing Methods

Automakers are allowed to apply two basic methods for measuring exhaust gas emissions. The full-flow dilution method (CVS method) involves diluting the entire exhaust mass flow with ambient air. Measuring the gaseous exhaust components is normally done using exhaust gas bags into which some of the diluted exhaust gas is sampled during the test; it is also possible to sample the diluted exhaust gas continuously and directly. When testing diesel engines, it is mandatory to measure the total hydrocarbons (THC) and nitrogen oxides (NO_x) in the diluted exhaust gas. Particulate emissions (mass and, depending on the relevant regulations, also particulate number) are also measured in the diluted exhaust gas.

A second permitted method involves the determination of the gaseous emissions in undiluted (raw) exhaust gas and particulate emissions by using a partial-flow dilution system. When using this approach, the concentrations of gaseous exhaust components are measured in the raw exhaust gas and the mass emissions determined mathematically based on the engine's exhaust mass flow which is generally derived from fuel consumption and intake air mass. The individual measurement quantities must be sampled with sufficient time resolution to be able to calculate correct results. For measuring the particulate mass and particle number, a proportional partial flow is sampled from the engine's exhaust mass flow, appropriately diluted and drawn through the particulate analysis filter to the particle number counter. The space and power requirements of the partial-flow dilution method

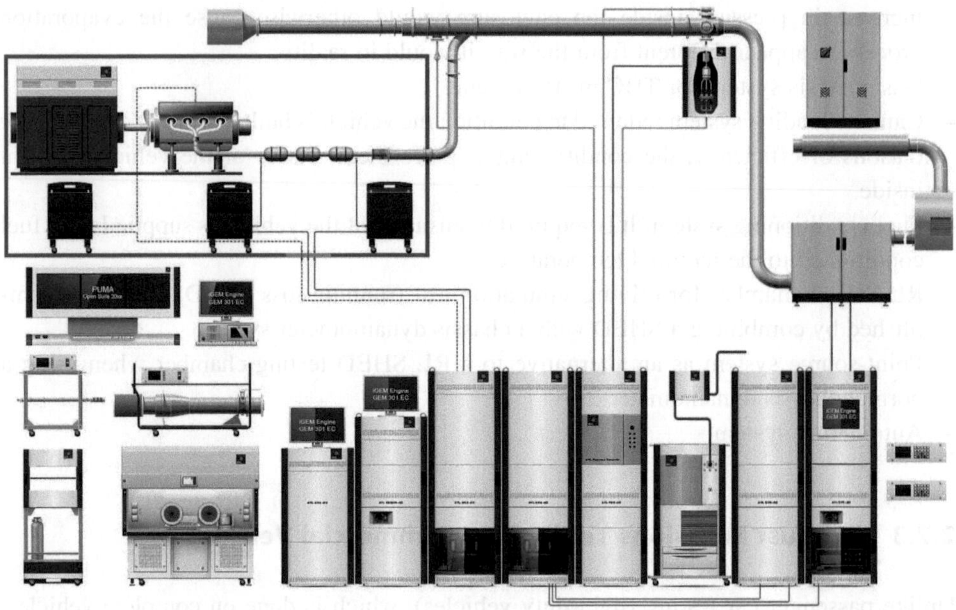

Fig. 2.86 Commercial vehicle engine test bed with raw exhaust analysis and diluted exhaust emissions measurement system

are much lower, although dilution control and result calculation are more time-consuming. The way in which it is applied makes the partial-flow dilution method more like an undiluted measurement method than a full-flow dilution method (Fig. 2.86).

2.7.3.2 Test Cycle

According to the currently applicable standards, exhaust emissions testing is conducted either only in transient engine test runs or, as is the case in Europe, both in steady-state and transient test runs. The transient tests are executed twice: once by starting with a cold engine and once with a hot start. The results are averaged after being adjusted by weighting factors. The employed speed and load profiles—these are normalized and defined in the regulations—are converted into a speed and torque profile that fits the unit under test based on the tested engine's full load curve.

The currently most important test runs are:

- WHTC (World Harmonized Transient Cycle) in accordance with UN-ECE GTR-4: two phases, 1800 s each with a 10-min soak period between phases
- WHSC (World Heavy Duty Stationary Cycle) in accordance with UN-ECE GTR-4: 13 different operating points (modes), total duration: 1895 s
- USHDTC (US Heavy Duty Transient Cycle) in accordance with 40 CFR Pt.86: two phases, 1200 s each with a 20-min soak period between phases. There are different test profiles for gasoline and diesel engines.

To convert the demand speed and demand torque curve defined as the percentage of a reference value, the engine's latest full-load curve is used. With the help of the full-load curve's different characteristics, the normalized driving profile is converted into a speed/torque curve. Upon completion of each phase, the demand and actual speed, torque and power values are validated statistically by applying a regression analysis. This establishes whether the engine was operated in compliance with the standard's requirements and whether the test is valid or invalid.

2.7.3.3 Certification Testing Procedure

All of the mentioned test cycles have the following fundamental emissions certification procedure in common:

- Full-load curve recording
- Engine and exhaust system conditioning (generally more than 6 h)
- Conditioning to start temperature (engine standstill)
- Cold-start test (engine temperature at start: 25 °C)
- Pause (hot soak)
- Hot start

2.7.4 Exhaust Emissions Testing for Non-Road Engines

Similar to medium- and heavy-duty commercial vehicles, vehicles not designed for on-road use require only an analysis of their engine's exhaust emissions and exhaust aftertreatment system behavior. The testing is performed on the engine test bed and the exhaust emission limits are defined in mass of emitted gas per unit of energy (g/kWh).

Non-road engines (off-road engines) include internal combustion engines for the following applications:

- Tractors
- Mobile industrial machinery (e.g. excavators)
- Stationary power generators (all sizes)
- Lawnmowers, chainsaws and other portable/handheld equipment
- Engines for snow mobiles and other recreational vehicles
- Marine engines of all sizes
- Locomotives

Due to the broad scope of applications for non-road engines, the speed and load ranges in which the engines are operated differ very widely. It is therefore hardly surprising that both test cycles and test beds differ just as widely as the involved engine sizes.

2.7.4.1 Test Bed Equipment and Measuring Methods

For testing this engine category (with the exception of small and miniature engines), test beds are generally used that utilize emissions testing technology based on raw exhaust measurement for gaseous emissions and partial-flow dilution for particulate mass. If small and miniature engines require testing, CVS measuring technology is once again the best choice. The reasons are primarily down to the facts that the measuring system is less likely to become contaminated and the impact on the unit under test is minimized.

2.7.4.2 Test Cycles

Due to the applications' very different utilization characteristics, the following basic types of test cycles are distinguished:

Steady-state cycles consist of a different number of speed/load points that are measured separately. The measurement is conducted either for each load point separately with a subsequent mathematical weighting or as a single continuous measurement throughout the entire test with weightings for time. The choice of speed/load points and their weighting reproduces an engine's load profile in the form in which it occurs in each application. For example, the points of an engine for ship propulsion correspond to the speed/load of a ship's propeller. Also, large power generation engines, for example, are only measured in idle and at a single speed because such engines always only run at one speed which corresponds to the power line frequency (50 or 60 Hz).

A further test cycle that is used is the NRTC (Non Road Transient Cycle) in accordance with GTR-11. This one comprises two phases with 1238 s each and a 20-min soak period between the two phases. This cycle does not represent a realistic engine application, but covers phases for many different kinds of applications such as tractors, forklifts, etc.

To convert the demand speed and demand torque curve defined as the percentage of a reference value, the engine's latest full load curve is used. With the help of the full-load curve's different characteristics, the normalized driving profile is converted into a speed/torque curve.

2.7.4.3 Certification Testing Procedure

The test procedure is basically very similar to that described in the previous section for commercial vehicle engines. Depending on which driving cycle is to be executed, it might also be possible to omit the cold start and start the driving cycle with a warmed-up engine.

References

1. H. Eichelseder, M. Klell, Brennstoffzelle, in *Wasserstoff in der Fahrzeugtechnik* (Vieweg +Teubner Verlag|Springer Fachmedien, Wiesbaden, 2008), pp. 139–160
2. J. Schäuffele, T. Zurawka, *Automotive Software Engineering* (Springer, Wiesbaden, 2013)
3. K. Reif, *Bosch Autoelektrik und Autoelektronik* (Vieweg+Teubner Verlag|Springer Fachmedien, Wiesbaden, 2011)

4. ETAS GmbH, LABCAR-Komponentenübersicht: Modulares HiL-Testsystem für automotive Steuergeräte, 2014. [Online]. http://www.etas.com/de/products/applications_ecu_development-testing-labcar_component_overview.php. Accessed 28 Apr 2014
5. T. Dicke, M. Kreher, R. Schieber, R. Merl, W. Beer, *Hardware-in-the-Loop-Simulation als feste Größe im modernen Motorapplikationsprozess* (Haus der Technik, Munich, 2007)
6. ASAM e.V., ASAM Standards, 2013. [Online]. http://www.asam.net/nc/home/asam-standards.html. Accessed 28 Apr 2014
7. International Organization for Standardization, *ISO 15031-6—Road vehicles—Communication Between Vehicle and External Equipment for Emissions-Related Diagnostics—Diagnostic Trouble Code Definitions* (Beuth, Berlin, 2010)
8. S. Köhl, M. Rühl, J. Klahold, K. Krügel, Steuergerätetest im HiL-System. ATZ elektronik **7**(4), 290–293 (2012)
9. S. Albrecht, P. Decker, Schnelle Wege zur Restbussimulation. Automobil Elektronik **10**(3), 58–60 (2012)
10. T. Düser, R.V. Gravel, A. Haase, H. Olms, C. Schmidt, U. Schmidt, *Fahrzeugrollenprüfstände—Von der Zertifizierung bis hin zur mechatronischen Entwicklungsplattform* (Süddeutscher Verlag onpact GmbH, Munich, 2011)

3 Hardware Perspective

Following the presentation of the different types of test beds in Chap. 2, this chapter will continue by addressing each of the layers that make up the general architecture of a test bed (see Fig. 3.1).

We generally distinguish between the following four layers:

- Physical layer: sensors, actuators/conditioning systems, intelligent systems (e.g. control units, complex measuring systems, etc.)
- Connection layer between the elements of the physical layer and the automation system: analog connections, bus connections (CAN bus, PROFIBUS, EtherCat, PowerLink, ProfiNet, FlexRay, etc.—see Sect. 3.5), serial communication (RS232, RS485, EtherNet—see Sect. 3.6), parallel communication
- Automation layer (see Chap. 4)
- Data storage, data processing and process management layer across the test facility (see Chap. 5)

Test beds that are designed for testing the powertrain components of advanced vehicles face a variety of testing tasks. These differ very widely, depending on the development work performed by the engineers working on vehicle components. This is why the related automation systems must be capable of adapting to rapidly changing requirements. It must be possible to connect new components, such as control units or intelligent measuring devices, to the test beds (see Fig. 3.2) if the related interfaces were not yet available at the time of test bed commissioning.

Given their high investment costs, it is paramount that test beds be operated with maximum efficiency. Many development centers have decided to run their test beds in shifts because capacity utilization in close to 24/7 operation has shown to keep the costs charged to customers to a minimum.

M. Paulweber, K. Lebert, *Powertrain Instrumentation and Test Systems*, Powertrain,
DOI 10.1007/978-3-319-32135-6_3

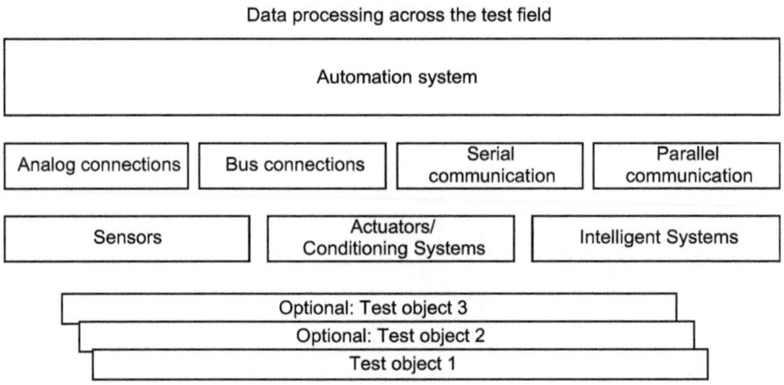

Fig. 3.1 General test bed architecture

The two key requirements outlined above result in the need for test beds with modular design. A test bed composed of standardized modules is easily expandable because, due to economies of scale, standardized modules are available at much lower cost than testing systems built for specific projects.

This chapter describes the two bottom layers in Fig. 3.1—the physical level and the connection level—above the unit under test. These layers establish the connection between the unit under test and the automation system. The first layer comprises sensors and actuators as well as intelligent systems for measuring the many different variables needed in the development process. The second layer connects these systems to the automation system layer via bus systems, analog connections or PC interfaces.

As a test bed comprises a multitude of different hardware modules, we have called this chapter Hardware Perspective. The hardware components measure the required variables in the unit under test(s) and supply them to the automation system. Hardware components (actuators or stimuli), in turn, transfer the set values from the automation system back to the unit under test.

3.1 Test Bed Mechanics

A test bed typically requires multiple mechanical components that fulfil a variety of functions. This section describes the mechanical components employed in the various test bed configurations such as internal combustion engine, powertrain, transmission, racing or exhaust emission test beds.

These components include:

- Test bed bases
- Unit-under-test mounting systems and pallet systems
- Shaft connections and safety covers

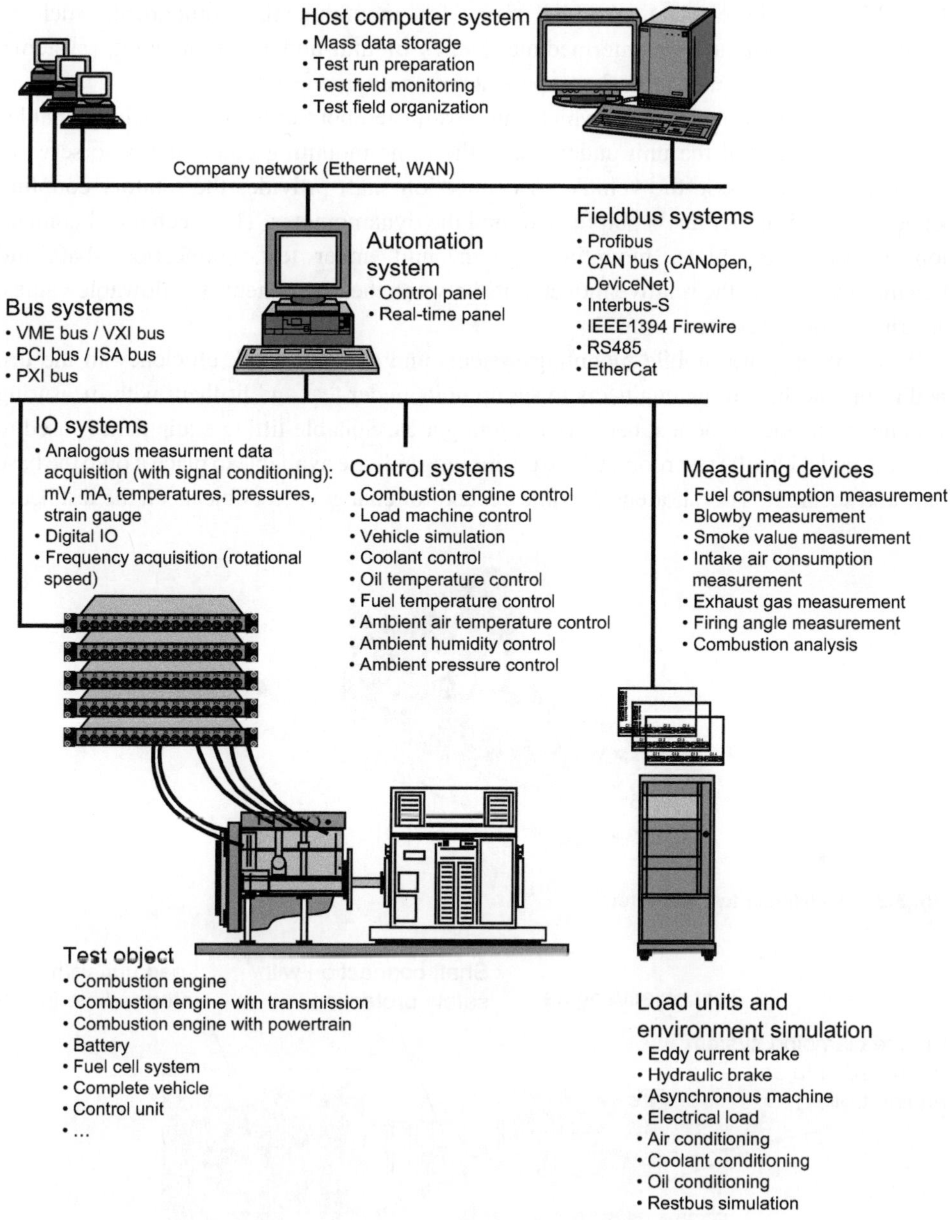

Fig. 3.2 Setup of a test bed

The test bed base is the lowest part of the test bed's substructure. The base plate, which is installed inside the test room, rests on a decoupling system that separates the test bed setup from the rest of the building, isolating test bed and building against vibration

(Fig. 3.3). The base plate provides room for mounting the components such as dynamometers mounted on intermediate frames or unit-under-test mounting systems. Figure 3.4 shows the mechanical components of an engine test bed.

The engine mounting system, which can be adjusted both in width and height, enables precise positioning of the unit under test at the same mounting points that also serve to install the engine inside the vehicle. A connection shaft provides the rotatory coupling between the unit under test's output shaft and the dynamometer. The mechanical connection between base plate, mounting system, unit under test, connection shaft and dynamometer limits the relative movements between the components to allowable values, ensuring a stable test bed setup.

Pallet systems are mobile mounting systems and enhance work efficiency in the test facility in that they allow engineers to set up units under test and fit them with measuring equipment outside of the test bed, in the rigging area. Suitable lifting equipment is used to bring the pallet into the test room, where positioning aids are available to place it onto the base plate and secure it. This shortens the time needed for changing test objects on the test bed.

Fig. 3.3 Mechanical test bed setup

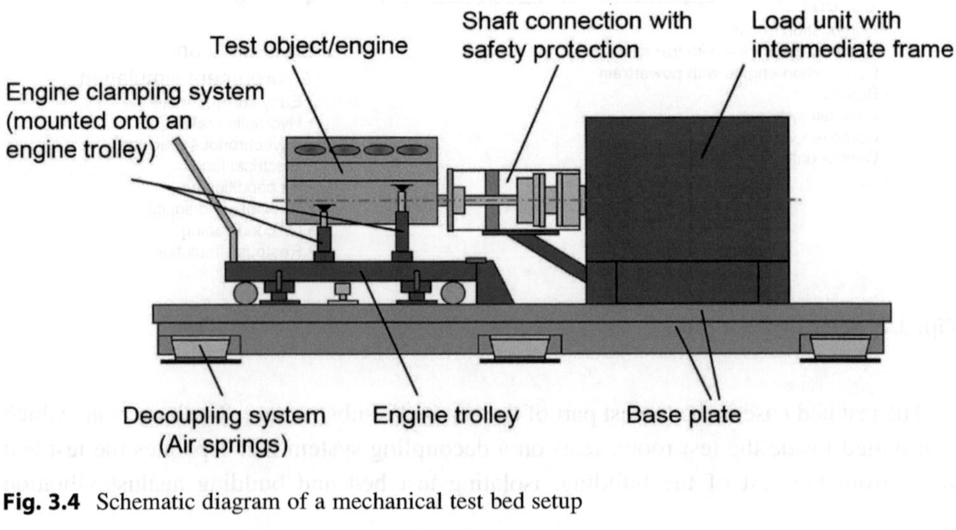

Fig. 3.4 Schematic diagram of a mechanical test bed setup

3.1.1 Isolated Base Plate

A test bed base consists of an isolated base plate and a decoupling system. The base plate allows for fast and easy setup of the test bed components. The components are mounted using T-slots or threaded screw holes that are incorporated into the base plate. The staff can walk on the base plate surface and use it to move and handle the equipment. Threaded screw holes have the advantage that they scarcely affect the surface, as only a small number of rather small holes are required. T-slots, on the other hand, are able to increase flexibility for component setup. By adding transversal T-slots, T-slots achieve an even higher level of flexibility.

A foundation pit inside the test cell provides the room needed for integrating the base plate. The pit is slightly larger than the actual base plate, allowing sufficient room for clearance. This in-between mounting clearance serves as a room for technical facilities where pipes and tubing for the media, and cable lines are installed. It also allows the staff to do commissioning and maintenance jobs on the decoupling system. Upon completion of the installation work, the in-between mounting clearance is covered up with cover plates, creating a level test bed working surface.

Any liquid media escaping the unit under test, such as oils, coolants or fuels, are collected in a drain channel running around the foundation plate and led away from the test bed for appropriate disposal.

3.1.1.1 Types of Base Plates

The choice of size and type of the base plate depends on the type and size of the unit under test and the kind of test to be conducted. The following plate types are available (Fig. 3.5):

- Gray cast-iron base plates consist of cast iron and nodular graphite (also referred to as spheroidal graphite cast iron or ductile cast iron). They are manufactured by using a sand casting technique. After the material has cooled down and solidified under controlled conditions, the surface is machined and planed before T-slots are machined or screw holes drilled.

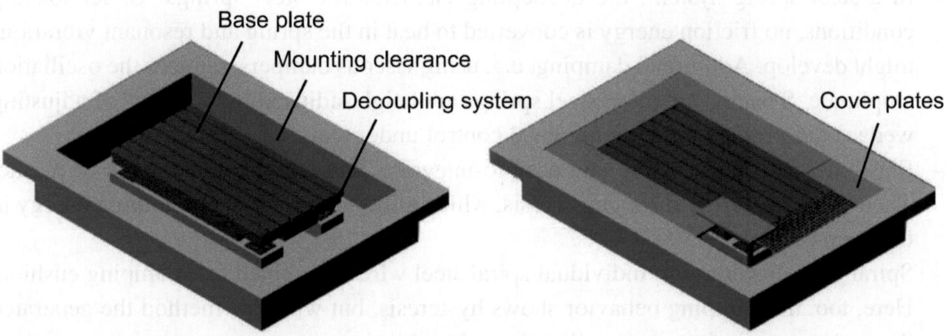

Fig. 3.5 Schematic diagram of an isolated-base plate setup

- Base plates of all-steel welded construction are produced by welding together steel frames to form a skeleton structure. Due to their low weight, this solution is simple and economical in view of manufacture and transportation.
- Concrete block base plates (seismic blocks) consist of a block of reinforced concrete with incorporated elements, e.g. steel profiles to accommodate mounting rails.
- Compound base plates are a combination of gray cast iron or all-steel welded base plate and a concrete block base plate. The substructure consists of a concrete block with a large mass. The components are set up on top of a gray cast iron plate or an all-steel welded plate. A high degree of damping (due to the large mass) and the possibility of simple component setup distinguishes this variation.

All plates made of cast iron or steel are either oiled, pickled or galvanized to protect the plate surface from corrosion.

3.1.1.2 Decoupling Systems

Due to the combustion process taking place inside the engine and the rotatory motion of individual test bed components, the test bed setup is vibrationally excited. Although these vibrations can neither be avoided nor suppressed, the appropriate measures allow them to be decoupled from the rest of the building. To this end, systems are used with isolating components that have damping properties, e.g. air spring elements or steel spring systems. As the generated vibrations and oscillations are significantly reduced, allowing only a very small portion to be transmitted to the building, their amplitude is considered unproblematic for the structures surrounding the test cell.

Depending on the intended tasks, various types of decoupling systems are used:

- Apart from having a vibration isolation function, air spring systems additionally feature an automatic level adjustment system for the base plate. Varying load distribution, moving loads or torque reversals can lead to level variations in the base plate. Level control makes sure that the base place remains in a constant, load-independent horizontal position.
- In a steel spring system, the decoupling elements are steel springs. Under loading conditions, no friction energy is converted to heat in the spring and resonant vibrations might develop. Additional damping, e.g. using viscous dampers, reduces the oscillation amplitude. Systems based on steel springs are height-adjustable by means of adjusting wedges but provide no dynamic level control under loading.
- Elastomers are inexpensive and easy-to-integrate damping elements made of plastic. Their damping effect shows hysteresis, which allows the release of damping energy in the form of heat.
- Spiral steel cushions are individual spiral steel wires combined to a damping cushion. Here, too, the damping behavior shows hysteresis, but with this method the generated thermal energy is far more easily released to the air.

3.1.1.3 Considerations Regarding Vibration

A vibration is produced by supplying energy to a system capable of oscillation. Vibration isolation means the reduced transmission of vibrations from one body to another (e.g. from a test bed to its environment or to a building). Damping refers to a reduction in vibration amplitude in an oscillatory system.

The test bed systems described above are all mass-spring systems. They form a system capable of vibration that oscillates freely at natural frequency ω_0, the mass set in motion having a single degree of freedom. The square of the natural frequency is inversely proportional to the mass.

$$\omega_0 = \sqrt{\frac{c}{m}} \tag{3.1}$$

ω_0 Natural frequency of the oscillating system
c Stiffness [N/m]
m Mass [kg]

Equation 3.1 Natural frequency

The efficiency of vibration isolation is significantly determined by η, the ratio of exciting frequency to natural frequency of the oscillating system. If the system's mass is large, η is small.

$$\eta = \frac{\text{exciting frequency}}{\text{natural frequency of oscillating system}} = \frac{f_{err}}{f_0} \tag{3.2}$$

Equation 3.2 Ratio of exciting frequency to natural frequency

Figure 3.6 shows the amplitude response (magnitude of the transmission function) of a single-mass oscillator for different damping values. The transmission coefficient Vg is plotted against the frequency ratio.

The magnification function Vg underlying the vibration isolation calculation is:

$$Vg = \sqrt{\frac{1 + 4d^2\eta^2}{(1-\eta^2)^2 + 4d^2\eta^2}} \tag{3.3}$$

Vg Transmission coefficient by taking into account damping
η Ratio of exciting frequency to natural frequency
d Damping

Equation 3.3 Transmission coefficient by taking into account damping

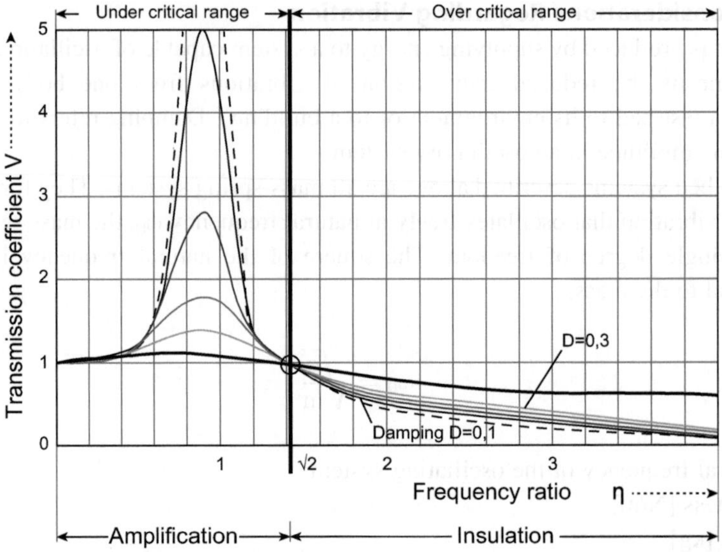

Fig. 3.6 Transmission coefficient Vg versus frequency ratio

Critical to the transmission coefficient is the natural frequency of the body to be isolated. Vibration isolation only occurs if frequency ratio $\eta > \sqrt{2}$ (overcritical range). Another factor for isolation behavior is damping. A high damping level also means a lower isolating effect. With $d = 0$, a low transmission coefficient is obtained in the isolation range, but in the magnification range a high magnification is achieved, which at $\eta = 1$ is practically infinite. This means that it is necessary to seek suitable damping for a sufficiently small transmission coefficient.

Where test beds are concerned, vibration isolation is obtained when the mechanical test bed setup is operated above its natural frequency. This means working in the overcritical range or insulation range (see Fig. 3.6).

3.1.2 Mounting Systems and Pallet Systems for Units Under Test

3.1.2.1 Engine Mounting Systems

An engine mounting system is a special version of unit-under-test mounting system. It serves to mount the engine onto a base plate (see Fig. 3.7) or a pallet. It essentially consists of a crossbeam with two stands. The crossbeam can be positioned by sliding it along the T-slots incorporated into the base plate. The fact that the stands are able to be moved along a slot in the crossbeam allows precise lateral positioning and height adjustability to fit the engine's original mounting points.

When mounting the engine onto the engine mounting system, the original engine suspension elements are used (the same ones that are used inside the vehicle). These

Fig. 3.7 Engine mounting system

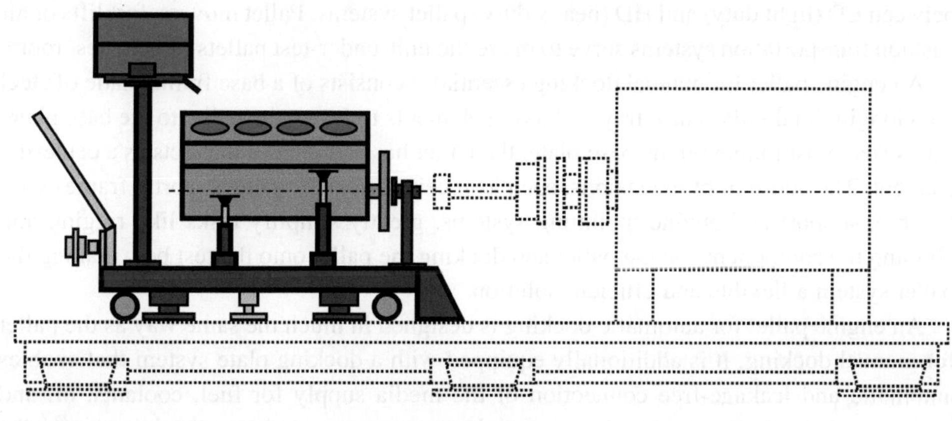

Fig. 3.8 Schematic diagram of a pallet system setup

engine suspension elements consist of metal plates with rubber parts vulcanized onto them. These rubber parts serve as initial damping devices to absorb the engine oscillations. Different types of engines require the use of engine suspension elements with varying damping properties. The engine mounting system itself must be mechanically designed to suit the unit under test's weight and the forces the dynamics induces.

3.1.2.2 Pallet Systems

Pallet systems allow you to pre-mount the unit under test (e.g. an internal combustion engine) outside of the test bed in a dedicated rigging area where the unit under test and the measuring equipment are assembled and connected on a pallet (Fig. 3.8). As pallet systems shorten the time needed for exchanging the unit under test on the test bed and make it easier to move the test object (e.g. an engine or a transmission) around the test environment, they make a major contribution toward efficiency enhancement throughout the testing operation.

Fig. 3.9 Engine trolley (on the *left*) and engine pallet (on the *right*) for manual docking

The purpose of working with pallet systems is to bring the unit under test onto the test bed in a condition ready for operation. End-of-line test beds, on which many different types of engines are tested, are consequently the main application areas. Pallet systems are additionally frequently used for testing certain units under test that take long to set up.

Pallet systems are divided into engine pallets and engine trolleys. There are different designs available for manual and automatic docking (see Fig. 3.9). Pallet systems vary by the maximum weight or size of the test object they are able to manage. We distinguish between LD (light duty) and HD (heavy duty) pallet systems. Pallet movers, forklifts or air cushion transportation systems serve to move the unit-under-test pallets into the test room.

An engine pallet for manual docking essentially consists of a base frame made of steel with longitudinal rails, a drip tray and fixing elements to fasten the pallet to the base plate. For correct positioning on the base plate, the pallet has a guide rail that acts as a centering element. The addition of extensions, such as receivers, center gauges, portal frames with test box support and engine mounting systems, greatly simplify tasks like rigging and aligning the components on the pallet and docking the pallet onto the test bed, making the pallet system a flexible and efficient solution.

An engine pallet for automatic docking is designed in much the same way as the pallet for manual docking. It is additionally equipped with a docking plate system that enables automatic and leakage-free connection of the media supply for fuel, coolants, oil and compressed air. An automatic clamping and support system helps to fix the pallet to the base plate. A control unit to process individual tasks complements the system. When docking automatically, the engine pallet is transported, e.g. by forklift, to the base plate on which it is placed. The pallet is aligned by pre-centering and then pushed into its final position by hand. Having reached its final position, it is automatically secured by pneumatic clamping elements after activating the relevant controls on the control unit. Pushing another button on the control unit automatically connects the media. Next, the connections to the shaft and the exhaust gas system are established.

Using an engine trolley, i.e. a mobile engine pallet, is a convenient way of transporting the unit under test from the rigging area to the test bed. It is designed for manual docking. The engine trolley has smooth-running rollers underneath. A handle permits easy pallet navigation, even by a single person.

3.1.2.3 Advantages of Pallet Systems as Shown by the Example Engine Test Beds

Conventional testing involves tasks like rigging the engine, executing test schedules and analyzing test results on the test bed. In an advanced test facility (also referred to as test factory) the tasks are distributed among several test beds, engine and parameter preparation stations and data evaluation sites (Fig. 3.10) (see also Sect. 4.7). While these individual sites might all be found at the same place or spread out among different locations, all of them are connected to a central data server. Compared with a rigging room or an office, the test bed is the most expensive resource. By distributing the tasks among different sites, the test bed is exclusively used for running tests. This means that the test bed is never unnecessarily blocked by tasks that can be done just as well elsewhere.

Pallet systems are an indispensable part of modern test facilities as they help to reduce unit-under-test changing and setup times on the test bed. Test objects completely assembled on the pallet are stored intermediately in the buffer zone and are available for flexible test-bed scheduling. Quick and easy unit-under-test changing on the test bed along with flexible test-bed scheduling guarantees optimal test bed utilization.

3.1.3 Shaft Connections and Safety Covers

A shaft connects the unit under test's output to the dynamometer and is designed for a specific combination of a unit under test and a dynamometer. That said, it is advisable to

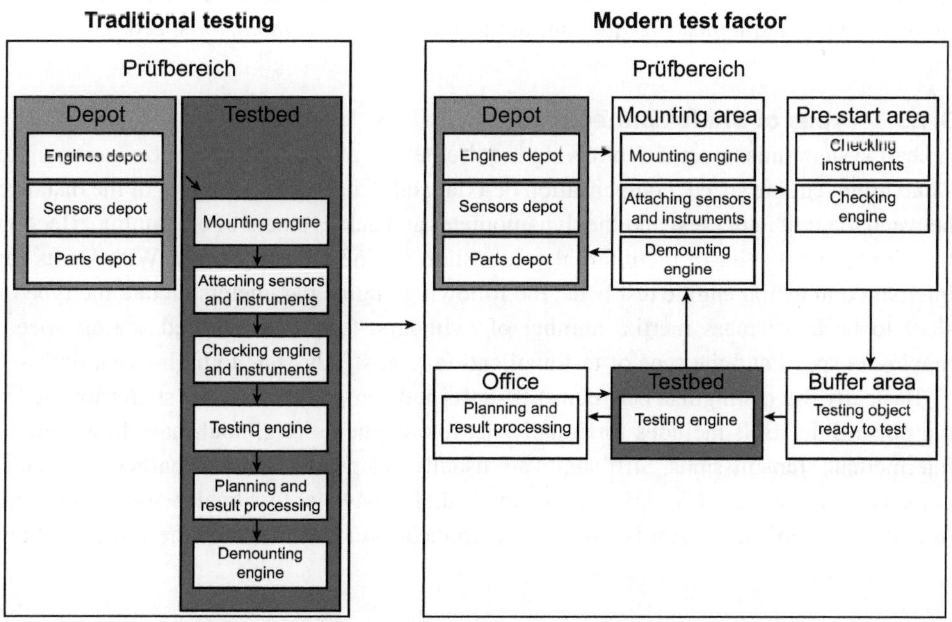

Fig. 3.10 Setup of a modern test system illustrated by the example of an engine test center

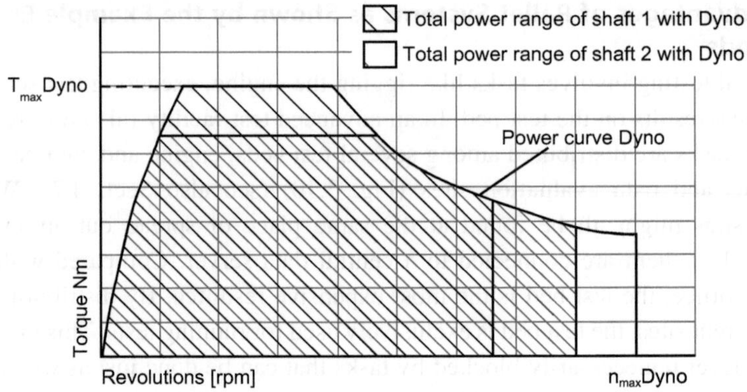

Fig. 3.11 Test bed testing range of shaft with dynamometer

have a single shaft that can be used for multiple types of test units across a broad operating range and for a variety of test applications. In practice, the task of preparing shafts with the necessary dimension in adequate quantities for a test bed will always be an optimization process. As the shaft system, among other things, determines the overall system's vibration behavior, incorrect shaft calculation may lead to excessive wear of components, more specifically, to bearing damage in the unit under test and in the dynamometer and—in extreme cases—even to their destruction. Together with the unit under test and the dynamometer, the shaft defines the admissible testing range (maximum speed and maximum torque). Figure 3.11 shows a valid testing range illustrated by the example of an engine test bed. The chart shows that the setup with Shaft 1 enables test bed operation up to maximum dynamometer torque, but not to maximum dynamometer speed.

3.1.3.1 Types of Shaft Connections

A shaft system must meet the following requirements: transfer of torques, operation up to a maximum engine speed, compensation of axial and radial offset, bridging of the distance between the unit under test and the dynamometer and achievement of a damping effect on the vibration behavior of the internal combustion engine's load systems. With shafts for internal combustion engine test beds, the following engine parameters dictate the type of shaft to be used: mass inertia, number of cylinders, torque, idle speed, starter speed, maximum speed and the type of test application (e.g. steady-state to high-dynamic).

Basically, we distinguish between stiff shafts and damped shafts. Stiff shafts are used if the powertrain itself includes torsionally flexible elements (e.g. dual-mass flywheel) or intermediate transmissions. Stiff shafts are usually designed as cardan shafts or constant-velocity joint shafts (CV joints). Damped shafts have a torsionally weak coupling (e.g. rubber elements or Hardy disks). The application areas in which these are used are steady-state or transient testing where higher degrees of cyclic irregularity or high alternating torques are encountered. High-elastic shafts provide damping on both sides and are distinguished by excellent vibration isolation properties in relation to the dynamometer.

3.1.3.2 Shafts for Electric Motor Test Beds

Electric motor test beds are generally characterized by higher speeds and lower excitation levels. While the shaft calculation procedure remains basically the same, the dimensioning diagrams and parameters will differ.

3.1.3.3 Safety Covers

Safety covers protect the staff and the test bed equipment from accidental contact with the rotating shaft and reduce the effects in the event of shaft breakage. The safety cover consists of two pivot mounted halves, which are interlocked by a quick-release lock. The upper half can be opened, while the lower part is either connected to the intermediate frame by a support or to the base plate by a stand. Movable sheet-metal panels are adjustable to suit the shaft length. An insert is mounted inside the body of the safety cover to keep the rotating shaft's range of free movement as small as possible in order to reduce any consequential damages to a minimum in the case of shaft rupture. This insert also serves as a place to put the unflanged shaft when changing the unit under test (Fig. 3.12).

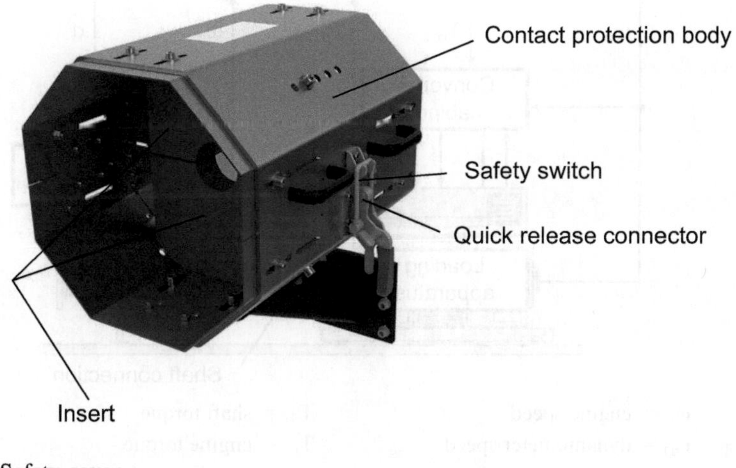

Fig. 3.12 Safety cover

During operation (shaft is rotating) the safety cover must be closed. A safety switch is provided to check if the safety cover is closed properly. This ensures that a test with a running engine only starts if the safety cover is closed.

3.1.4 Shaft Dimensioning

A shaft connection acts as a torsional-vibration damper that connects two masses with each other. To this end, based on defined force transmission properties (mass inertias, torsional spring rates and damping levels), a setup is obtained that meets the limits for torques, torsional vibrations and load surge resistances needed for test operation. The following shaft calculation results from the example of an internal combustion engine test bed (Fig. 3.13):

To obtain a suitable shaft, torsional vibration analysis is applied. For shaft connections, the calculation considers the main excitation, i.e. the engine combustion process in its dominant order (Fig. 3.14).

The prevalent dimensioning limit is the critical speed ($n_{critical}$) at which the system goes resonant. This speed must be below the smallest engine speed capable of being tested. The exciter amplitude depends on the cyclic irregularity of the engine speed and the mass ratio

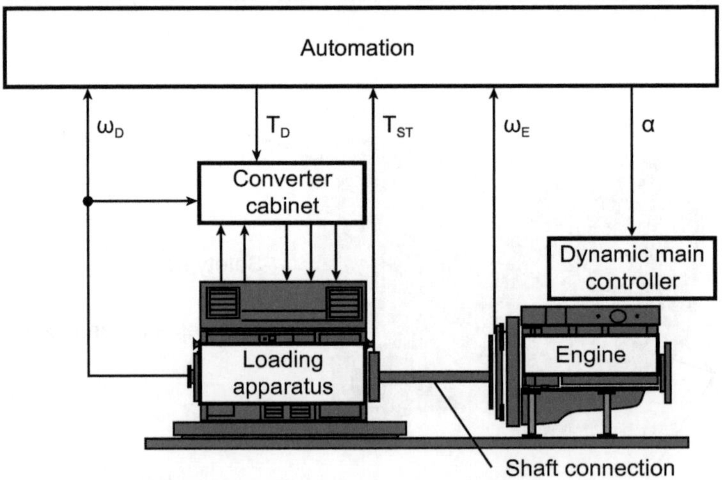

ω_E = engine speed
ω_D = dynamometer speed
T_D = dynamometer torque
T_{ST} = shaft torque
T_E = engine torque
Δ_φ = twisting angle of shaft

Fig. 3.13 Schematic diagram of a shaft setup

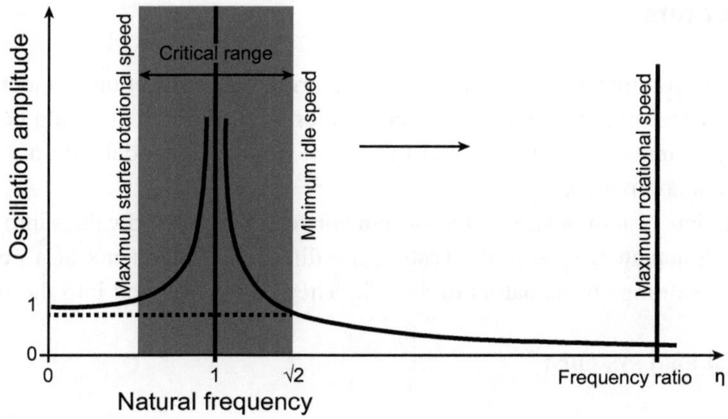

Fig. 3.14 Safe shaft range

of engine and dynamometer. The amplitude of vibration under resonance conditions is additionally dependent on the damping element's damping factor.

$$
\begin{aligned}
n_{min} &> \sqrt{2} * n_{critical} \\
n_{critical} &= \frac{60}{2pi * j}\sqrt{c * \left(\frac{1}{\theta_{ICE}} + \frac{1}{\theta_D}\right)} \\
d_0 &= \frac{d * f_0 * pi}{c}
\end{aligned} \tag{3.4}
$$

c	Torsional stiffness of shaft
d	Shaft damping
θ_{ICE}	Inertia of engine
θ_D	Inertia of dynamometer
j	Dominant harmonic of internal combustion engine (e.g. four-stroke engine with four cylinders $j = 2$ and for six in-line cylinders $j = 3$)

Equation 3.4 Critical speed

A shaft is usually always calculated specifically for one test bed and one engine. As the market requires that a single test bed be used to test several engines and run different types of test applications, it has become standard for shaft suppliers to offer catalog shafts. Diagrams show different types of catalog shafts, dynamometers and engine types. Shaft selection is based on the internal combustion engine's inertia, the minimum idle speed and the maximum starter speed.

3.2 Actuators

To be able to run automated tests, it is necessary to bring the inputs of units under test into defined states. These may be required by a certain test procedure or, as would be the case in HiL or XiL systems, calculated in real time from simulation models of only virtually available system components.

The automation system or simulation system converts the demand values into physical quantities using actuators or stimuli. Testing the different components of a powertrain requires a diverse range of actuators or stimuli. They can be divided into the following main groups:

Mechanical load systems:

- Devices capable of producing a certain torque or speed on a shaft (asynchronous dynamometers, PMM dynamometers, eddy-current dynamometers,. . .)
- Throttle actuators that set a defined throttle position
- Gear shift robots that shift gears, actuate the clutch, etc.

Electric load systems: These systems set the electrical loads for batteries, power electronics components or fuel cell systems. They include:

- Battery emulators
- Electrical loads
- E-motor emulators

Climate/media conditioning systems: These are used to set the climatic conditions for units under test such as temperature, pressure or humidity. They include:

- Conditioning systems for intake air
- Conditioning systems for cooling fluids
- Conditioning systems for oil or fuel
- Conditioning systems for boost air
- Conditioning systems for hydrogen or other reformate gases

ADAS stimuli: Control systems for advanced driver assistance systems (ADAS) or partially or fully automated vehicles have a number of additional sensors, such as video cameras, LIDAR sensors, radar sensors, ultrasonic sensors, that receive information via vehicle-to-vehicle communication. Automation or simulation systems have to create the necessary physical input values for these sensors to perform in component tests. Stimuli for this particular purpose are currently being developed by the industry. They include:

- Video sensor stimuli that produce images for a video sensor in a vehicle. Complex road, traffic and vehicle simulation systems create these video streams (similar to video games).
- Radar sensor stimuli
- Lidar sensor stimuli
- V2V and V2I stimuli produce V2V or V2I communication flows that correspond to a simulated traffic situation (V2V stands for the communication between vehicles [vehicle to vehicle] and V2I stands for the communication between vehicle and infrastructure [vehicle to infrastructure]).
- GPS stimuli use the GPS coordinates of a simulated road to produce matching satellite signals for a GPS receiver in a vehicle under test.

3.2.1 Mechanical Load Systems

3.2.1.1 Overview of Dynamometers

Dynamometers (in short also referred to as "dynos") for power measurement based on the quantities torque and rotational speed have been used on test beds since the beginning of the twentieth century. In order to apply the required load to the crankshaft, early test beds for measuring power in internal combustion engines initially used belts that were mounted manually with variable force to a belt pulley fastened to the crankshaft. The friction torque generated by this process was transmitted via levers to a mechanical indicator that displayed the braking force.

Dynamometers together with control systems today are used for high-precision setting and measuring of speed and torque. The latter is transmitted via a shaft connection to the dynamometer. Based on these two quantities—and taking into account any gear transmission ratios—it is possible to determine the exact power transmitted via the shaft. On advanced test beds, dynamometers make it possible to achieve defined load points as well as load curves within the shortest period of time. In addition, unit under test and dynamometers can be operated in different control modes, depending on what is required.

Basically, we distinguish between two types of dynamometers:

- **Passive (or absorption) dynamometers** are only capable of applying a load to the unit under test, i.e. they apply a braking force to it against its rotational direction.
- **Active (or universal) dynamometers** are capable of both decelerating the unit under test against its direction of rotation and accelerating the unit under test in its direction of rotation, as would be the case with an internal combustion engine installed in a vehicle driving downhill or uphill, for example.

With regard to dynamometer use, we distinguish between different time sequences in which the load conditions are set at the mechanical connection between unit under test and dynamometer (see also Sects. 2.1.3 and 2.1.4):

1. **Steady-state**: In steady-state tests, the load points are executed step by step. The load levels are defined as desired, and the measurement of speed and torque is conducted in steady state condition at each load point.
2. **Transient**: With transient tests, predefined curves of speed and torque points are provided. During tests, typically speed, torque and other sensor signals are measured continuously. In transient tests, transitional states, as they would be encountered in a vehicle under real load conditions, are only reproduced approximately.
3. **Dynamic**: In dynamic tests, the speed and torque curve on the test bed should match as realistically as possible the signal traces measurable in vehicles under real road conditions. Sometimes synthetic dynamic load curves are used. Therefore, it must also be possible to achieve fast changes of speed and torque with high gradients. One has to accept, though, that the kind of transition states encountered under real-life loads cannot be reproduced with absolute precision. Differences between real operations and test-bed tests might be caused by pulsations from the internal combustion engine's firing sequence or high speed gradients during clutch engagement/disengagement.
4. **High-dynamic**: High-dynamic tests also take into account transition states as they occur at the crankshaft of internal combustion engines during the gear shift process, or the impact of torque pulsations caused by the internal combustion engine's firing sequence.

Consequently, the demands on the speed of dyno load changes are critical to the dynamometer's structural design. Typical load ranges for dynamometers in different applications are listed in the Table 3.1 below.

3.2.1.2 Passive Dynamometers

A passive dynamometer is a dynamometer which exclusively applies a braking force to the unit under test against its own direction of rotation. The most important passive dynamometers still in use today are described below.

Table 3.1 Torque and speed ranges for typical dynamometers

		Torque T	Speed ω_D
Tests on internal combustion engines	Passenger cars and small commercial vehicles	100 Nm < T < 750 Nm	750 rpm < ω_D < 10,000 rpm
	Heavy-duty commercial vehicles	900 Nm < T < 5000 Nm	500 rpm < ω_D < 2500 rpm
Tests on transmissions and powertrains	Wheel dynos for passenger cars	0 Nm < T < 2500 Nm	0 rpm < ω_D < 3000 rpm
	Wheel dynos for commercial vehicles	0 Nm < T < 40 000 Nm	0 rpm < ω_D < 1200 rpm

Eddy Current Dynamometers

Eddy current dynamometers consist of a toothed rotor made of ferromagnetic material (iron). The stator consists of magnetically conductive material too and a coil (also referred to as 'winding') to which a variable DC voltage source is connected. Excited by the DC current, sections with small and large air gaps are created in the magnetic field near the rotor teeth (Fig. 3.15).

The stator's magnetic field generated by the coils greatly magnetizes the rotor in the area of the teeth (small air gap), creating pronounced magnetic poles on the rotor tooth that match the direction of the magnetic field created by the coils. If the unit under test rotates the rotor, the magnetic field changes at a defined point on the stator surface because of the air gap change caused by the rotor's tooth spacing. In the rotor itself, the magnetic field remains constant because, in relation to any point of the rotor surface, there is no change in the air gap to the stator surface, which means there is no change in the magnetic field either.

The changing magnetic field in the stator induces eddy currents in the stator material. These, in turn, also produce magnetic fields initiating forces of attraction and repulsion acting on the magnetic poles at the rotor teeth and, in accordance with Lenz's Law, oppose the rotor's direction of rotation.

Because of the eddy current losses, the braking energy is converted into heat in the stator. To dissipate this heat, the stator material requires cooling. For this purpose, water is used as a coolant because of its good heat transfer properties and its high specific heat capacity. The braking force increases in proportion to the DC voltage applied to the coil and the speed. Both variables have an effect on the magnitude of the induced eddy currents in the stator and, as a result, on the magnitude of the forces of attraction and repulsion acting on the rotor's magnetic field.

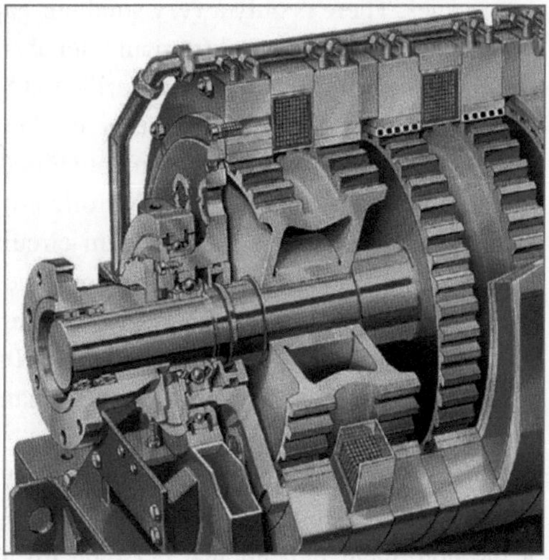

Fig. 3.15 Eddy current dynamometer

Eddy current dynamometers are easy to control and capable of producing high torques even at low speeds. This is why they are often used for applying loads to slowly rotating diesel engines, largely in the area of heavy-duty vehicles. The generated high braking energy can be dissipated at low cost via a cooling water supply system.

One of the drawbacks of eddy current dynamometers is that they can only act as a load to the unit under test and not as a drive unit. Also, attributable to the high inductance of the exciting coil, it takes a relatively long time to achieve the magnetization process in the material and the current build-up and current reduction in the case of load changes. This is why eddy current dynamometers are only suitable for steady-state tests.

Today, efforts are additionally being made to put the unit under test's braking energy to use. Utilizing the energy dissipated into the cooling water circuit, however, is only possible to a limited extent because of the low difference in temperature between the cooling water supply and return line. Usually there are no suitable consumers requiring heating power anywhere near the dynamometers, so the water is dissipated to the environment by means of cooling towers or air/water heat exchanging devices without being used. That is why nowadays eddy current dynamometers are being used less and less.

Eddy current dynamometers can be used for test bed applications with power ratings ranging from several kW to the MW range and torques ranging from several Nm to several tens of thousands of Nm. The increase in torque is normally achieved by arranging several coils and rotors on one rotor shaft. The torque is almost exclusively measured via a cradle-mounted stator. A load cell measures the counterforce (see page 168).

Hydraulic Dynamometers

These dynamometers consist of a rotor and stator into which semi-circular (or pocket-) shaped vanes are cast. The spinning rotor causes the rotor vanes to move past the fixed stator vanes. There is only a very small air gap between the rotor and the stator vanes. Via the stator, water is supplied from an inlet above and passed into the inside of the hydraulic dynamometer. It drains off by an orifice at the bottom of the stator. The water inlet and outlet have to be decoupled from the cradled stator so that no forces are introduced that could impact the torque measurement (Fig. 3.16 and Fig. 3.17).

The load unit's torque is controlled by varying the water fill level inside the dynamometer. Because of their semi-circular shape, the rotating rotor vanes create a flow transversal to the axis of the rotor. Due to shearing caused by the flow between the vanes, friction forces are created in the water, which convert the rotation energy supplied by the unit under test into thermal energy. Its operation principle is the same as that of a hydraulic clutch. Control valves at the dynamometer's water inlet and outlet regulate the fill level.

Due to the fact that it takes relatively long to change the fill level inside the dynamometer, hydraulic dynamometers are used for steady-state testing only. Because of the non-linear dependence between fill-level, speed and produced torque, control is more complicated than with eddy current dynamometers. In addition, the water inlet and

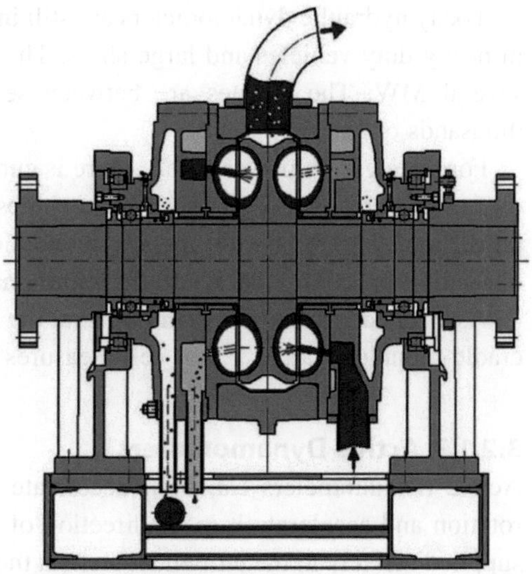

Fig. 3.16 Schematic diagram of a hydraulic dynamometer

Fig. 3.17 Hydraulic dynamometer

outlet valves are typically controlled to act in opposition to each other. This leads to improved dynamic behavior.

One of the advantages of hydraulic dynamometers is their low mass inertia due to the small rotor diameter even at high braking power. As with eddy current dynamometers, torque increase is accomplished by connecting several, typically two, pairs of vanes in series.

Constant water supply pressure and an even distribution of water across all vane pairs is necessary to ensure sufficiently good control. The water exiting the system must drain off freely without counter pressure. This is the reason why hydraulic dynamometers cannot be directly integrated into a closed cooling water circuit.

Another disadvantage is their high susceptibility to wear because of the cavitation occurring in hydraulic dynamometers. Cavitation occurs when gas bubbles forming in the water implode and cause extremely high pressure waves that can cause massive damage to the rotor and stator material.

Today hydraulic dynamometers are still in use, primarily for testing large engines used in heavy-duty vehicles and large ships. The power range covers several hundred kW to several MW. The torques are between several hundred Nm to several hundreds of thousands of Nm.

For testing high power levels, there is currently practically no alternative to hydraulic dynamometers. Although electric machines are available that manage the respective power ranges, they are not always economically feasible due to the missing possibility of feeding the energy back into the supply network.

As with eddy current dynamometers, the torque is almost exclusively measured via a cradle-mounted stator. A load cell measures the counterforce (see page 168).

3.2.1.3 Active Dynamometers

Active dynamometers can both decelerate the unit under test against its direction of rotation and accelerate it in its direction of rotation. For this purpose, energy has to be supplied actively to the unit under test via the mechanical coupling. A brief description of the most widely used active dynamometers follows below.

Asynchronous Machines

Today, asynchronous machines (also referred to as induction machines) are by far the most widely used dynamometers available (see Fig. 3.18). An asynchronous machine consists of a stator and a rotor, each consisting of layers of highly permeable metal sheet elements that are insulated from one another. This is necessary to reduce the eddy currents in the material, which would otherwise lead to an impermissible temperature rise in the material.

The stator contains a three-phase winding, which results either in one magnetic pair of poles around its entire circumference or two magnetic pairs of poles around half its circumference and so forth. The ends of the three-phase windings are routed to a terminal

Fig. 3.18 Example of an asynchronous machine—AVL DynoRoad

Table 3.2 Speeds in asynchronous machines dependent on the number of pole pairs and supply frequency

Number of pole pairs	Speed in 50 Hz supply system (rpm)	Speed in 60 Hz supply system (rpm)
1	3000	3600
2	1500	1800
3	1000	1200
4	750	800
5	600	720
6	500	600
7	428	514
8	375	450

board and connected to the three-phase power supply either via a star connection or a delta connection. This produces a rotating field inside the machine. Where there is one pair of poles, it moves around the stator's entire circumference in one period of the supply frequency. Where there are two pairs of poles, the rotating field moves around half the stator's circumference and so forth. Depending on the supply current frequency, this results in rotational speeds for the rotating field inside the machine that are shown in Table 3.2.

The rotating field induces high eddy currents in the conductors of the rotor, which, following Lenz's Law, produce a magnetic field against the cause of induction, specifically the speed difference between the rotating field and the rotor speed. This produces a force on the rotor. As a result, the difference (slip) between the rotor speed and the speed of the rotating magnetic field of the stator is a measure for the magnitude of induced rotor current, and thus for the force acting on the rotor, which is made available at the end of the shaft as usable torque. So the higher the load at the end of the shaft, the more the rotor is decelerated against the stator's rotating field and the higher the torque that is available at the shaft of the rotor.

The speed is controlled via a frequency converter. This basically consists of a grid-side PWM (pulse-width modulation-controlled) converter, which transforms the input voltage from the grid (AC voltage) into DC voltage, a DC intermediate circuit (DC link) that smoothens the DC voltage, and a motor-side PWM (pulse-width modulation-controlled) inverter that reconverts the DC voltage into a three-phase, sinusoidal AC voltage with variable frequency. This technology allows the energy to flow either as propulsion energy from the power supply to the asynchronous motor, allowing it to drive the unit under test via its rotor shaft in its direction of rotation or, vice versa, as braking energy from the unit under test back to the power supply.

In doing so, the asynchronous machine applies a counter torque to the test unit's shaft, decelerating the unit under test against its direction of rotation. In this case, the asynchronous machine acts as a generator and transmits its braking energy to the power supply via the DC link and the PWM converter. The rotation frequency of the stator's rotating field increases linearly with the frequency of the applied supply voltage. The rotor speed increases correspondingly, so that the speed is generally controlled by adjusting the motor voltage frequency.

The converter also controls torque. To increase torque, a current must be generated in the rotor that matches the required torque. To accomplish this, the PWM inverter sets the frequency at the stator windings in such a manner that a sufficient speed difference between the stator field and the rotor speed (slip) is obtained. As the asynchronous machine basically acts like a transformer short-circuited on the secondary side, the required current also has to be produced at the stator winding. To do this, the PWM inverter changes stator terminal voltage. For the purpose of control improvement, an electric model of the asynchronous machine is stored in the converter and parameterized according to the machine actually connected.

Torque control, however, is only possible up to a maximum value defined by the magnetic saturation of the laminated core. A measure for the magnetization is the ratio of terminal voltage to the applied frequency (U/f). A machine achieving maximum saturation at e.g. 400 V and 50 Hz corresponds to a stator field speed of 3000 rpm in a machine with one pair of magnetic poles. Up to this point, the frequency and voltage may vary, depending on the maximum allowable U/f ratio, so that maximum torque can be demanded within the entire speed range, from 0 rpm to 3000 rpm. Due to the winding design, it is impossible to increase the voltage any further beyond this threshold speed, and the torque drops at a rate of 1/f (field weakening range) when the speed increases.

It is vital that the PWM inverter provides the asynchronous machine with current largely without harmonics to limit power dissipation due to high-frequency eddy current losses in the dynamometer. The shorter the switching elements' switching periods in the PWM inverter are, the lower the harmonics content will be in the current supplied to the machine. This implies that the harmonics content is directly dependent on the switching frequency. Modern frequency converters have switching frequencies ranging from 2.5 kHz to 10 kHz. However the higher the switching frequency, the higher the losses in the frequency converter, which means that a reasonable compromise must be found between the machine and the converter in terms of maximum service life of the overall system.

Asynchronous machines are deployed in almost all test applications, such as transmission testing, powertrain testing and internal combustion engine testing, and even on vehicle test beds. In these applications they cover speeds ranging from several hundred revolutions per minute to up to 16,000 rpm and torques from 100 Nm to 100,000 Nm. Power outputs in testing applications range from 20 kW to 5.5 MW.

There are two methods to measure torque (see also Sec. 3.3.6):

Cradle-mounted stator and measurement of the counterforce using a load cell

When using asynchronous machines in transient or dynamic tests, the part of the torque that is used to accelerate the rotor has to be compensated for in order to achieve accurate measurements. A fast torque calculator computes the accelerating torque and makes the necessary corrections. This requires an exact model of the AC machine and sufficient computing capacity. This method has the advantage of measuring the precise counterforce of the AC machine against the cradle, largely unaffected by temperature-related factors, resulting in a dynamic torque measurement with high accuracy. The disadvantages are the additional costs for friction-free support bearings and online torque correction calculation.

Torque measuring flange

By this method, a torque measuring flange is mounted in the shaft between the dyno and the unit under test. It measures the desired torque in the shaft with high dynamics. However, the measurement is subject to high temperature impacts, which cannot be fully compensated for. So if the machinery (and thus the shaft) reaches high temperatures, this can lead to a serious error, which is in the magnitude of the measuring flange's steady-state accuracy. The low overall costs, however, are an advantage because in this case the dynamometer can be designed as a simple foot-mounted machine.

Synchronous Machines

On the stator side, synchronous machines are similar in their structure to asynchronous machines. Here, too, a three-phase winding is distributed evenly around the stator's circumference and the ends of the windings are routed to a terminal board. The windings can be either wye (star) or delta connected. Unlike asynchronous machines, the excitation field is produced by the rotor. The excitation is achieved in two ways:

Externally excited synchronous machine

In these machines, a winding is placed in slots along the circumference of the rotor's laminated core. This winding is energized externally with DC voltage supplied to the rotor through slip rings, resulting in a magnetic field with a pronounced north and a south pole on the rotor. The magnitude of excitation is adjusted by varying the DC voltage. Because of the additionally required direct voltage supply and the high speeds occurring in test bed applications, externally excited synchronous machines have failed to catch on in test bed applications.

Permanent-magnet synchronous machine

In permanently excited synchronous machines (see Figs. 3.19 and 3.20), permanent magnets made of rare-earth metals are distributed around the machine's circumference. The way these are arranged once again produces a pronounced magnetic field with a north and a south pole on the rotor. The magnets must keep their magnetic properties even if rotor temperatures exceed 100 °C. They must also remain firmly affixed to the rotor surface even when the speeds and the associated centrifugal forces are high. This

Fig. 3.19 Example of a permanent-magnet synchronous machine—AVL DynoWheel

Fig. 3.20 Further example of a permanent-magnet synchronous machine—AVL DynoSpirit

is accomplished by using special adhesives or bandages, or by embedding the magnets in slots in the rotor surface. One of the advantages of this technique is that the energizing winding, along with the related DC voltage supply, is no longer needed. A drawback, on the other hand, is that over recent years the cost of permanent magnets has been increasing steadily. Today, dynamic test bed applications almost exclusively use permanent-magnet synchronous machines as dynamometers.

In synchronous machines, the rotor spins at exactly the same frequency as the stator, but offset from it by a fixed angle referred to as rotor displacement angle. If this load becomes too great, the rotor can come to a complete standstill. Nowadays such risks are fully mastered by the frequency converter which detects the rotor position and adjusts the stator current as required. In the event of a line voltage breakdown, e.g. as a result of supply voltage failure in the frequency converter, the unit under test (e.g. internal combustion engine) could at worst drive the synchronous machine to maximum speed. In such a case, the rotor would induce a high voltage in the stator winding that could result in damage to the frequency converter. For this reason, dynamometers are designed in such a way that the maximum amount of voltage the rotor induces in the stator winding in the event of failure will never exceed the permissible DC link voltage.

Synchronous machines are operated using frequency converters, as are asynchronous machines. Similar to asynchronous machines, synchronous machines are able to either decelerate the unit under test against its direction of rotation or accelerate the unit under test in its direction of rotation. In the first case, the braking energy is fed into the power supply system, and in the second case the energy is drawn from the power supply system. Synchronous machines with permanent excitation are easy to control.

As the level of excitation remains the same across the entire operating range, the speed is solely controlled by variation of the stator voltage frequency through the PWM inverter. Unlike the asynchronous machine, the synchronous machine's torque forms in the stator by an increase in stator current. Using a machine model stored in the frequency converter, a corresponding active and reactive current is calculated that is impressed onto the machine by imposing a pulse-width modulated voltage on the stator. Synchronous machines are primarily distinguished by the fact that the rotors can be manufactured with small diameters, which helps to keep the rotor's mass inertia relatively small compared with asynchronous machines. This circumstance makes synchronous machines particularly suitable for use as load units for test beds with high-dynamic applications.

They are suitable for high-speed applications too. This is because synchronous machines have good regulation characteristics, even in the field-weakening range and at high speeds. One of their disadvantages, however, is that, compared with asynchronous machines, they have higher power requirements, particularly in the field-weakening range. This mainly results from the fact that the field produced by the rotor needs to be weakened at speeds exceeding the threshold speed in order to prevent excitation from reaching magnetic core saturation as speed increases. This requires that the stator winding is supplied with current to produce a field that weakens the rotor field. The higher power requirements make it necessary to ensure sufficiently sized frequency converters, a circumstance that makes the overall system more expensive in comparison to asynchronous machines.

Despite this drawback, synchronous machines are much more dynamic, enabling realistic reproduction of loads on the internal combustion engine to be tested, for example, or realistic internal combustion engine simulation. Synchronous machines today are therefore mainly used for simulating internal combustion engines on powertrain test beds or as wheel machines for applications where the wheel's inertia is simulated. They are also used for tests on internal combustion engines requiring realistic loading of the crankshaft. Power outputs range between close to 100 kW to 500 kW and torques of a few 100 Nm to 3000 Nm when used as wheel machine in light-duty vehicle (passenger car) testing.

Direct Current Machines

Today direct current (DC) machines are practically no longer used as test bed dynamometers. Reasons for this trend are that they are more expensive to buy than asynchronous machines and have higher maintenance requirements. The advancements in frequency converter technology allowing equivalent control quality in asynchronous and synchronous machines are another critical factor. In individual cases, DC machines are still being used in brake test beds or, in some rare cases, in vehicle test beds. For such test bed applications, DC shunt-wound machines are mainly used. In this technology, DC current windings are distributed around the stator's circumference as exciter winding. The ends of the winding are routed to a terminal board. Unlike asynchronous or synchronous machines, the stator housing is made out of solid material and not laminations. This is possible because the stator has no alternating fields and, consequently, no eddy currents can be induced which would lead to heat generation in the material.

There is also a coil wound around the rotor surface, the ends of which are connected section by section to conductive contact surfaces on a cylindrical element mounted on the rotor shaft (see Fig. 3.21 and 3.22). The contact surfaces are isolated from each other. Slip rings (typically carbon brushes) isolated against each other connect the winding sections to the DC voltage supply. The slip rings are therefore also connected to the terminal board at the stator housing. In direct current machines, the rotor is also referred to as armature.

Fig. 3.21 Schematic diagram of direct current machine

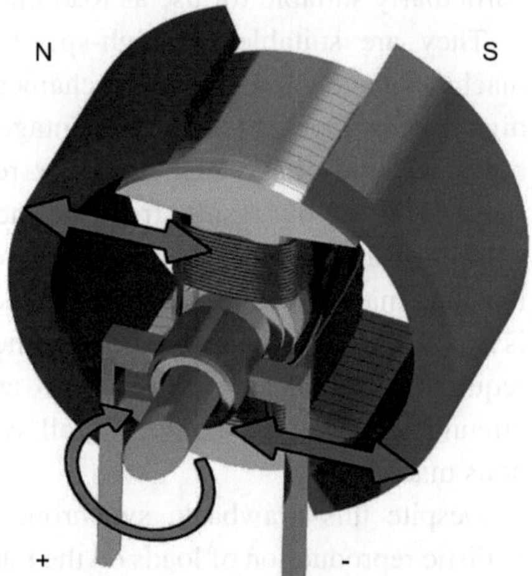

Fig. 3.22 Direct current machine

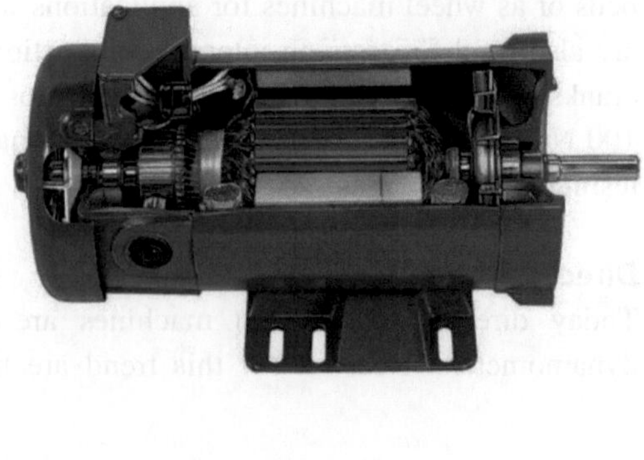

Accordingly, the rotor winding is also referred to as armature winding and the rotor current as armature current.

The rotor winding is connected in such a way that the current passes through the winding sections and produces a strong north and south pole on the spinning rotor. Since the direct current has to change direction ("commutate back and forth") each time the coil rotates through one of its sections, the cylindrical element with the contact surfaces is called a commutator.

In a DC shunt-wound machine, the exciter winding in the stator is connected in parallel (or "shunted") to the armature winding. Adjustable (slip ring) resistors permit varying the voltage at the excitation winding continuously, allowing the excitation field to be changed too, regardless of the armature voltage. The repulsive forces exerted by the poles in the exciting and rotor winding cause the rotor to spin. The increasing speed induces a voltage known as counter-electromotive force (CEMF) in the rotor winding that opposes the rotor (armature) voltage. In DC shunt-wound machines, it is therefore possible to adjust the speed in the armature control range at constant excitation voltage by varying the armature voltage between 0 and the rated armature voltage in the rotor winding, at constant torque between 0 and rated speed.

Above the rated voltage, the speed can be increased beyond the rated speed by reducing the excitation field and thus magnetic flux. This is referred to as field weakening range. In this area, a decrease in excitation increases the speed at constant load until an armature current is produced, leading to a dyno torque which is equivalent to the external, requested torque. The armature current is caused by the counter-electromotive force induced in the armature. In the field weakening range, the motor torque decreases inversely proportional to the speed because the armature voltage cannot be increased beyond the rated voltage. This behavior makes DC shunt-wound machines very easy to control; torque and speed can be controlled almost linearly by varying the armature voltage and excitation voltage. Torque is measured either by using a load cell in a cradle-mounted stator housing or by using a torque measuring flange in foot-mounted machines.

Hydrostatic Systems

Much like DC machines, hydrostatic dynamometers today have no real significance any more as load units in test bed applications. In hydrostatic dynamometer systems, a hydraulic pump brings a fluid (hydraulic oil) inside an intermediate circuit and pressurized tank to a high pressure level (around 300 bars). The energy stored in the intermediate circuit is then converted to torque via an impeller with adjustable vane position. The produced torque can be continuously controlled between the impeller's idle position (no effective vane surface in the fluid's flow direction) and the rated torque (maximally effective vane surface in the fluid's flow direction) by varying the impeller's slope angle.

Significant benefits of hydrostatic dynamometers are their extremely low moment of inertia and their high dynamics. Disadvantages include the high costs, the complexity of the overall system, as well as safety issues arising from possible leaks stemming from the high pressure level in the intermediate circuit. In the past, hydrostatic dynamometers were used for testing internal combustion engines up to 250 kW. Torque is measured using a torque measuring flange connected to the dynamometer's coupling flange via a short shaft with intermediate bearing.

Tandem Dynamometers

Tandem dynamometers consist of a passive (absorbing-type) dynamometer (hydraulic or eddy-current dynamometer) and an active (motoring) dynamometer (almost exclusively an asynchronous machine) connected in series (Fig. 3.23). The rotor shafts of both machines are connected to each other by a rigid coupling. Design consideration should be given to ensuring that the shaft between the unit under test and the dynamometer—as well as the rotor shaft of the first (usually passive) dynamometer—can withstand the maximum required torque at the unit under test's connection point.

Tandem dynamometers find their application in the following tasks:

- To simulate motoring, i.e. active acceleration of the unit under test (usually an internal combustion engine) in its direction of rotation
- To dissipate braking energy via a cooling water circulation system if the braking energy cannot be completely fed back into the electrical supply grid
- To increase the dynamics of the passive dynamometer in case it is incapable of setting the load conditions required by the test cycle within the defined short period of time. This is frequently the case with legal exhaust emission cycles for heavy-duty engines.

In tandem combinations it is usually the passive dynamometer that produces the far greater amount of the required braking torque (approx. 70–80 %). The active dynamometer can therefore be sized much smaller because the torque required to

Fig. 3.23 Tandem dynamometer

simulate motoring operation is much lower than what is needed to decelerate the unit under test.

Apart from that, the active dynamometer additionally only needs to produce an additional torque of about 20 % of the dynamometer's entire torque to achieve the required dynamics. However, to increase the dynamics, the passive and the active dynamometers are mechanically braced against each other by way of electrical control. This means that the active dynamometer exerts a certain amount of torque against the passive dynamometer's torque. In dynamically critical cases, i.e. sudden variations from high torque to low torque, this allows the active dynamometer to instantly provide a torque at the passive dynamometer's shaft that speeds up the reduction in torque at the tandem dynamometer unit's connection point. This is a major improvement over a test bed with an entirely passive dynamometer unit. This compensates a disadvantage of both eddy current dynamometers and hydraulic dynamometers, which are slower when changing from high to low torque than in reverse direction.

In order to achieve the above advantage, the current in the exciting coil of eddy current dynamometers needs to be reduced quickly. This is normally not possible given the high inductance and the simple type of bridge circuits commonly used in the power units. An improvement is achieved by using bridge circuits that allow bidirectional current flow, so that the energy present in the coil inductance is reduced more quickly by means of negative current.

In hydraulic dynamometers, a lowering of the water level inside the dynamometer reduces torque. This process is slow too because it has to be accomplished in an unpressurized state against the atmosphere. To improve dynamics, some hydraulic dynamometers are equipped with air vents in their water compartment housing that can be opened to the atmosphere. This allows air to flow into the water compartment inside the hydraulic dynamometer, acting against the negative pressure and helping the water to drain off more quickly.

Tandem dynamometers are almost exclusively used for testing diesel engines of heavy-duty vehicles and off-road vehicles with power outputs up to 800 kW and torques up to 5000 Nm. A torque measuring flange acquires the torque. Measurement of braking torque can be accomplished via the cradle-mounted passive dynamometer and a load cell. This torque is used to control the complete system and to distribute torque between the active and the passive dynamometer.

3.2.2 Other Mechanical Load Systems

3.2.2.1 Throttle Actuators

To be able to run internal combustion engines in automatic operation, the control, simulation or automatic system must be capable of adjusting the engine's "accelerator pedal". The actuators used for this purpose are throttle actuating systems that consist of small electric motors that adjust throttle position (accelerator pedal position) by way of precise and dynamic position control according to the required demand value. Moreover, sophisticated powertrain concepts today even allow sending engine torque requests directly via a CAN message to the engine control unit.

3.2.2.2 Driving Robots and Driving Simulators

There are a variety of tasks performed on testbeds for combustion engines, electric or hybrid vehicles that require the reproducible adjustment of different vehicle conditions. In most cases, synthetic test cycles are executed in an automatic mode, but there are more and more applications where real road tests are being transferred to the test bed ("road to rig"). For all of these applications, actuators are needed that are capable of controlling the vehicle's powertrain operating elements. There are various ways in which this can be accomplished:

The simplest interface to the vehicle is a hybrid connection to the accelerator pedal (or a gearshift button). It is also possible to access the control unit directly, but this requires a special, open access control unit (e.g. via the CAN bus). In advanced vehicles, such interfaces allow a wide range of functionalities such as the influencing the accelerator pedal, the brake or the gearshift in automatic transmissions. Another possibility would be to use pedal actuators (generally for the accelerator pedal and brake; please refer to the section above). One major advantage of these actuators is the degree of flexibility they provide in the integration into the vehicle. This is due to the fact that they require no special interfaces to the vehicle.

The most sophisticated actuators needed to perform automated vehicle tests are driving robots (see Fig. 3.24). They not only actuate the complete set of pedals (including the clutch pedal in manual transmission vehicles), selector levers and shift levers in automatic

Fig. 3.24 Driving robots for different vehicle types (*source*: Stähle GmbH)

or manual transmissions; but they also operate the ignition key or switch on the air conditioning or other loads.

3.2.3 Electric Load Systems

The battery emulator is a key part of a variety of component and system test beds for the development, validation and end-of-line testing of the following units that are typical of advanced electric and hybrid vehicles:

- Electric motors
- Drive inverters
- Inverters for auxiliary units such as air-conditioning compressors or electric superchargers
- DC converters for coupling the voltage of the high-voltage energy bus to the 12-V on-board supply
- On-board charging devices
- Fuel cells

The system additionally serves to characterize, verify and validate vehicle drive batteries and other energy storage systems such as lithium-ion, nickel-metal-hydride and lead batteries, as well as supercapacitors.

A battery emulator is a power-electronics-based emulation system that mimics electrochemical storage units (e.g. lithium-metal-polymer, lithium-ion, Na/NiC12-, Ni-MH packs). It uses demand values, for example provided by a battery model, to emulate the real battery's terminal behavior. The load applied by the units under test matches what would be required in real operation with a high-voltage battery, i.e. the output terminal voltage and the battery emulator's charge and discharge current match those of a real battery. Many battery emulators are regenerative, i.e. the energy discharged from the battery is fed back into the three-phase power supply, giving the system a high degree of operational efficiency exceeding 95 %.

Based on modular system components and an expandable control system, commercially available systems can be optimally adapted to meet customer-specific requirements (see Fig. 3.25).

To ensure a battery emulator's broad scope of usage scenarios on the test bed, the system should satisfy the following technical requirements:

- High control dynamics and stability despite transient load changes
- Very small level of passive output impedance
- Highest possible accuracy for the adjustable output voltage (<0.1 % full scale)
- Broad adjustment range for output voltage (8–800 V)
- Low residual ripple in output voltage and output current

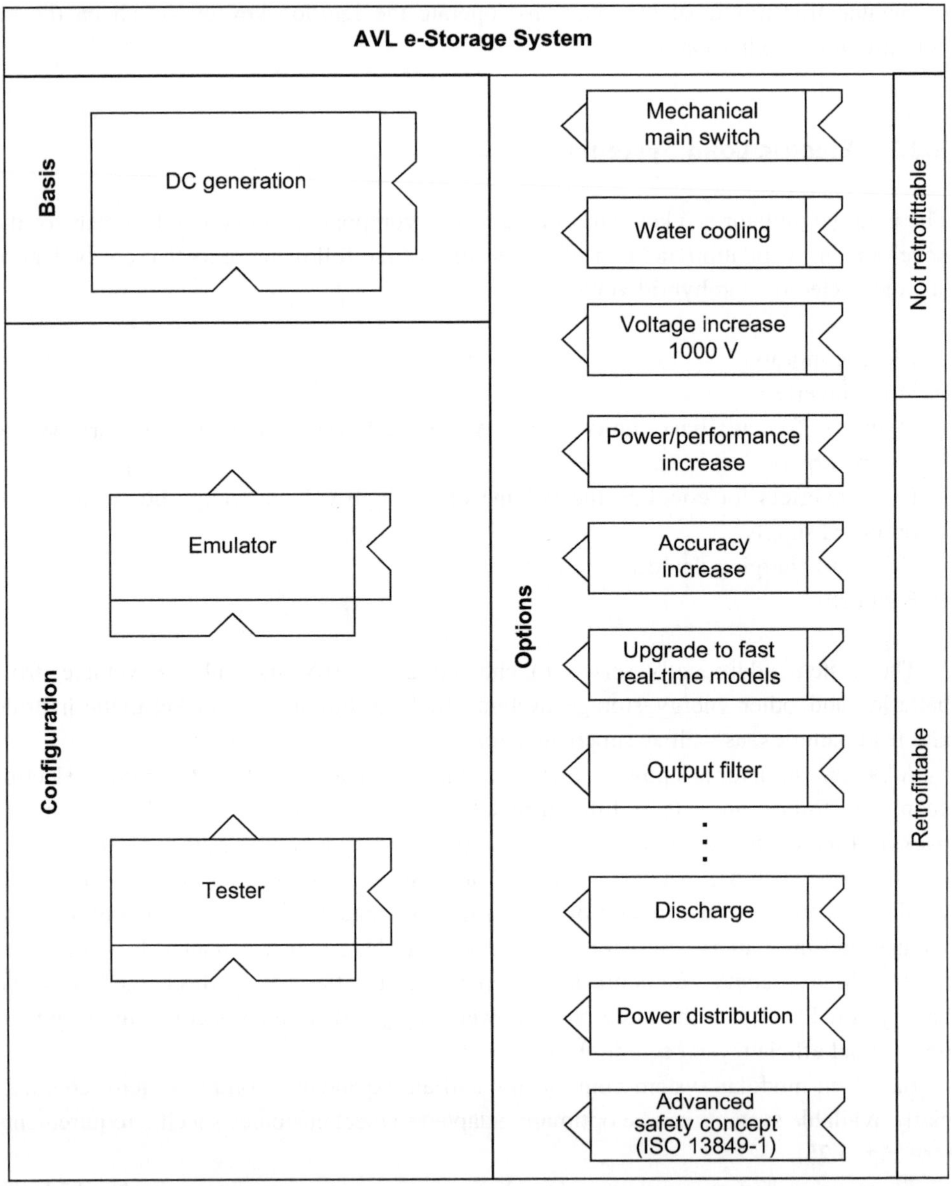

Fig. 3.25 Modularity and customer-specific expansion options for a battery emulator illustrated by the example of the AVL e-Storage Emulator

- Electrical isolation from the power supply
- Insulation (ground fault) monitoring
- Functional safety with optional upgrade to comply with ISO 13849-1 (SIL 2, performance level d, optional PLe)
- Fully automated test runs

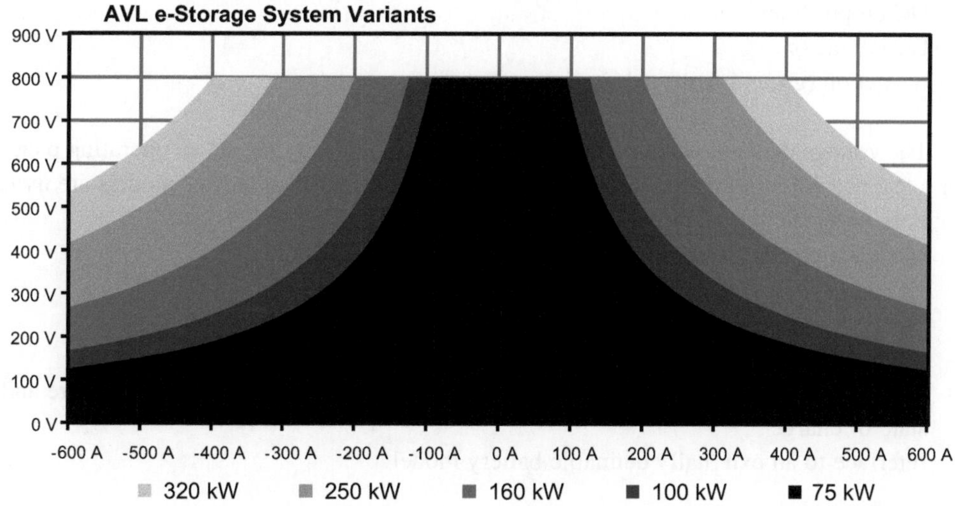

Fig. 3.26 Example of standardized system power classes and nominal DC output current

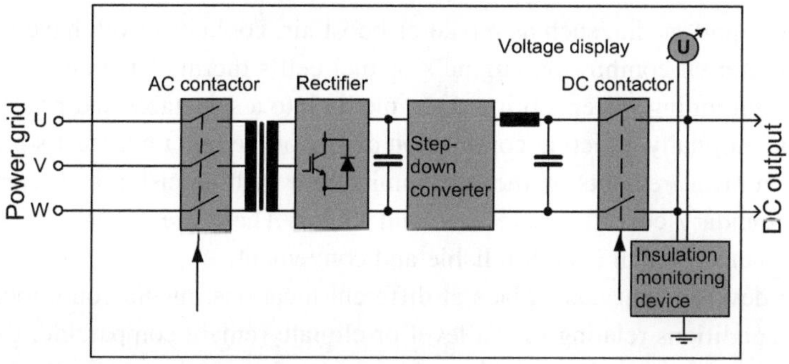

Fig. 3.27 Main components of a battery emulator

Figure 3.26 shows an example of system power and nominal DC output currents of battery emulators. These battery emulators may also be designed as multiple-channel systems or have switching facilities for connecting multiple emulators in parallel to produce higher output currents.

The main power path components of a battery emulator are shown in Fig. 3.27 and listed below:

- Power supply disconnection device
- Power transformer including line filter and EMC filter
- Power rectifier (AC/DC conversion)
- DC-link capacitor
- Buck converter DC output stage

- DC output filter
- DC contactors
- Insulation (ground fault) monitoring

The voltage demand value for battery emulation is typically set via an operating panel or a test bed automation system. Alternatively, parameter-driven battery models are used that are implemented on a real-time system or a HiL system.

Models are typically available with different levels of detail:

- Freely definable internal resistance models
- Internal resistance model based on temperature and state of charge
- Expandable battery model as an electrical equivalent circuit based on temperature and state of charge
- Interface to an externally definable battery model

3.2.4 Climate/Media Conditioning Systems

Gaseous and liquid media, such as intake or boost air, coolants or oil, have a substantial effect on an internal combustion engine's or fuel cell's thermodynamic and mechanical behavior. Conditioning systems bring these media into a selectable and predefined range without unintentionally affecting combustion or any of the unit under test's processes. To be able to run measurements on these test units with high precision and reproducibility, stable test boundary conditions are a crucial factor. These are ensured by conditioned media, in a manner which is both reliable and convenient.

If engine development takes place at different locations, media conditioning ensures that testing conditions relating to sea level or climate remain comparable. Conditioning systems are therefore an indispensable and key part of test bed equipment when it comes to the development and certification of internal combustion engines or fuel cells.

Another application area for conditioning units involves the simulation of extreme driving conditions on the test bed. For such usage, conditioning systems also have to be capable of producing extreme media conditions. To enhance test bed efficiency, it is also vital to establish initial conditions very rapidly in order to shorten testing periods. Oil and coolant conditioning systems allow the unit-under-test media to cool down fast, which helps essentially to raise the number of tests per time unit, e.g. starting tests.

3.2.4.1 Intake Air Conditioning

The ever increasing pressure to shorten development times and the growing complexity of internal combustion engines and fuel cells (see Sect. 2.2.8) is making it more and more important to optimize test runs.

The condition of combustion air, for instance, has a key impact on the operating behavior of an internal combustion engine. Engine torque increases accordingly by

Fig. 3.28 Intake-air conditioning system

approx. 0.12 % per hPa as air pressure rises. Power changes by approx. 0.5 % per Kelvin of temperature change. The humidity content of intake air has no major effect on engine power, but the effects on exhaust emissions, particularly on nitrogen oxides, are notable. The water content of intake air additionally affects the knocking limit, which has to be taken into account when tuning engine management systems. For this reason, it is essential to rule out ambient factors and maintain predefined values. This is made possible by employing intake air conditioning systems. The resulting reproducible, high-precision measurements help significantly to raise product quality (Fig. 3.28).

The required air conditions are achieved by cooling (drying) and subsequent heating or moisturizing. The mentioned processes are usually controlled centrally by means of a control system and compared with a demand value. This prevents air from drying (dehumidifying) or cooling down too strongly compared to the demand value. In doing so, the system helps to adapt the energy requirements optimally to the relevant specifications (Fig. 3.29).

Dependent on the device configuration, pressure can be controlled at steady-state engine conditions as well as in dynamic operation. The challenge, however, is to keep the pressure upstream and downstream of the engine at a constant level and almost without any time lag, even during dynamic engine behavior (and, consequently, varying combustion air consumption). This is usually achieved by installing an air bypass around the engine or by allowing the air to flow off in a controlled manner upstream of the engine (see Fig. 3.30). So, at least where combustion air is concerned, different ambient pressure levels (and therefore altitudes) can be set—even in highly dynamic engine operation. A separate exhaust back pressure valve, on the other hand, simulates pressure loss in the vehicle's exhaust system.

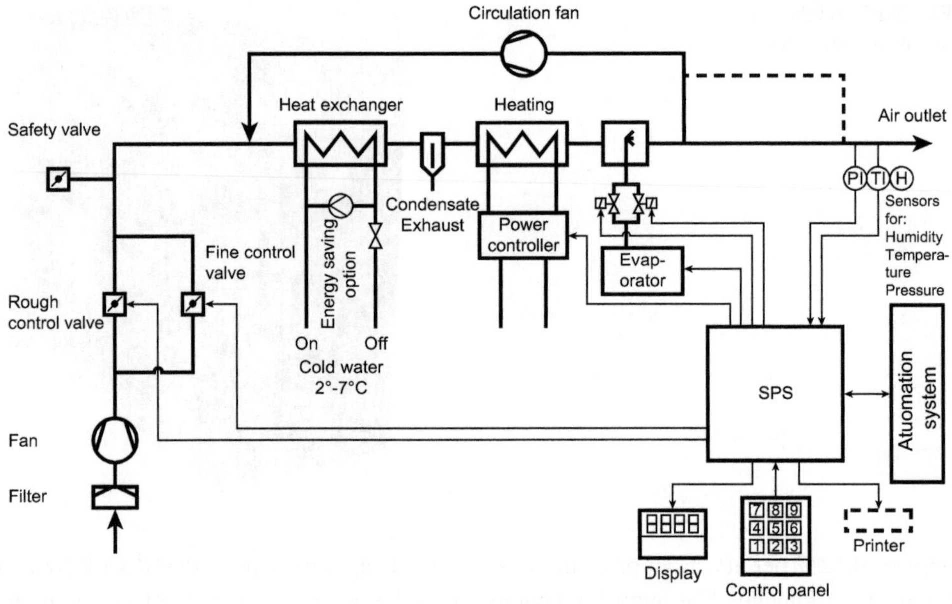

Fig. 3.29 Schematic diagram of an intake air conditioning system

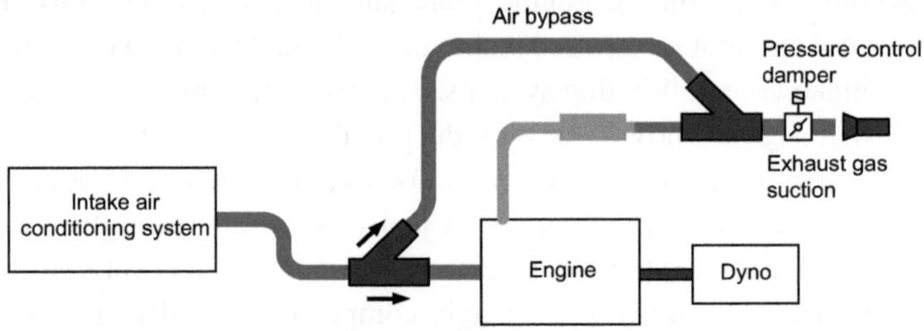

Fig. 3.30 Air bypass for stable combustion air pressure control, independent of engine behavior

Specifications for combustion air on engine test beds are typically:

- Humidity range: 7–20 g H_2O/kg of dry air
- Temperature range: 15–70 °C
- Pressure range (abs): 600–1050 mbar (60–105 kPa)

In humidity ranges below 6 g H_2O/kg of dry air, a pre-drying facility is needed to prevent frost from damaging the system. This is true at least for systems in which the air is cooled with cold water.

The use of special additions, for example exhaust gas vacuum systems, extends the pressure control range, even in dynamic engine operation. As a result, altitudes of up to

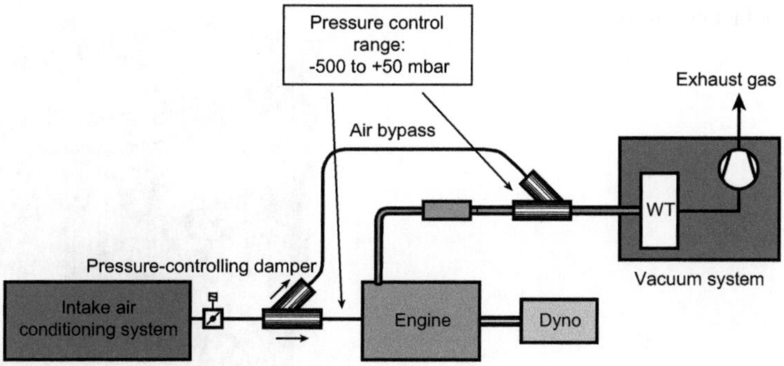

Fig. 3.31 Schematic diagram—altitude simulation

5000 m above sea level can be simulated, no matter where the test cell is located (Fig. 3.31).

3.2.4.2 Coolant Conditioning

Coolant conditioning systems primarily control the coolant temperature in engines on the test bed and keep them at a constant level. Such systems replace the vehicle's radiator that is cooled by the headwind (Fig. 3.32). A similar system is needed to keep the PEM fuel cell stacks cool.

The simplest method would be to use a water/coolant heat-exchanger whose cooling water flow is controlled. But the problem with such systems is that control behavior is heavily dependent on the unit under test's behavior.

Integrated heating systems, on the other hand, not only bring the coolant to the demand temperatures very quickly but also provide for more stable control in dynamic engine operation. An additional circulation pump compensates for test-bed related pressure losses (piping, docking plate couplings in pallet systems, etc.).

Systems with circulation pumps also find applications in areas such as:

- Preheating with and without engine
- Fast cooling for stopped engines (starting tests, climate chamber applications)
- Replacement of the engine's own circulating pump (research engines)
- Control of dynamic temperature profiles
- Enhancement of control performance in idle or partial-load operation

One of the most challenging tasks a coolant conditioning system has to accomplish is to simulate dynamic temperature curves on engine test beds of the kind that are measured on chassis dynamometers, for example, in the course of exhaust emission cycles.

A mixed control system usually serves to set the required temperature. For this purpose, one part of the coolant passes from the engine via the heat exchanger (cooling), whereas the other part is routed from the engine via an electric heating system (heating)

Fig. 3.32 Coolant conditioning

using a bypass. The mixing valve adjusts the required coolant temperature. Figure 3.33 shows the dynamic demand value curve for temperature in an NEDC test run.

Optionally, one may also connect the electric heating system in series with the heat exchanger, although this method will lengthen the response time somewhat (Fig. 3.34). Control valves that additionally regulate the coolant flow rate can help to broaden the application range (e.g. in view of engine size). Simple systems as shown in Fig. 3.35 consist merely of heat exchanger and control valve and are usually employed in steady-state test runs.

Coolant pressure control can be of particular advantage in development tasks:

- System pressure control for temperature compensation
- Pressure loss tests on the radiator
- Tests on the engine's own circulating pump

3.2.4.3 Oil Conditioning

Oil conditioning systems set and maintain oil temperature and oil pressure at precisely definable levels. To achieve fast preheating of oil and improved control accuracy in dynamic engine operation, the device may have an integrated electric heating system. Once again, the required application dictates which functions the system must satisfy.

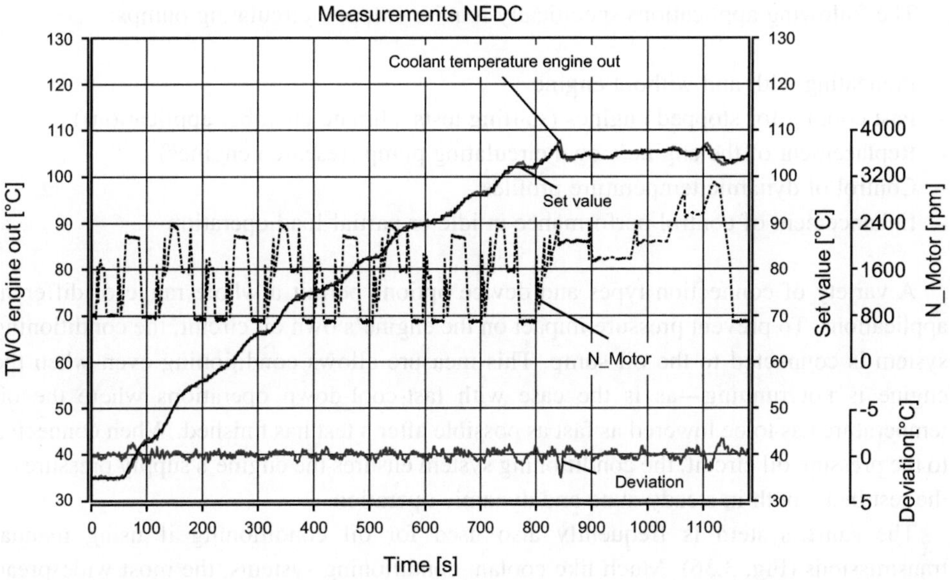

Fig. 3.33 Simulation of dynamic temperature profiles on the engine test bed

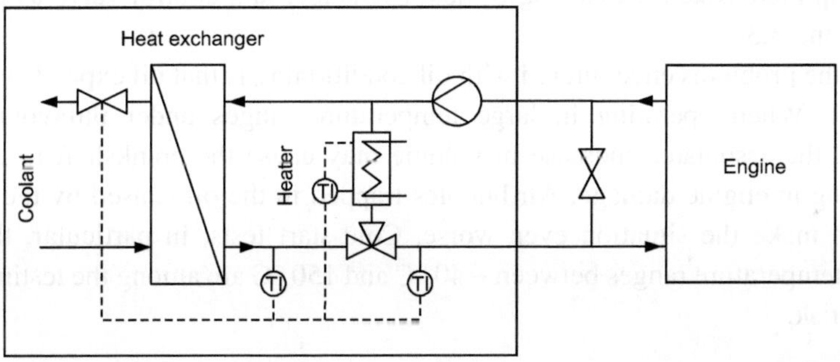

Fig. 3.34 Schematic diagram of a coolant conditioning system with broad application range

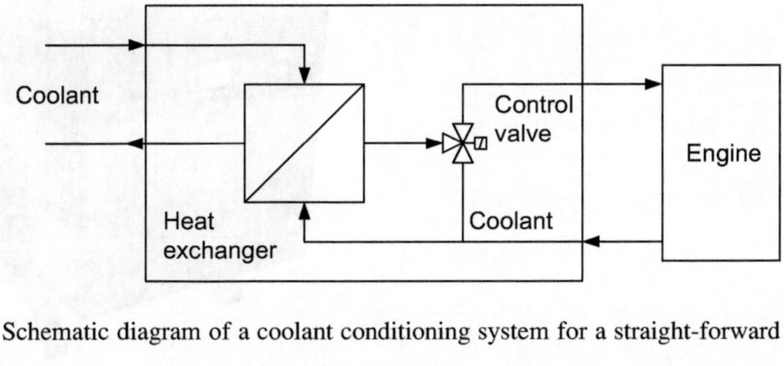

Fig. 3.35 Schematic diagram of a coolant conditioning system for a straight-forward steady-state application

The following applications specifically require external circulating pumps:

- Preheating with and without engine
- Fast cooling for stopped engines (starting tests, climate chamber applications)
- Replacement of the engine's own circulating pump (research engines)
- Control of dynamic temperature profiles
- Enhancement of control performance in idle or partial-load operation

A variety of connection types and device options permit a whole range of different applications. To prevent pressure impact on the engine's own oil circuit, the conditioning system is connected to the oil sump. This measure allows conditioning even when the engine is not running—as is the case with fast-cool-down operations where the oil temperature has to be lowered as fast as possible after a test has finished. When connected to the pressure oil circuit, the conditioning system ensures the engine's supply pressure on the test bed—both in steady-state and dynamic operation.

The same system is frequently also used for oil conditioning if using manual transmissions (Fig. 3.36). Much like coolant conditioning systems, the most widespread models are based on a mixed control system. Once again, where simple applications (generally steady-state test runs) are concerned, it is perfectly sufficient to employ a greatly simplified system consisting of heat exchanger and control valve as indicated in Figs. 3.37 and 3.38.

One of the problems encountered with oil conditioning is that oil expands when heated (Fig. 3.39). When operating in large temperature ranges under unfavorable engine conditions, the associated increase in volume may cause the crankshaft to dip into the oil, resulting in engine damage. Air bubbles trapped in the oil caused by the circulating pumps can make the situation even worse. Cold-start tests, in particular, that involve executing temperature ranges between −40 °C and 150 °C are among the testing scenarios at highest risk.

Fig. 3.36 Oil conditioning system

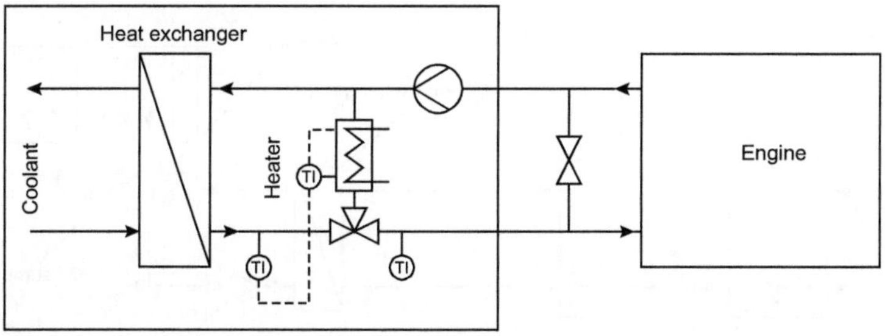

Fig. 3.37 Schematic diagram of an oil conditioning system for a broad range of applications

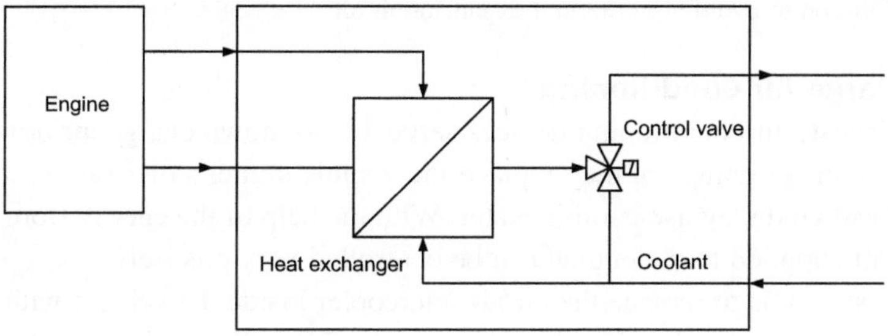

Fig. 3.38 Schematic diagram of a simple steady-state test run

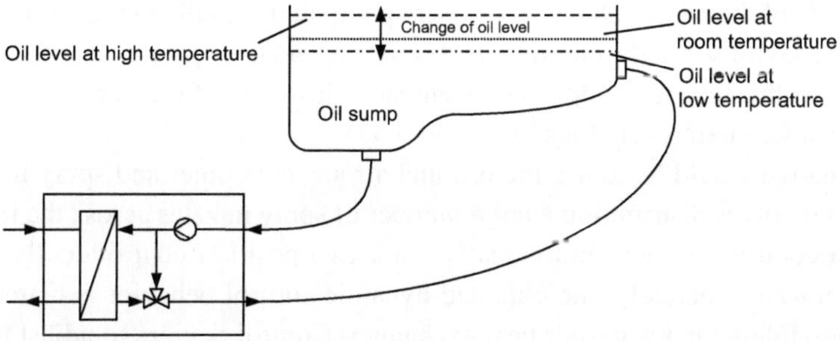

Fig. 3.39 Oil expansion problems in high temperature ranges

To solve the problem with these applications, the oil circuit, and thus possible thermal volumetric expansion, has to be kept as small as possible. This is achieved by separating the conditioning component and the oil circuit as shown in Fig. 3.40 using a heat exchanger. The oil circuit with the pump module, which has to be placed close to the engine, is reduced to a circulating pump and a heat exchanger. A coolant conditioning device serves as a conditioning circuit. The improved thermal properties of the coolant mixture in comparison to the oil compensate for any control-related disadvantages arising from the additional heat exchanger.

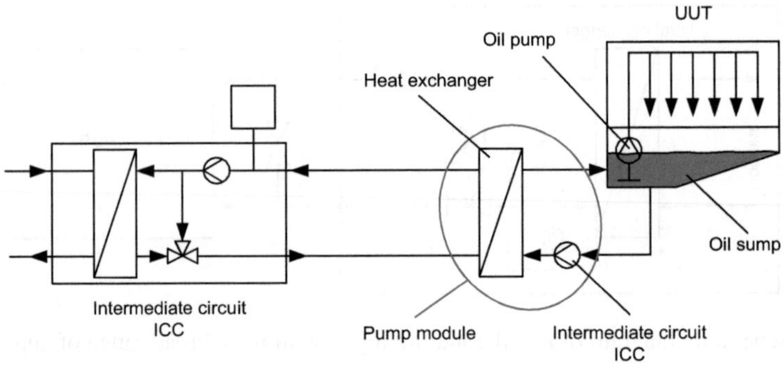

Fig. 3.40 Solution to avoid high thermal expansion in oil

3.2.4.4 Charge Air Conditioning

Charge (or boost) air conditioning devices serve to cool down charge air downstream of the turbocharger. In doing so, they replace the vehicle's intercooler on the test bed, but instead of headwind they use cooling water. With the help of the energy from the hot air, the charge air supplied to the engine can be controlled very precisely.

The easiest way is to replace the air/air-intercooler inside the vehicle with a water/air heat exchanger that has to be placed as close as possible to the engine to optimize control behavior. The turbocharger's boost air, which is transferred via a heat exchanger, is conditioned indirectly to a constant temperature by controlling the cooling water on the secondary side of the heat exchanger. An extra pump in the cooling water circuit improves the control behavior when the engine speed is low. Manually operated or automatic valves installed inside the air circuit additionally enable adjustment of pressure loss in relation to the engine's own intercooler (Figs. 3.41 and 3.42).

An alternative would be to use the original air/air intercooler and spray it with water. The method involves distributing a large number of spray nozzles across the intercooler's surface and controlling them individually via a two-position controller. By controlling each spray nozzle separately, the obtained dynamic control behavior is better than what would be possible with a water/air heat exchanger. Control devices to adjust the pressure loss on the air side are not required in this case because the engine's own intercooler is being used (Fig. 3.43).

3.3 Measuring

3.3.1 Temperature Measurement

Temperature is a physical state variable. Its unit of measurement is Kelvin (K). Other common temperature scales are degree Fahrenheit (°F) in Anglo-Saxon areas and degree Celsius (°C).

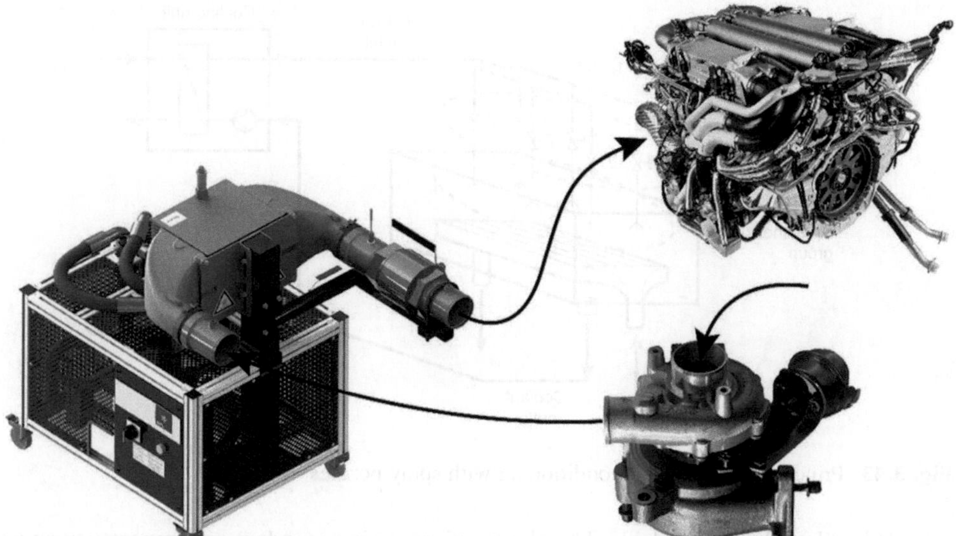

Fig. 3.41 Charge air conditioning

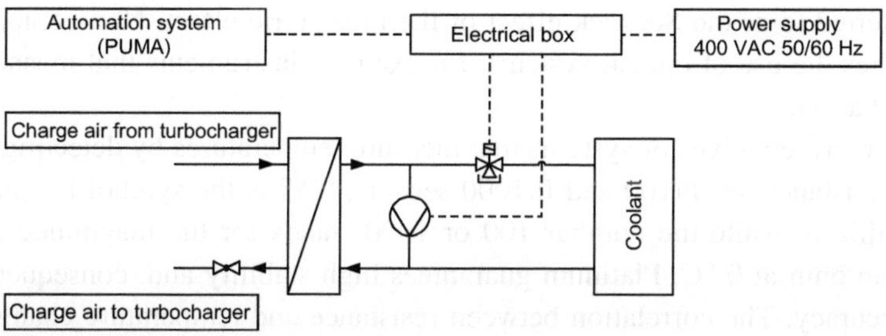

Fig. 3.42 Schematic diagram of a charge air conditioning system with water/air heat exchanger

$$\begin{aligned} t\,[^\circ C] &= T[K] - 273.15 \\ 1\,^\circ C &= \left(\frac{9}{5} + 32\right)\,^\circ F \end{aligned} \tag{3.5}$$

Equation 3.5 Conversion of units of temperature

The measurement of temperature at various positions of a unit under test or unit-under-test environment is of vital significance, as it very often gives an indication of the state of unit under test or test environment. We distinguish between a number of temperature ranges. Operation frequently takes place in ranges between −40 °C and 180 °C. Examples are oil and water temperatures, or even battery temperatures in electrically powered vehicles. An engine's exhaust gas system produces exceedingly higher temperatures of

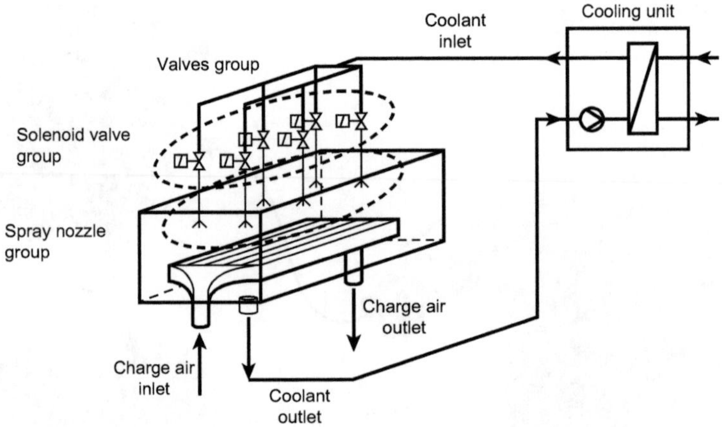

Fig. 3.43 Principle of charge air conditioning with spray nozzles

several hundred degrees Celsius. The choice of sensor is dependent on temperature range, accuracy specifications, the required dynamics and the possible installation site.

Normally, we measure temperatures by taking advantage of the temperature's dependence on electrical resistance, the thermal expansion of solid matter, gases or fluids, or what is referred to as the Seebeck effect or thermoelectric effect. Non-contact measurement requires the use of optical systems, for example instruments that measure infrared thermal radiation.

Typical representatives of systems that measure temperatures by detecting changes in electrical resistance are Pt100 and Pt1000 sensors. "Pt" is the symbol for platinum, the material utilized, while the number 100 or 1000 stands for the magnitude of the used resistance in ohm at 0 °C. Platinum guarantees high stability and, consequently, a high level of accuracy. The correlation between resistance and temperature change is in fact linear over a certain temperature range. The use of nickel (Ni100) is usually less costly. These sensors achieve higher levels of sensitivity, yet within a narrower temperature range.

Thermistors (high-temperature conductors) provide another way of measuring temperature by correlating electrical resistance *R* with temperature. Thermistors are classified into PTC (positive temperature coefficient) and NTC (negative temperature coefficient) type thermistors. Heat increases the number of active charge carriers in the materials in use (usually metal oxides), resulting in a change in resistance that is expressed in approximation equations.

$$\frac{1}{T} = a + b\ln(R) + c\ln^3(R) \tag{3.6}$$

Equation 3.6 Steinhart-Hart approximation equation for NTC thermistors

Thermocouples are particularly well suited for use in high-temperature ranges. The underlying measuring principle is the Seebeck effect. In a metal conductor, an electric voltage U is generated whenever there are different temperatures (T_1 and T_2) at its ends.

$$U_{th} = c\,(T_2 - T_1) \tag{3.7}$$

Equation 3.7 Thermoelectric power dependent on the material-specific Seebeck coefficient

Thermocouples are designed by combining materials that are capable of generating maximum thermoelectric power. Thermocouples are standardized in accordance with DIN IEC 584-1 and DIN 43710 and are identified by a letter for the particular pairing of thermocouple materials. The letter J, for example, stands for iron-constantan and the letter K for chromel-alumel (chromel is made of 90 % nickel).

Liquid-based thermometers rely on materials that have the property of expanding when heated. Either the volume or the pressure on a membrane is then measured. Today, depending on the temperature range needed, liquids like pentane, ethanol or toluol are used.

For further details on measurement principles or usage of temperature sensors in the automotive field, please refer to the relevant publications (e.g. [1, 2]).

3.3.2 Measuring Electrical Quantities

Voltage indicates how much energy is needed to move an electric energy charge within an electric field. The symbol for voltage is U and it is indicated in units called volts (V). One volt is defined as the potential difference between two points of a metal conductor that will result in one joule of energy per coulomb of charge that passes through it. In a constant electric current (denoted by I) of one ampere, this corresponds to one watt of power (denoted by P).

$$P = U \cdot I \tag{3.8}$$

Equation 3.8 Relationship between electric voltage, current and power

Voltage measurement generally requires drawing energy from a voltage source and then converting it into some other form that is measurable. This usually involves transforming the energy into an electromagnetic field, heat or current.

In pointer instruments, spring systems, for example, usually transform the electromagnetic force into a corresponding mechanical counterforce. Typical representatives are moving-coil instruments. When converting energy into heat, the temperatures in heating elements are used. If the drawn amount of energy must be low for measurement purposes, Ohm's law is applied to transform the voltage U to be measured via a defined resistance (denoted by R) into a current which is proportional to the voltage. The current is integrated over a certain interval of time and used to determine the voltage.

Electric current (denoted by I) is the movement of charge carriers and is expressed in units called amperes (A).

$$I = \frac{dQ}{dt} \tag{3.9}$$

Equation 3.9 Definition of electric current

Electric current is usually determined by measuring the drop in voltage when using an internal or external measuring resistance (i.e. shunt resistance). For measuring larger currents, a not uncommon method is to use current dividers that take advantage of Kirchhoff's first law. With the help of this law, which is also called Kirchhoff's junction rule (or nodal rule), currents are defined as branches. Current flow also has a thermal effect, which means that temperature-dependent quantities can be used to measure current too.

Conductors with current flowing through them produce a magnetic field around the conductor. By taking advantage of this physical effect it is also possible to measure current. Magnetic sensors or special sensors called Hall sensors, which are available as integrated circuits, enable the measurement of magnetic fields. The action of force that accompanies a magnetic field is also a measure for electric current. Using a mechanical counterforce it is relatively easy to indicate amperage using analog instruments.

3.3.3 Strain Measurement

Strain ε describes a body's change in length induced by an external load. It has no unit. It is usually expressed as a percentage. The equation below analyzes strain (stretching or compression in the length of a body) caused by an external force. Section 3.3.1 Temperature Measurement deals with thermal strain used as a measuring principle for temperature measurement.

$$\varepsilon = \frac{\Delta l}{l_0} \tag{3.10}$$

Equation 3.10 Definition of strain based on change in length

Where elastic deformation is concerned, most materials show a linear relation between force and strain, which follows Hooke's law. This is the reason why strain is also useful for measuring the acting force (see Sect. 3.3.4). In addition to altering a body's length, strain proportionately also affects its diameter. The ratio is expressed using the Poisson number.

The most well-known technique for measuring strain is a strain gauge (also: strain gage). A change in length or cross-section of an electrical conductor results in a change in electrical resistance. Strain gauges are made of materials in which the change in resistance is high. Meanwhile, a large variety of designs have become available, in the form of foil, wire or semiconductor strain gauges. Frequently, four strain gauges are mounted in a radial configuration, each separated by 90°, to be able to record both changes in length and in circumference when a load is applied in a certain direction. There are even prefabricated commercial solutions available for these characteristic arrangements.

Another strain measuring method involves evaluating the power output of an optical fiber using what is termed as fiber Bragg grating (FBG). If light is fed into an optical fiber, and if the wavelength of the light and that of the interference filter match, the light wave is reflected, allowing the detection of only a small output of power at the end of the optical fiber. A strain creates a change in the grating period, causing the reflection of less light. This results in the measurement of a higher power output at the end of the fiber.

3.3.4 Force and Pressure Measurement

According to the principle of linear momentum for objects with constant mass (Newton's second law), force (denoted by F) acting on an object with constant mass (denoted by m) is proportional to the object's acceleration (denoted by a). Force is measured in units called newtons.

$$\begin{gathered} F = m \cdot a \\ 1\,N = 1\,\frac{kg\;m}{s^2} \end{gathered} \tag{3.11}$$

Equation 3.11 Force and its unit newton

Force always acts in one direction. This is of particular relevance when locating the positions to install force sensors. Physical principles used for measuring force include piezoelectric, piezoresistive or electromagnetic effects (see [2]). On test beds, particularly piezoelectric and piezoresistive sensors are in use.

Piezoelectric sensors transform a force directly into a measurable voltage. Due to an external influence, the sensor's crystal lattices suffer a distortion and charge carriers move outward. The resulting voltage is directly proportional to the force. If a force is allowed to act on a metal conductor, it stretches. This mechanical deformation changes electrical resistance. This piezoresistive effect allows evaluation using strain gauges (see Sect. 3.3.3). As expressed by Hooke's law, force can be derived from the deformation (strain).

An application example for this particular measuring technology is the load cell, which, among other things, is used on chassis dynamometers and cradle-mounted asynchronous machines (see also Sect. 3.3.6).

Pressure (p) is defined by a force (F) applied perpendicular to a surface (A). It is measured in units called pascals (Pa). But there are also other units commonly used to express pressure that vary by country.

$$\begin{aligned} &p = \frac{F}{A} \\ &1 \ \text{Pa} = 1 \frac{\text{N}}{\text{m}^2} = 1 \frac{\text{kg}}{\text{m}\,\text{s}^2} \\ &1\,\text{bar} = 1 \cdot 10^6 \ \text{Pa} \\ &1\,\text{bar} = 14.5 \ \text{psi} \\ &1 \ \text{Torr} = 1 \ \text{mm} \ \text{Hg} = 133.3 \ \text{Pa} \end{aligned} \tag{3.12}$$

Equation 3.12 Pressure in the units pascal, bar, psi and mm of mercury

On test beds, aside from temperature, pressure is one of the most important indicators for assessing the condition of the unit under test and the test bed environment. Examples to illustrate this fact are the pressures in the intake manifold or in an engine's exhaust gas system, pressures in hydraulic or pneumatic components, braking pressures, positive or negative pressures, e.g. in refueling systems, or even combustion chamber pressures that are measured using indicating measurement systems (see Sect. 3.3.16). Pressure measurement is primarily needed when dealing with liquids and gases. With solid materials, the most widespread requirement is force measurement.

Direct pressure measurement is only considered useful if pressures are excessive. All of the known resistances also exhibit pressure dependence. However, the high dependence on temperature and the difficulties presented by ensuring a sealed lead-out of the connections from the pressure medium are aspects that do not support using direct pressure measurement.

The most widely used instruments are indirect pressure sensors that measure the deformation of a diaphragm (i.e. membrane) by determining the difference in pressure between its two faces. Depending on what measuring range is required, the sensor is selected by its material, diaphragm diameter and diaphragm thickness. Most widespread materials are steel, ceramics and semi-conductor materials. Depending on the installation position, different kinds of pressure measurements are performed. With differential pressure measurement, the pressures requiring measurement act on either side of the diaphragm. In a relative (gauge) pressure measurement, the pressure to be measured acts on one side of the diaphragm. The other side refers to the reference pressure (often this is the ambient air pressure). An absolute pressure measurement is a measurement in which a defined pressure is applied to one face of the diaphragm at constant volume. As a result, the reference pressure is known and the absolute pressure in the volume to be measured is determinable.

The measurement of the membrane's deformation is possible using a variety of effects. If the pressure is indicated the conventional way using a manometer, the diaphragm's deformation is directly indicated via a linear torsion spring. Apart from the direct mechanical method, the deformation can also be transformed into an electrical signal first. In many cases this is done by using piezoresistive methods and strain gauges (see Sect. 3.3.3). Another useful principle is the change in a capacitance due to the deformation in the diaphragm. According to this technology, a capacitor's ability to store an electric charge changes depending on the distance that separates the diaphragm from a fixed reference surface. This distance can also be measured using inductive sensors. Another way would be to detect the force applied to the diaphragm using a force sensor.

For further details on measurement principles or usage of temperature sensors in the automotive field, please refer to the relevant publications (e.g. [1, 2]).

3.3.5 Acceleration Measurement

As with force measurement, Newton's second law of motion can be used to describe the relationship between force (denoted by F) and acceleration (denoted by a) for a constant mass (denoted by m).

$$\begin{aligned} F &= m \cdot a \\ [a] &= \frac{m}{s^2} \end{aligned} \tag{3.13}$$

Equation 3.13 Relationship between force and acceleration for a constant mass

This law shows that, similar to force, acceleration is a vector quantity (it has a certain magnitude and direction). As is the case when measuring force, acceleration is also determined by using systems that measure displacement and mechanical stress.

In displacement-measuring systems, a defined seismic mass is elastically attached to a body whose acceleration is supposed to be measured. The coupling is described by using the model of a spring-damper system. Using the seismic mass along with the spring and damping constant, the measuring range and the level of sensitivity are adjustable. Now either piezoresistive, piezoelectric or capacitive methods (see Sects. 3.3.2 and 3.3.4) can be used to measure the mass's displacement. If the seismic mass is position-controlled, on the other hand, the restoring force necessary to hold the position constant can be used as a measure for acceleration. Integrated acceleration sensors are now available for all space coordinates, which even take into account gravitational acceleration, allowing their installation in any desired position. Thanks to MEMS technologies (micro-electromechanical systems) surprisingly small models are available.

Aside from obtaining acceleration by displacement measurement, it is also possible to determine acceleration by measuring mechanical deformation induced by a force. This task once again requires the use of strain gauges (see Sect. 3.3.3).

3.3.6 Torque Measurement

The torque produced by an internal combustion engine or an electric motor is the most important output parameter in these automotive powertrain components. Precise measurement of torques in powertrain development is consequently of fundamental significance. The following section addresses different methods of torque measurement.

3.3.6.1 Measuring Torque Using a Torque Measuring Flange

Figure 3.44 shows the setup of a test bed with a torque measuring flange (torque flange for short). The torque flange consists of a rotor integrated in the shaft system on the dynamometer and a stator firmly installed at the dynamometer's bearing support. Inside the rotor, strain gauges capture the torsion caused by the torque exerted from the outside. Telemetry transfers the torsion to the stator from where the torsion undergoes direct conversion into torque, based on the known rotor geometry and the material properties.

The advantage of using a torque flange is that the unit under test's torque is measured directly, without being influenced by the acceleration torque from the dynamometer

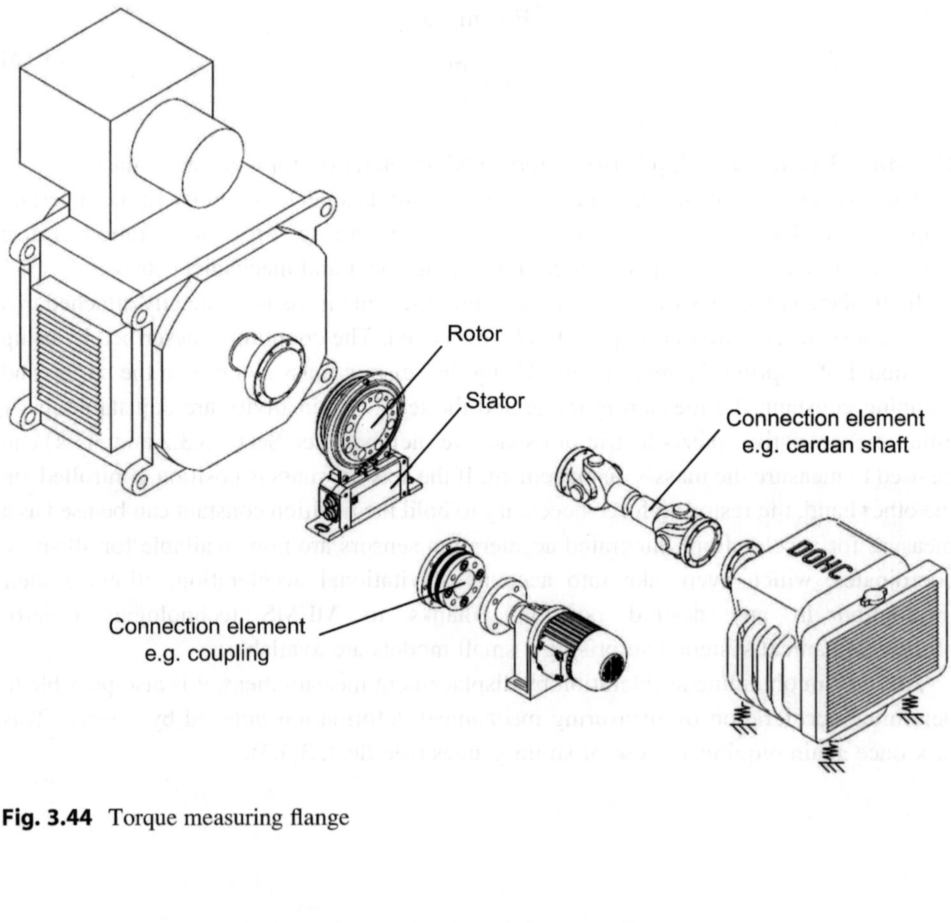

Fig. 3.44 Torque measuring flange

rotor's rotating mass. One of its downsides is the torque flange's great temperature impact on measurement accuracy, as it is directly exposed to the high temperature of the dynamometer's rotor. During operation, temperatures in the rotor shaft, where it is connected to the torque flange, can reach up to 60 °C.

3.3.6.2 Measuring Torque Using a Cradle-Mounted Dynamometer and a Load Cell

With this method, the dynamometer's stator is cradle-mounted with as little friction as possible. Cradle mounting is provided in addition to the rotor shaft bearing system (see Fig. 3.45). The cradle bearing serves to decouple the power flow and so ensures that the electrical torque present in the air gap of a machine is counterbalanced via the load cell. This internal electrical torque matches the mechanically effective torque produced via the rotor shaft.

Given what has been said, the counterforce measured with the load cell and the known effective lever arm is now used to directly calculate the mechanical torque acting on the rotor shaft.

A disadvantage associated with this method, however, results from the circumstance that, during rotor acceleration processes, the sensor also measures the acceleration torque. To achieve dynamically correct measurements, this requires computational compensation in real-time.

One of the benefits of this arrangement, on the other hand, is that, as a rule, the force measurement is only subjected to a low and steady temperature impact, so the counterforce itself can be measured very precisely. Figure 3.46 shows a load cell. Its application in test bed dynamometers can be seen in Fig. 3.47.

3.3.6.3 Calibrating Torque Measuring Chains

For calibrating torque measuring chains, the methods most commonly applied today (which we will address in further detail below) are as follows:

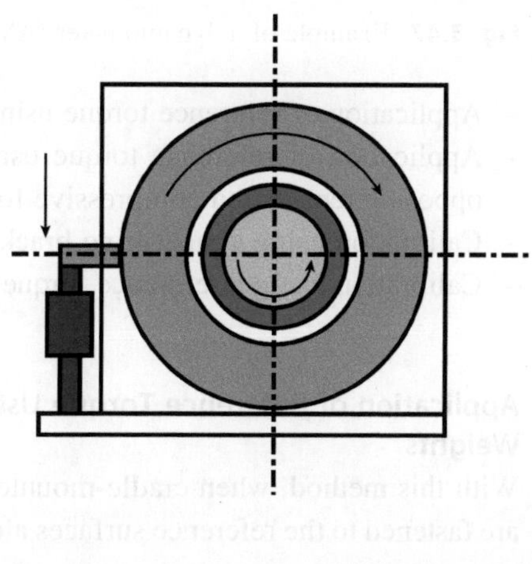

Fig. 3.45 Schematic diagram of load cell

Fig. 3.46 Load cell

Fig. 3.47 Example of a dynamometer (AVL DynoExact) with load cell

- Application of reference torque using lever arm and force application using weights
- Application of reference torque using lever arm and force application by means of opposing tensile and compressive forces via reference load cells
- Calibration using a measuring bracket
- Calibration using a reference torque flange

Application of Reference Torque Using Lever Arm and Force Application Using Weights

With this method, when cradle-mounted dynamometers are being calibrated, lever arms are fastened to the reference surfaces along the stator's circumference on either side of the

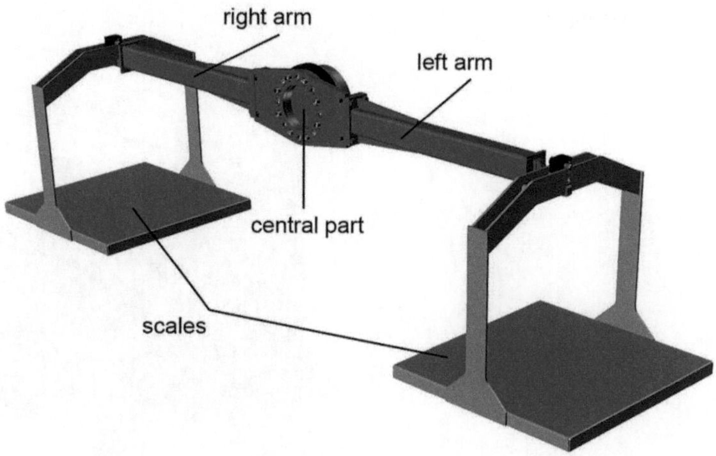

Fig. 3.48 Calibration using a lever arm

stator. If torque flanges are to be calibrated, the lever arm is symmetrically screwed on to the torque flange. For calibrating torque flanges, the rotor shaft has to be blocked so that the machine's stator counterbalances the transmitted torque.

For calibrating cradle-mounted dynamometers, the lever arms are fastened to the reference surfaces on the left and right side of the stator housing (see Fig. 3.48). Calibrated weights that are placed on scales on the left and the right side of the stator housing apply the force, so that the calibration is performed in both directions. The reference torque is now determined based on the known lever arm and the weight force.

The disadvantage of this method is that the distance between the force application point and the rotation point is subject to manufacturing and component tolerances. Also, the bearing clearance in the cradle bearings and friction bearings along with the bending and shearing forces caused by the mounting of the levers and resulting from force transmission are possible error sources. When dealing with cradle-mounted dynamometers, possible errors caused by breakaway and friction forces in the cradle bearing system must also be considered. Another factor that we must know for calculating the force is the gravitational acceleration at the calibration location. All things considered, this method is suitable only to a limited extent for the exact calibration of high-precision torque flanges and load cells.

Application of Reference Torque Using Lever Arm and Force Application by Means of Opposing Tensile and Compressive Forces Via Reference Load Cells

When applying this method, the setup is fundamentally the same as the technique described above. It differs, however, in that a high-precision reference load cell is integrated at each of the force application points. Via this load cell, using e.g. air cylinders, a tensile force is transmitted to the one side, whereas an equal magnitude of compressive force is transmitted to the other. This helps to reduce the effects of shearing forces. Moreover, the force measured at the reference load cells is captured in high

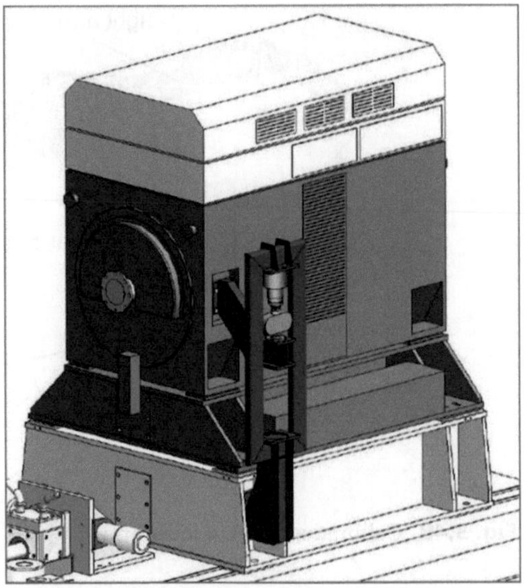

Fig. 3.49 Calibration using a measuring bracket

precision, eliminating the need to take into account the acceleration of gravity when calculating the torque.

Calibration Using a Measuring Bracket

This method is exclusively used when calibrating cradle-mounted machines, where extremely high loads are needed in order to perform calibration. As a rule, this applies to large hydraulic dynamometers (Fig. 3.49).

The measuring bracket consists of a steel bracket with a bend that has been very accurately measured with regard to its elasticity. For calibration purposes, the measuring bracket is connected in series to the load cell in the power flow. Force transmission is initiated, for example, hydraulically or using a spindle. A dial gauge records the measuring bracket's deformation. The deformation is then converted directly into a reference force matching the deformation, based on the bracket's deformation/force curve, which is established beforehand. Apart from possible error sources associated with weight-based calibration, the dial gauge's reading accuracy and the interpolation of the measured deformation over the measuring bracket's deformation/force curve have a critical impact on the calibration's overall accuracy.

Calibration Using a Reference Torque Flange

With this method, a high-precision reference torque flange is connected directly to the dynamometer's torque flange requiring calibration. The dynamometer's shaft is blocked against the bearing support to counterbalance the transmitted torque. Force transmission must be free of shearing and bending forces. The reference torque flange transmits the

Fig. 3.50 Calibration using a reference torque flange

force exerted by levers, spindles, compressed air or other force generating devices (Fig. 3.50).

This method has the advantage of providing a direct comparison with a high-precision measurement standard, which also makes it suitable for achieving an exact calibration of high-precision torque measuring flanges. If the measuring flanges are positioned very precisely with respect to each other and the connection is elastically decoupled in transverse direction at the same time, the effects of the shearing and bending forces can be reduced to a minimum. The advantage of this method is that the process can be automated, reducing the time required for calibration very significantly in comparison to conventional methods.

3.3.7 Speed Measurement

Speed is generally measured using incremental encoders. The following measurement principles are applied:

3.3.7.1 Optical Encoders

With these sensors, light (markings 1, 2 and 3 in Fig. 3.51) passes through an optical aperture (4) to a disk that has a number of tracks of equally spaced slots (A and B track).

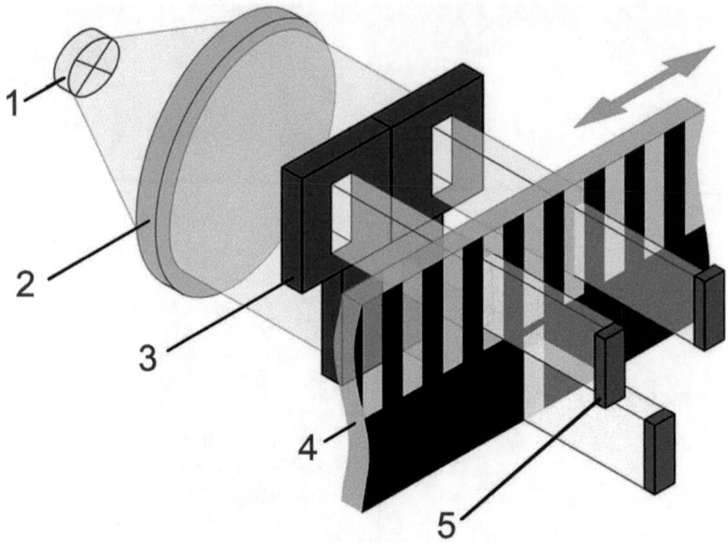

Fig. 3.51 Optical encoder

The light detectors (5) opposite the tracks record the light pulses they receive when the disk rotates.

Using two tracks with slots positioned out of phase, signal curves are produced in which the light pulses received from one of the two tracks lead the other track, depending on the direction of rotation. This makes it possible to determine both speed and direction of rotation.

A further aperture and an additional incremental track on the disk, which only has one slot per signal period, makes it possible to determine an additional zero position and thus the position in relation to the circumference.

3.3.7.2 Magnetic Encoders

Magnetic encoders (see Fig. 3.52) require the use of a disk made of magnetic material, upon which a certain magnetized pattern serves as an incremental track. Here Hall elements act as receivers.

3.3.7.3 Gear Tooth Encoders

These types of sensors (see Fig. 3.53) use the tooth pitch of a gear as an incremental encoder. Inductive sensors record the pulses and subsequently count them. These somewhat simpler encoders are commonly used to determine the speed only and not the direction of rotation or the zero position.

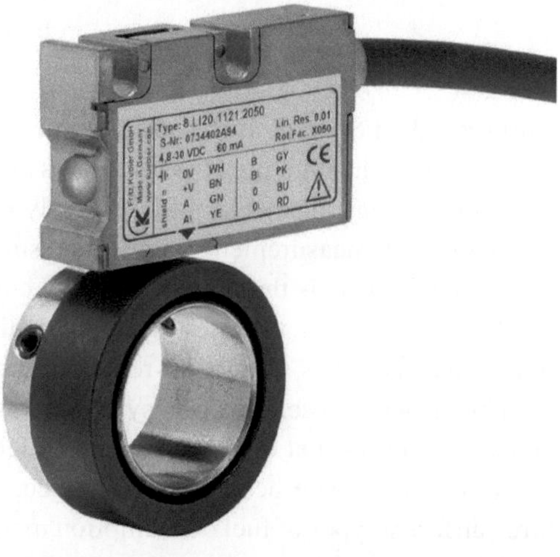

Fig. 3.52 Magnetic encoder

Fig. 3.53 Gear tooth encoder

3.3.8 Fuel Measurement

3.3.8.1 Goal of Fuel Consumption Measurement

The goal of fuel consumption measurement is to determine the amount of fuel injected into the cylinder. This measurement quantity is a key value, for example, if you wish to determine, for example, the engine's efficiency. Fuel consumption is a fundamental target variable for the optimization of present-day combustion engines.

As engine efficiency is defined by power, the target variable of fuel consumption is the mass of fuel supplied within a time unit—usually indicated by kg/h. For determining efficiency directly, typically specific fuel consumption (usually expressed as g/kWh) is measured, which is essential for optimizing the engine's degree of efficiency. Specific fuel consumption b_e is calculated from the mass of fuel B injected per time unit and the effective power output by the engine P_e:

$$b_e = \frac{B}{P_e} \tag{3.14}$$

Equation 3.14 Specific fuel consumption

Because of the high injection pressures and the immense temporal dynamics of modern injection systems, the mass of fuel actually injected into the cylinder is technically barely accessible for measurement without causing a serious impact on the unit under test's functionality. This is the reason why fuel consumption is generally measured in the feed line to the engine's injection system. In this respect, we must establish whether the injection system has a function-related return line, e.g. for cooling purposes. Especially in vehicles with diesel injection systems, this line is fed back into the vehicle's fuel tank. When measuring fuel consumption, this means that consumption is determined by measuring the difference between the fuel feed flow and the fuel return flow. Accordingly, three different types of fuel consumption metering systems have become established in the market.

3.3.8.2 Measuring Systems for Determining Fuel Consumption

Measuring System for Engines Without Return Line

This application usually uses simple sensors which require installation directly inside the feed line to the engine because of a slight pressure drop. Particularly the PLU measurement principle and simple volumetric sensors, such as four-piston meters or spindle displacement meters, have proved useful. The disadvantage of such sensors is that they rely on an indirect mass measurement method, according to which mass flow determination is only possible with an additional density measurement or by means of tabulated temperature-related density values (see also Sect. 3.3.1 and Fig. 3.54).

Differential Measurement System Using an Integrated Hydraulic Circuit

If the engine has a return line, or if the engine's fuel feed line requires a certain fuel temperature to ensure reproducible engine conditions, a fuel conditioning system is mounted between the flow sensor and the engine. The system is integrated as an independent hydraulic circuit downstream of the flow sensor and in close proximity of the engine using a bypass or H-piece (see Fig. 3.54).

One or more heat exchangers within the measuring circuit condition the fuel temperature. An additional fuel pump maintains a permanent circulation flow in the circuit. A bypass hydraulically delivers the required difference between the engine's fuel feed line and return line. The amount of fuel that continues to flow into the circuit is equivalent to the engine's consumption under the boundary condition that the mass within the measuring circuit is constant. With this type of system setup, maximum accuracy is obtained. Due to the large volume in the measuring circuit, the attainable accuracy is directly dependent on the temperature stability achievable within the measuring circuit.

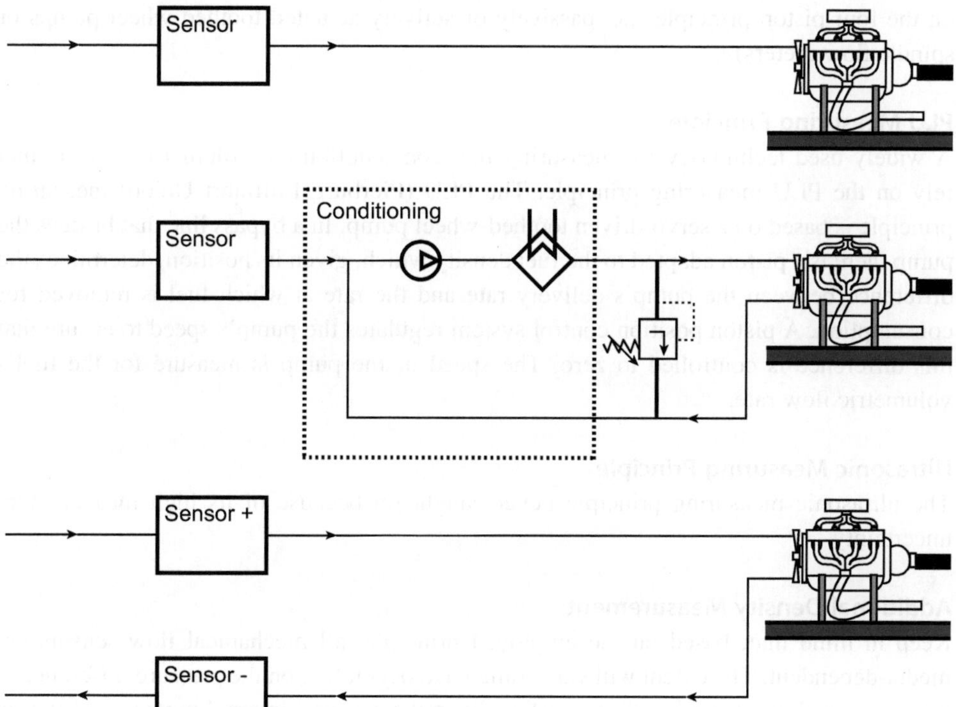

Fig. 3.54 Schematic diagram of a measuring system for engines without fuel return *line* (*top*), differential measurement system using a hydraulic circuit (*middle*) and measuring system with difference obtained by calculation (*bottom*)

Measuring System With Difference Obtained by Calculation

With large engines and stationary engines, a widely established method involves obtaining the difference by calculation. To accomplish this, two volumetric flow sensors are mounted in the engine's fuel feed line and return line, each with low pressure drop and in the proximity of the engine. The consumption measurement requires calculating the difference between the two signals. The disadvantage of this method is that two large flows are involved in calculating the difference. Particularly where consumption levels are low, this method of obtaining the difference must be viewed critically, as a very high absolute measurement uncertainty is generated. While this method is fine for engines in steady-state operation, high measurement uncertainties are generated in engines designed for transient operation.

3.3.8.3 Sensors

In principle, we can distinguish between volume-based and mass-based sensors. With mechanical volumetric sensors, the translational flow is always transformed into a rotary motion, which is often output as pulses per unit of volume. Widespread sensors are based on the four-piston principle (i.e. passively or actively actuated toothed-wheel pumps or spindle flow meters).

PLU Measuring Principle

A widely used technology for measuring fuel consumption are volumetric sensors that rely on the PLU measuring principle. The PLU (Pierburg Luftfahrt Union) measuring principle is based on a servo-driven toothed-wheel pump. In a bypass line that bridges the pump there is a piston adapted to the fuel density which, given its position, determines the difference between the pump's delivery rate and the rate at which fuel is removed for consumption. A piston position control system regulates the pump's speed to ensure that this difference is controlled to zero. The speed of the pump is measure for the fuel's volumetric flow rate.

Ultrasonic Measuring Principle

The ultrasonic measuring principle never caught on because of its high measurement uncertainty.

Additional Density Measurement

Keep in mind that, based on the employed principle, all mechanical flow sensors are media-dependent. The extent will vary somewhat, depending on the pressure difference at the sensor and the viscosity of the media. Given the sensors' internal moments of inertia and friction, the measuring range is restricted, making measurements in the lower range possible to a limited extent only.

The calculation of the mass flow rate of interest requires not only the volumetric measurement but also the determination of density. For this purpose, process density meters are very often used that are positioned at a suitable position within the fuel circuit and allow determining the density and the density-related temperature. An additional temperature sensor in the supply line to the volumetric flowmeter is used to calculate the mass flow rate, assuming the fuel's volumetric expansion coefficient is known. Depending on the media temperatures and the design of the sensors, further factors contribute to the overall measurement uncertainty.

Among the currently available methods that measure the mass flow directly are methods that measure continuously (e.g. Coriolis sensor) and discontinuously (e.g. scales).

Fuel Balance

A fuel balance is technically a fuel tank that ensures minimum impact on the unit under test (the combustion engine) with regard to engine operation. The mass of this tank is measured continuously using a balance, which permits an accurate measurement of the mass of fuel consumed by the engine. One of the drawbacks of this method is that during the intermittent refilling of the balance, no readings are available. A further principle-related disadvantage lies in the fact that a transient flow is only obtained by determining the difference between balance results weighed at two different times. However, when time differences are short, the procedure is obviously susceptible to scatter.

Coriolis Sensor

Thanks to further developments by renowned manufacturers in the past 20 years, the Coriolis sensor has become very well suited for consumption measurement in a vast measuring range. The Coriolis sensor's sensitivity (and thus accuracy) is directly proportional to the medium's flow rate in the measuring tube and, consequently, indirectly proportional to the pressure drop across the sensor. To ensure high accuracy, it is therefore necessary to compensate for the pressure drop occurring in the measuring system.

The simplest and most common way of accomplishing this involves building up the maximum required pressure upstream of the sensor during operation and using a mechanical pressure regulator downstream of the sensor to remove the residual pressure, depending on the flow rate. The disadvantage encountered with this method is that such mechanical pressure regulators act as "hydraulic diodes," having the effect that the medium is prevented from flowing back through the sensor. For the overall application, this tends to entail negative side effects because it is usually no longer possible to keep the pressure within the limits needed by the unit under test.

Alternative methods secure a pressure build-up upstream of the sensor by ensuring that there is always a constant outlet pressure downstream of the Coriolis sensor. This allows the medium to flow back through the sensor. Methods like these are usually referred to as "open system" methods and are distinguished by being just as simple to use in engine applications as fuel balances are.

Apart from its dependence on temperature, which results from the variable rigidity of the measuring tube, a Coriolis sensor's dependence on the medium is minimal. Manufacturers have come up with type-specific technologies to compensate for these two dependences in order to ensure the required reproducibility. Beside this behavior, the employed measuring principle also entails a dependence on vibration, which only becomes noticeable in a narrow frequency band around the sensor's resonance frequency. Unbalanced vibration behavior in an individual sensor will lead to a zero-point stability problem which becomes apparent as a full-scale error. This also means there is always a measurement signal at low flow rates, but it will add a significant uncertainty to the measurement.

3.3.8.4 Measurement Uncertainty Factors in Fuel Measurement

The flow sensors currently available—whether they measure volume or mass—feature minimal measurement uncertainty ($<0.1\ \%$ of the measured value). In the case of small flow rates, measurement uncertainty is significantly higher due to scatter effects (either of a mechanical nature or due to the zero-point stability problem), which means that the mentioned level of measurement uncertainty is only achieved at high consumption levels. In commonly used sensors, the aforementioned scatter effects amount to a few grams per hour, so that due to the sensors alone, idling fuel consumption measurements in a range of

0.5 kg/h should be associated with an uncertainty of at least 0.5 %. From a consumption of ~1 kg/h onward, the sensors alone may exhibit a measurement uncertainty of ~0.2 %.

With volumetric flow sensors, there are additional contributing factors relating to density measurement. On the one hand, these result from the density sensor's measurement uncertainty, but they are also the result of temperature measurement uncertainty in the density sensor and in the volumetric sensor. Typical process density sensors feature a measurement uncertainty of ~1 kg/m^3, which results in a relative measurement uncertainty of ~0.15 % for typical fuels.

The error contribution is much higher due to the uncertainty associated with temperature measurement. In this respect, absolute temperature might not be relevant, but the uncertainty of the differential temperature between volumetric measurement and density measurement may easily amount to 5 °C due to dynamic effects. Should fuels have an average volume expansion coefficient of 0.1 %/°C, this can lead to an uncertainty in mass measurement of as much as 0.5 %.

The most significant contribution to measurement uncertainty is attributed to the thermal expansion of the measuring circuit volume. A rising average fuel temperature between the flow sensor and the injection nozzle leads to a volumetric expansion of the fuel amounting to 0.1 %/°C. In this case, the expansion of the engine geometry, tubing and piping is practically negligible. As the temperature in the engine allows conditioning only to a very limited extent, stabilizing the temperature in the measuring circuit is the only way to minimize this influence. It must be noted that neither the volume in the measuring circuit nor the absolute temperature of the fuel has any influence on this measuring error. Responsible for the error is exclusively the power that is input during the measuring process.

The following simple calculation illustrates this. Imagine the entire measuring circuit volume concentrated between the flow sensor and the injection nozzle in a volume V_0. Now should this volume undergo a change in temperature, it will either expand or contract. If it expands, a flow sensor will identify it as negative flow. If it contracts, it will be identified as an inflow into the measuring circuit and, consequently, as consumption. This is referred to as "apparent" flow. The question now is to establish the relationship between the input power and the apparent flow. Depending on the current temperature T, the volume trapped in the measuring circuit is expressed using the thermal expansion coefficient α (T_0—temperature at the time t = 0):

$$V(T) = V_0(1 + \alpha(T - T_0)) \tag{3.15}$$

Equation 3.15 Measurement circuit volume

After obtaining a derivative with respect to time, it is possible to calculate the change of volume per time unit, which the flow meter detects:

$$\frac{dV}{dt} = V_0\alpha\frac{dT}{dt} \tag{3.16}$$

Equation 3.16 Change in flow volume

Multiplication by the density at the time t = 0 yields the change in fuel mass:

$$\frac{dM}{dt} = M_0 \alpha \frac{dT}{dt} \tag{3.17}$$

Equation 3.17 Change in fuel mass

The internal energy Q of the medium altered by the temperature change can be expressed by the second law of thermodynamics (c_p is the specific heat capacity at constant pressure):

$$dQ = M_0 c_p dT \tag{3.18}$$

Equation 3.18 Change in internal energy

An analysis of the change in the internal energy per time unit obtains the power introduced to the medium:

$$P = \frac{dQ}{dt} = M_0 c_p \frac{dT}{dt} \tag{3.19}$$

Equation 3.19 Introduced power

This finally results in the apparent flow detected by the flow sensor:

$$\frac{dM}{dt} = -\frac{\alpha}{c_p} P \sim 1.8\ [\text{kg/hkW}]P \tag{3.20}$$

Equation 3.20 Apparent flow

Consequently, it is solely the instantaneous power that is introduced into the measuring circuit which is responsible for the apparent flow. Following this estimation, a power impact of as little as 5 W is capable of producing an apparent consumption of 0.009 kg/h, causing an error of as much as ~2 %, assuming the idling fuel consumption mentioned previously. Accordingly, it may be noted that a good fuel consumption measurement system is not only distinguished by having an accurate flow sensor but, above all, by the quality of its temperature conditioning system.

The power transferred into the measuring circuit comprises the fuel flowing in and out, thermal influences from the environment and the engine and the power impact (positive or negative) resulting from the conditioning. The goal of the temperature conditioning is to even out the instantaneous power impact at any point in time. To accomplish this, the fuel temperature is measured at one or more suitable points, with the temperature being controlled using heat exchangers. Admittedly, this only helps to regulate the average temperature under certain conditions, but currently no further options are available.

Besides the power impact, volume variations in the measuring circuit have an effect on the measuring signal too. In particular, the variable pressure-related volumes of trapped gas bubbles have a major impact, an issue that can only be prevented by ensuring an adequate line routing in the measurement application. Inevitable volume changes result from pressure regulator membranes and damping vessels that are sometimes needed for installation purposes. Due to the diversity of possible error impacts addressed above, fuel consumption measurement will never be accurate without the correct choice of measuring device and a precisely planned and designed hydraulic system for the entire measuring circuit.

3.3.9 Air Flow Measurement

3.3.9.1 Goal of Air Flow Measurement

Air flow or air mass measurement is, among other things, needed for the modal determination of exhaust gas emissions. At this point, we would like to limit the discussion to direct air mass measurement, as this is the main method nowadays applied in the area of diesel engine development.

Having a specifically designed intake tract, most gasoline engines are not suited for mounting additions as these would impact the engine, and its behavior would no longer be representative. So the determination of air mass in gasoline engines requires an indirect approach, either by using the fuel mass alone or based on lambda measurement (air-fuel ratio) and the fuel. Of particular interest to engine developers is the air-mass flow at the cylinder's intake valve, which, for structural reasons, is usually difficult to access or not accessible at all. This is the reason why nowadays the average intake air quantity is measured almost exclusively either upstream or just slightly downstream of the air filter. The high pulsation rates in the engine's intake system, which may be as much as 100 % of the average value, pose a particular challenge to intake air volume measurement. These pulsation rates may be associated with backflow. Only in recent times have devices based on ultrasonic transit time measurement become capable of determining such flows correctly in terms of time.

3.3.9.2 Air Flow Measurement Methods

Hot-Wire Measurement

Direct air mass measurement generally relies on the use of hot film anemometers. The sensor system consists of a heating electrode applied to a silicon substrate, which is kept at a constant temperature by means of a control loop. The necessary energy is directly proportional to the mass of air flowing past the electrode. The calibration of the sensor also requires the medium's absolute temperature, which an additional temperature sensor determines, usually located on the same silicon substrate.

The disadvantage of this measuring method is firstly the sensor system's high sensitivity to contamination, which makes frequent calibration a necessity. Secondly, there is the fact to consider that the method not only fails to correctly identify backflows, it even interprets these as positive flows. Over the last decades, development efforts have increasingly been focusing on compensation methods for on-board measurement. The attempts to correct the measuring signal involve flow control or the integration of additional detection mechanisms. Despite these efforts, such measures have not yet found their way into standard measurement and testing equipment.

Rotary Piston Gas Meters

Another way of measuring the intake-air quantity is to use a rotary piston gas meter. With this method, volumetric measurement is achieved by two rotary pistons arranged opposite each other while almost always remaining in contact with each other. The piston bearings are designed to allow a rotary motion at the slightest of pressure. The rotation is then converted into a frequency. For evaluating the air mass, the density is determined via a table stored in the metering device based on temperature and pressure measurement and applied in a further calculation using the measured volume. Due to the rotary pistons' inertia, this type of measurement permits determining only an average flow rate. The flow's transient behavior remains unmeasurable.

Ultrasonic Flow Measurement

A technique that was introduced to the market only recently is ultrasonic flow measurement that allows dynamic determination of the engine air intake quantity. The measurement is based on an ultrasonic, transit time-based process that is utilized for volumetric measurement. To this end, a transmitter sends two ultrasonic signals to a receiver, one in the direction of flow and one against the direction of flow. Due to the pulling effect, the transit time decreases in the direction of flow and increases against the direction of flow. Using the two transit times, it is then possible to determine the sound velocity at rest and the flow velocity averaged over the propagation path. For mass measurement, the calculation also takes media pressure into account.

The measuring method itself is nothing new, yet the high pulsation rate involved in intake air flow measurement demands an ultrasonic signal repetition rate of at least 500 Hz for the averaging results to be correct. It took years of work before this issue could be solved and the breakthrough was eventually achieved with this novel measuring technique. The ultrasonic measurement technology now allows determining the flow of intake air (at frequencies of up to several kHz) with the benefits of low measurement uncertainty, no restrictions (such as sensitivity to contamination), no damping of pulsations, minimum pressure drop and high dynamics in relation to time.

3.3.10 Oil Consumption Measurement

3.3.10.1 Goal of Oil Consumption Measurement

For a variety of reasons, engine oil consumption has become a major concern to engine developers. One critical factor, for example, is its significance to the engine's emissions in case of oil combustion. Unburned or partially burned oil has a harmful effect on the emission control components. Apart from that, oil consumption has become a major quality factor for automobile customers. While only 10 years ago all that mattered was oil consumption itself, engine developers today still focus on oil consumption, but are also extremely interested in the oil's quality. Due to the separately controlled pre- and post-injection of fuel in modern injection systems, the fuel dilution and oil foaming behavior have gained additional significance.

3.3.10.2 Oil Consumption Measuring Methods

Volumetric Methods

The fundamental oil consumption measuring methods are volumetric methods. Among these methods are those based on the oil dip stick. These measure the oil level in the crankcase either via additionally welded-in overflow drains in the area of the engine's maximum oil level or by means of receptacles provided alongside the oil pan. These techniques also include refill methods that determine the fill level based on the volume of oil refilled.

The most important factor for determining oil consumption correctly is to reproduce the engine state prior to the measurement. Factors, such as the thermal expansion of oil or different volumes filled into the engine, ultimately lead to large errors if the consumption is small, which it usually is. Another significant factor is the geometry of the oil pan. This should ideally have an even surface at all oil levels. To determine the actual oil consumption, it is necessary to take an additional sample in order to be able to distinguish between consumption of oil and quantities of fuel contaminating the oil.

Mass Determining Methods

As an alternative to volumetric, level-based methods, measuring devices are available that determine the level using pressure sensors. The advantage of methods like these is that thermal expansion has little or no significance because the determined pressure is a function of the product of density and fill level. Similar to volumetric level determination, however, the geometry of the crankcase has a considerable impact on the measurement.

The most widespread method of measuring oil consumption is still the drain-and-weigh method. This method involves draining and weighing the unit under test's entire oil quantity, or a defined portion of it, at regular intervals. This can be done either on an automated basis (by pumping) or manually. In view of the procedure's reproducibility, the human factor is of utmost significance. A possible contamination with fuel can only be established by running an additional laboratory test on the oil's composition. A disadvantage of these conventional methods is that the engine has to be operated in a particular demand state for a relatively long time before the detection limit required by the measurement method is reached.

Tracer Methods

The use of tracer methods speeds up oil consumption measurement considerably. With this technique, the oil is marked by adding a tracer. In use are radioactive tracers, such as bromine or tritium, or elemental or molecular tracers such as sulfur or pyrene. Oil consumption measurement is achieved by determining the concentration in the exhaust gas as well as the exhaust gas mass flow, either from fuel/air measurement, air/lambda or air/carbon balance.

The tracer method has the advantage that entire maps are determined in relatively short periods of time. It is equally possible to study the transient oil consumption mechanisms. One of the disadvantages of radioactive methods is the fact that the local authorities demand the implementation of specific safety precautions for the handling and disposal of the entire testing material. There are also the high costs of disposal to consider. This also explains why there are only a few highly specialized laboratories worldwide that provide this measuring technique as a service.

One particular method is the pyren tracer method. Pyren is a comparably stable organic substance that shows good miscibility with oil and is relatively easy to detect in the exhaust gas by using mass spectroscopy. Unless the tracer enters the combustion flame, the tracer remains stable. This means that this particular tracer only detects oil consumption mechanisms that do not come in direct contact with the flame.

In view of oil fractions, the mass spectroscopic analysis of exhaust gas composition is considered an elemental tracer method. With this technique, the oil composition first undergoes an analysis of its mass numbers. One or more representative mass numbers are chosen as tracers. A mass spectrometer subsequently detects these in the exhaust gas during engine operation. The problem associated with this method is that thc oil consumption does not have the same effect on all mass fractions, a fact which complicates the calibration process considerably. Furthermore, where diesel engines are concerned, fuel and oil have overlapping mass numbers, a circumstance which makes the results ambiguous.

3.3.11 Ignition Timing Measurement

In many cases, the measurement of events in relation to a time value belongs to the standard tasks on a variety of test beds. The time value is frequently measured at angular intervals of a rotating shaft. In case of combustion engines, this is often the crank angle or cam angle.

Examples of measurements like these are:

- Injection point(s)
- Ignition angle
- Peak pressure
- Timing of valve opening

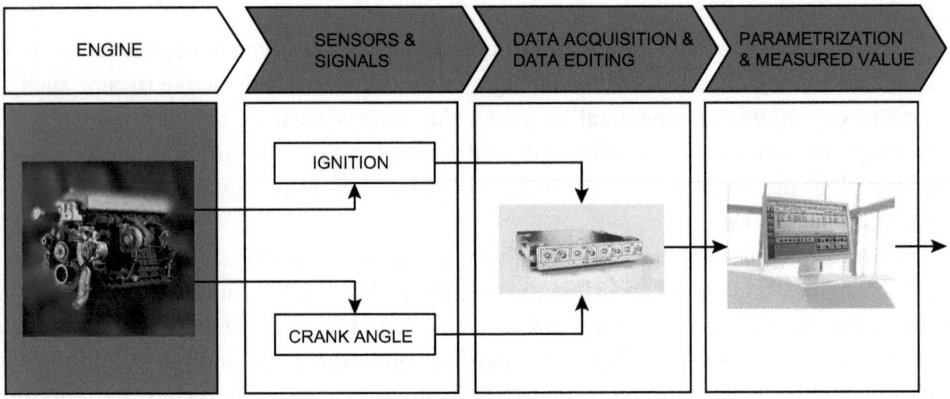

Fig. 3.55 Setup of an ignition timing measuring system

As an example of this type of measurement, we will address the ignition angle in further detail. Its position is generally referred to as °CA relative to ITDC (crankshaft degrees relative to ignition top dead center).

There are both special ignition timing measuring devices and combined measuring systems (e.g. indicating systems) that are able to provide this measuring signal. The setup of such timing measuring systems is basically always the same (Fig. 3.55). Using sensors that convert the measured values as required, these are then evaluated in a special data acquisition system, so that the ignition angle can typically be output in °CA relative to ITDC.

The effort required for this measurement depends on the availability of the relevant input signals. One way of measuring these involves the use of special sensors. Alternatively, the signals are picked up on the ignition side of the ignition system's primary or secondary circuit or as a logic signal, e.g. by the engine control unit (ECU). The latter is also possible for the crank angle signal.

Choosing the simple option of collecting ignition timing by ECU signals delivers information on how the ignition is being controlled or how you want it to be controlled. But only a measurement of the real processes in the ignition system will reflect the conditions that have actually led to the combustion in question in the first place. This is the reason why the most widely applied method of acquiring the spark signal is to use a current transformer [3], i.e. ignition pliers.

If the investigation focuses on multiple cylinders of an engine, it is necessary to ensure proper assignment of the individual curves to the cylinders to enable their correct display and interpretation. This procedure requires specifying the firing order in the device and performing a "cylinder-1 detection". Depending on the device's functionality, multispark ignitions in one cylinder can lead to misinterpretation of the firing point. This, too, must be taken into account when setting the device's parameters for the evaluation of the characteristic values.

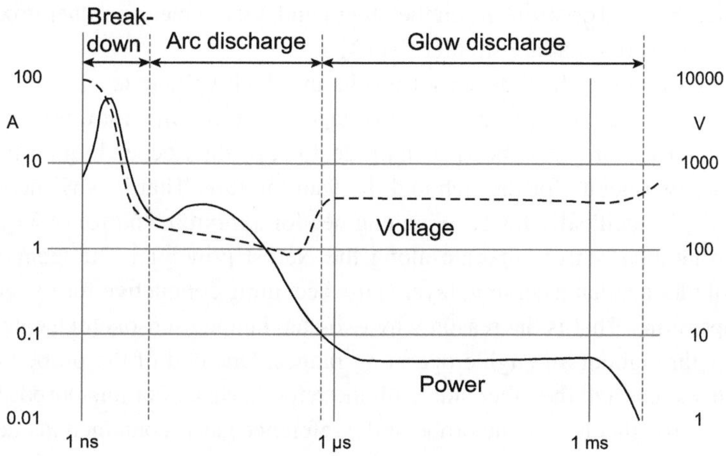

Fig. 3.56 Spark discharge phases [6]

For the sake of completeness, we should mention at this point that the number of required channels per cylinder may increase, depending on how many spark plugs are used and which tests are to be conducted. This makes it possible, for example, to exactly detect angular ignition timing irregularities between two ignitions.

The previous paragraphs addressed how to obtain characteristic values, e.g. the ignition angle in relation to a specific threshold value. Recording a curve of ignition voltage versus time (see Fig. 3.56), however, enables the acquisition of further meaningful information on how this parameter can be rated. In conjunction with other measured combustion values (e.g. from the indicating system, see [4]), one can evaluate this parameter in terms of its effects.

For the individual phases of an ignition voltage or ignition current curve, please see Fig. 3.56 or refer to [5]. If certain characteristic values, e.g. spark duration, are needed, these can be calculated using smart measuring systems (e.g. an oscilloscope) on the basis of the curves.

3.3.12 Lambda Probes

In vehicles, lambda probes (also referred to as oxygen sensors) are installed between the engine exhaust and the catalytic converter inside the exhaust pipe and allow the measurement of the proportion of oxygen remaining in the exhaust gas by comparison with a reference gas. This allows a conclusion to be drawn on the ratio between the combustion air and the fuel mixture in the combustion chamber. When the stoichiometric ratio is achieved, the amount of air added allows the entire fuel to be burned. This perfect ratio corresponds to a λ value of 1. Excess air would lead to a leaner mixture, resulting in λ values greater than 1. Too little air, on the other hand, would mean a richer mixture with λ values being measured that are smaller than 1.

The use of a two-step lambda sensor results in a high voltage level between 800 mV and 1000 mV for a rich mixture and a low voltage below 100 mV for a lean mixture. The characteristic curve for $\lambda = 1$ shows a sharp decline, so that, depending on the temperature, there are two points for the rich and the lean mixture. That is why these particular sensors are used specifically for gasoline engines for a mixture control of $\lambda = 1$.

Two-step lambda sensors operate along the Nernst principle. The technology takes advantage of the fact that a ceramic layer starts becoming conductive for oxygen ions at a certain temperature. That is the reason why two-step lambda sensors are heated to ensure functionality throughout all engine operating ranges. One end of the probe is in contact with the exhaust gas and the other side with the reference gas. For this purpose, either the ambient air is brought close to the probe or the reference gas is contained inside the probe. The difference in oxygen partial pressures leads to the transfer of oxygen ions across the outer layer.

However, control not only aiming at the stoichiometric ratio ($\lambda = 1$) but extending to other operating points requires a more exact measurement. Special probes called wideband lambda sensors are capable of measuring λ values of around 0.7 to ∞. If lambda equals infinite, the oxygen concentration in the exhaust gas matches that of the ambient air. This circumstance allows engine regulation that also satisfies lean concepts. These sensors are also appropriate for diesel and gasoline engines. The wideband lambda sensor combines a Nernst-type cell, which we are already familiar with from the two-step lambda sensor, with an oxygen pumping cell that transports oxygen ions. Simply said, a control mechanism ensures that the measurement chamber always contains the exact amount of oxygen necessary to yield a lambda value of 1 when combined with the Nernst cell. If the mixture is lean, oxygen ions have to be transported out, while a rich mixture will require that oxygen ions be transported in. The necessary pumping voltage is a measure for the oxygen concentration in the exhaust gas. For the exact setup and the operating principle of a wideband lambda sensor, please refer to the relevant publications (see [1]).

3.3.13 Exhaust Emission Measurement

3.3.13.1 Introduction

An important part of every engine development is the measurement of the exhaust composition. Somewhat simplified, any type of fuel used for combustion engines is primarily made up of hydrocarbon compounds. Depending on its chemical composition, fuel can be either gaseous or liquid and, depending on its base product, it is made from mineral oil, gas or biomass, or synthetically.

Critical for the combustion and the composition of the resulting exhaust gas is not so much the origin or the type of fuel, but rather more the ratio between the contained carbon and hydrogen atoms, the chemical energy carriers. During combustion in the engine, a chemical reaction occurs between the hydrocarbon compounds and the oxygen molecules present in the air. The air is chiefly composed of nitrogen and oxygen molecules with nitrogen molecules accounting for the largest proportion (approx. 80 %), most of which passes through the engine unaffected. A relatively small amount of nitrogen, however, burns and converts to nitrogen oxides (NO or NO_2).

During every chemical reaction, which includes combustion, there are two fundamental quantities that are conserved over time: the mass and the atoms (law of mass conservation by Lomonosov and Lavoisier). On the one hand, the total mass of air and fuel supplied to the engine is released again as exhaust mass, with total mass remaining unchanged. On the other hand, while both number and type of atoms remain unaffected, the chemical compounds of the involved atoms, along with the physical properties, such as temperature, pressure and density, change very fundamentally. The two aforementioned continuities provide the elementary basis for a great variety of exhaust emission measurement calculations.

All of the atoms listed in Fig. 3.57 are also present in the exhaust emissions, in the same number and of the same type. Combustion causes no change in the atoms but in the chemical compounds into which they unite. The best utilization of the energy contained in the fuel is achieved if all carbon atoms (C) are burned to carbon dioxide (CO_2) and all hydrogen atoms are converted to water (H_2O). Apart from making optimal energetic use of the energy contained in fuel (maximum efficiency), this kind of combustion produces no substances harmful to the environment—aside from global warming which is caused by an increase in CO_2. Optimal combustion is referred to as complete (perfect) combustion.

In reality, the technology today achieves results very close to optimal, but we are still left with some products that result from incomplete (imperfect) combustion (see Fig. 3.57). These particular exhaust components are harmful substances, as they present a danger to health and the environment. For this reason, the emissions legislation limits the maximum levels allowed to be emitted by combustion engines.

In most countries there are specific statutory emission regulations which an engine or vehicle manufacturer must adhere to. No engine or vehicle can be sold before compliance

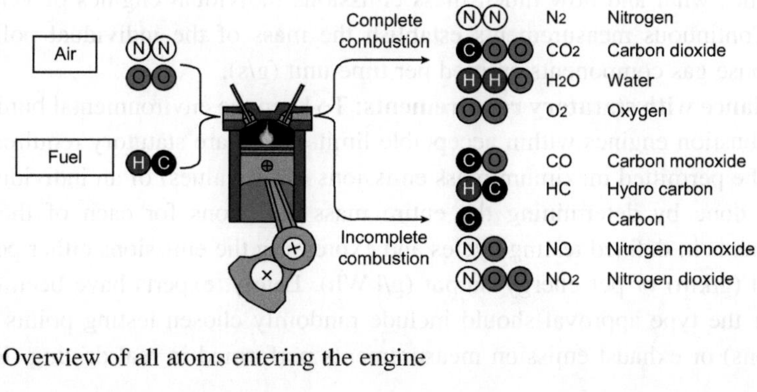

Fig. 3.57 Overview of all atoms entering the engine

is demonstrated (see Sect. 2.7). This type approval process involves testing and approving the type of engine or the vehicle (but not every single product). Manufacturers are additionally subjected to a conformity-of-production assessment based on random checks.

The emission legislation, the limits and the manner in which exhaust emission testing is carried out all take into account the specific engine application. Motorcycles and passenger cars, for example, have to undergo whole-vehicle testing by passing a defined speed profile on a chassis dynamometer (see Sect. 2.5.1). The emitted total mass of each pollutant component is determined and expressed per distance traveled (g/km).

With all other engine applications, such as buses, trucks, construction or agricultural machinery, trains and ships, the engine alone is tested in an engine test (speed and torque) that suits the application. The produced total emissions are expressed in mass of emitted gas per unit of energy output during the test (g/kWh).

Table 3.3 below shows the components limited in the legislations in Europe, U.S. and Japan for various combustion engine applications.

3.3.13.2 Applications in Exhaust Emissions Testing

The tasks of measuring the composition of exhaust emissions and determining the total mass of emitted exhaust gas components, particularly of pollutant and greenhouse gas components, are important but also very demanding and time-consuming parts of every engine and vehicle development. The motivation for measuring exhaust emissions is divided into three main areas [4]:

1. **Engine development:** The composition of exhaust emissions provides an insight into the engine's operating conditions (mixture formation, combustion, air/fuel ratio, etc.) and into the functionality of the exhaust aftertreatment system. To this end, the work focuses primarily on the emitted exhaust gas concentrations that are measured and evaluated together with the intake air and fuel masses (e.g. mass balances).
2. **Environmental relevance:** A sustainable and responsible use of combustion engines requires that the environmental impact resulting from the emitted pollutants and greenhouse gases be reduced to a minimum. Exhaust emissions measurement determines what and how much mass emissions individual engines or vehicles produce. Continuous measurements establish the mass of the individual pollutant and greenhouse gas components emitted per time unit (g/s).
3. **Compliance with statutory requirements:** To keep the environmental burden caused by combustion engines within acceptable limits, there are statutory requirements that define the permitted maximum mass emissions (limit values) of an individual engine. This is done by determining the entire mass emissions for each of the pollutant components in defined testing cycles and expressing the emissions either per distance traveled (g/km) or per energy output (g/kWh). Lately, experts have been discussing whether the type approval should include randomly chosen testing points (off-cycle emissions) or exhaust emission measurements performed in real driving (RDE—real

Table 3.3 Pollutant components limited by various emissions legislations (as of 2013)

Component	Abbr.	Two- and three-wheelers			LD vehicles			HD vehicles			Off-highway			Train			Ship		
		EU	US	JP	EU	US	JP	EU	US	JP	EU	US	JP	EU	US	JP	EU	US	JP
Carbon dioxide	CO_2		x		x	x	x		x			x			x			x	
Carbon monoxide	CO	x	x	x	x	x	x	x	x	x	x	x	x	x	x	x			
Nitrogen oxides	NO_x	x	x	x	x	x	x	x	x	x	x	x	x	x	x	x	x	x	x
Total hydrocarbons	THC	x	x	x	x	x	x	x	x	x	x	x	x	x	x	x			
Methane	CH_4							x											
Non-methane hydrocarbons	NMHC				x	x	x		x										
Particulate matter (mass)	PM				x	x	x	x	x	x	x	x	x	x	x	x			
Particle number	PN				x			x											
Ammonia	NH_3							x											
Alcohols						x			x										
Aldehydes						x			x										
Nitrous oxide	N_2O		x			x			x			x			x			x	

driving emissions) on the road (RDE testing) in addition to the test cycles already known beforehand.

With regard to engine-related information (Item 1) it is often sufficient to measure the pollutant concentrations only. When assessing environmental relevance and compliance with statutory requirements (see Items 2 and 3), the pollutant mass emissions are of significance. The concentrations of the individual exhaust gas components in engine development deliver valuable information about certain combustion properties. They also serve as a basis when inspecting vehicles already available on the market, for example for a special emissions test ("*Abgassonderuntersuchung*") required by law in Germany. The environmental relevance of emissions is only evident after identifying the mass of pollutants emitted. High concentrations in small quantities might be unproblematic, whereas small concentrations in great quantities might be harmful. That is the reason why the exhaust emission legislation also restricts the mass of pollutants emitted. However, a direct measurement of these pollutant mass emissions is not possible.

3.3.13.3 Determining Total Pollutant Mass Emissions

The emitted total mass of the individual pollutants is determined by measuring the concentration of the individual exhaust gas components, the engine's exhaust gas volumetric flow and the density of the individual components.

As shown in Fig. 3.58 there are three methods to determine mass emissions:

1. **Full-flow dilution and use of sampling bags:** The entire exhaust emissions of an engine are fed into a CVS (constant volume sampler) and diluted. The system keeps the entire flow (exhaust gas and dilution air) constant. With the CVS system's entire flow being constant, the exhaust gas is diluted to a smaller or greater extent, depending on the exhaust gas volumetric flow produced by the engine. An open inlet in the CVS draws in the difference between the exhaust gas flow and the CVS total flow as dilution air from the environment.

   So if the exhaust flow from the engine is low, a greater amount of dilution air is required, while an engine running at full load will require less dilution air to maintain the total flow rate. This process results in a mass-proportional weighting of the exhaust gas concentration in the diluted exhaust gas. The concentration of diluted exhaust gas is therefore proportional to the product of undiluted exhaust gas concentration and the total exhaust flow rate of the engine.

   Determining the diluted exhaust emission concentrations and the diluted total volume flowing through the CVS during the measurement permits calculation of the total mass of pollutant emissions. Although the diluted concentrations undergo continuous change, it is not necessary to measure them continuously. The final result calculation requires the average pollutant concentrations in the diluted exhaust gas. For this purpose, a small sample amount of the diluted exhaust gas is collected in an analysis bag. The concentrations in the bag are averaged using a pneumatic-mechanical

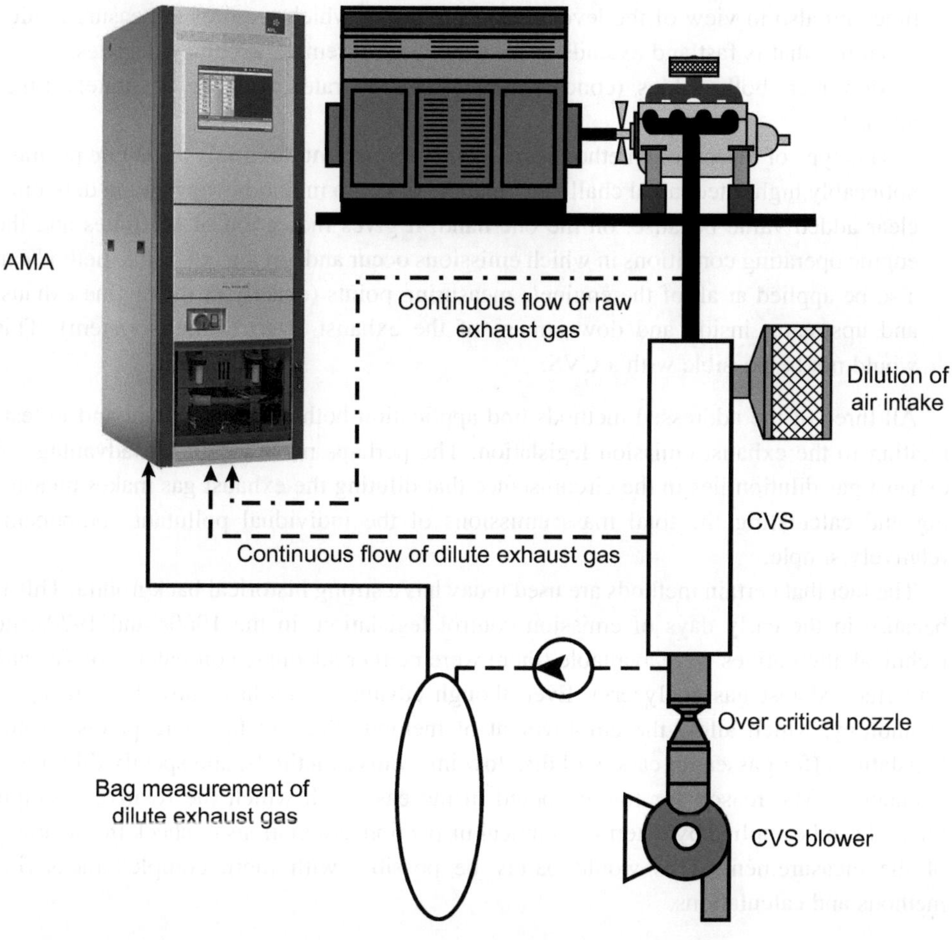

Fig. 3.58 Diagram of all three methods for determining mass emissions

process. After the measurement, the concentrations in the bag are analyzed and the total pollutant masses are calculated based on the total CVS volume.

2. **Full-flow dilution (CVS constant volume sampling) and continuous measurement of diluted exhaust emission concentrations:** Similar to Item 1, however averaging is not done with bags; instead, the diluted concentrations are measured on a continuous basis and mathematically processed with the CVS flow.
3. **Sampling undiluted exhaust emissions (raw exhaust) and engine exhaust volumetric flow:** With this method, the concentrations are measured directly from the engine's tailpipe. In addition, the current exhaust volumetric flows of the engine have to be known.

The continuously measured concentrations and flow rates are used to calculate the masses. In transient engine operation, these values are very dynamic, both in terms of time and also in view of the level of concentration, which requires a measured value dynamics that is fast and extends over several measurement accuracy decades.

Moreover, both values (concentration and flow rate) must be accurately time-aligned.

This type of measuring method is frequently termed **modal analysis**. While posing a noticeably higher technical challenge than the first two methods, the method delivers a clear added value because, on the one hand, it gives indication of the times and the engine operating conditions in which emissions occur and, on the other, the method can also be applied at all of the engine's measuring points (exactly at the engine exhaust and upstream, inside and downstream of the exhaust aftertreatment system). This would not be possible with a CVS.

All three of the addressed methods find application both in development and in tests relating to the exhaust emission legislation. The perhaps most significant advantage of exhaust gas dilution lies in the circumstance that diluting the exhaust gas makes measuring and calculating the total mass emissions of the individual pollutant components relatively simple.

The fact that certain methods are used today has a strong historical background. This is because in the early days of emission control legislation, in the 1960s and 1970s, no technical alternatives were available (there were neither adequate computers nor fast and drift-free exhaust gas analyzers). Even though advancements have now been made in technology, which allow the employment of methods that are far more precise, some legislations (for passenger cars) still disallow innovative methods, and specify dilution as mandatory. The reason for this is rooted in the ease with which the relatively simple method can be applied by external technical inspection associations to check the accuracy of the measurement. This would barely be possible with more complex measuring methods and calculations.

3.3.13.4 Exhaust Analyzers for Measuring Concentrations of Gaseous Emission Components

For measuring gaseous components, either individual analyzers are used, each measuring a certain exhaust gas component, or multi-component analyzers that are capable of measuring multiple components simultaneously (Fig. 3.59). Although there are a number of measuring principles (as shown in Table 3.4) that are capable of measuring the same gas component, the relevant exhaust emission legislations and standards specify—at least where the legally limited exhaust gas components are concerned—precisely which measuring principles have to be used for which component. Usually these measuring principles are not just used for the statutory measurements but already earlier on during engine development.

Fig. 3.59 Example of an exhaust gas analyzer AVL FID i60 HHD

Table 3.4 Typically used measuring principles; those marked with a (§) symbol are permitted for statutory emission measurements

Gas	Measuring principle	Abbreviation
CO_2	Non-dispersive infrared sensor Fourier transform infrared spectrometer	NDIR (§) FTIR
CO	Non-dispersive infrared sensor Fourier transform infrared spectrometer	NDIR (§) FTIR
HC	Flame ionization detector Fourier transform infrared spectrometer	FID (§) FTIR
NO_x	Chemiluminescence detectors Fourier transform infrared spectrometer	CLD (§) FTIR
O_2	Paramagnetic detector	PMD
NH_3	Laser diode spectroscopy Fourier transform infrared spectrometer	LDS (§) FTIR (§)
N_2O	Non-dispersive infrared sensor Fourier transform infrared spectrometer	NDIR (§) FTIR (§)

NDIR: Non-Dispersive Infrared Sensor

The light, which an infrared source emits, is directed through a measuring cell split into two compartments. The infrared radiation emitted by the infrared source (lamp) has a broad spectrum. Infrared radiation undergoes no change in its wave structure—a behavior known as "non-dispersive."

Gas detectors usually have two chambers, one for the measuring cell and one for the reference cell. Both chambers are filled with the gas requiring measurement. This means that in the detector the gas requiring measurement absorbs the same wavelengths as it does in the measuring cell. If there is no gas to be measured inside the measuring cell, the entire infrared radiation of this wavelength reaches the detector. The same applies to the reference cell. Both detector chambers contain the same amount of infrared radiation of this wavelength. This radiation is absorbed in the chamber, which leads to a rise in temperature and pressure inside the chambers. If the exhaust gas sample contains gas molecules that absorb infrared light, less infrared radiation will reach the detector from the measuring cell than it will from the reference cell. The detector measures the difference in intensity.

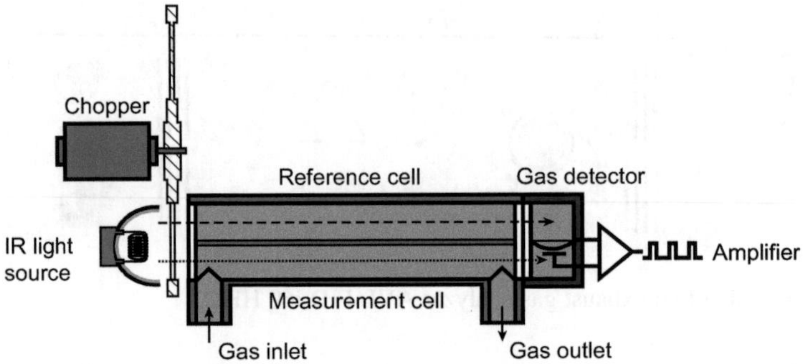

Fig. 3.60 Schematic diagram of an NDIR (non-dispersive infrared) detector

But if the measuring cell does contain gas to be analyzed, it will absorb infrared light. Less light will reach the downstream detector and less radiation will be absorbed in the detection chamber. This makes the temperature and pressure increase lower than on the reference side. The higher the concentration of the gas to be measured inside the measuring cell, the higher the difference in pressure will be between the two detection chambers. This difference in pressure is measurable as a deflection of a flexible diaphragm between the two chambers, for example. A chopper (a rotating perforated disk) serves to intermittently disrupt the infrared radiation (Fig. 3.60).

Interrupting (chopping) the infrared radiation intermittently makes it easier to amplify the detector signal in the electronics further downstream in the system. The larger the concentration in the exhaust gas to be measured, the bigger the detector's measurement signal will be. The correlation between the concentration and the measurement signal corresponds to the Beer-Lambert law, which is a nonlinear function. This explains why NDIR sensors always need to be linearized.

FID: Flame Ionization Detector

A flame ionization detector (FID) measures the concentration of hydrocarbons in the exhaust gas (see Fig. 3.61). There are a multitude of different hydrocarbon compounds in the exhaust gas, the composition of which depends heavily on the fuel being used. From an environmental and legislative point of view, the definition of meaningful separate limits for every single compound is not possible, so the limits commonly refer to total hydrocarbons (THC). The legally approved measuring technology is an FID which has shown to be the best suited instrument for measuring total hydrocarbons.

In the measuring cell a mixture of hydrogen (H_2) and helium (He) undergoes combustion with synthetic air. The reason for choosing hydrogen as combustion gas is the fact that the flame it produces involves no hydrocarbons. During the measurement, a sample of exhaust gas is fed to the flame. Introduced to the flame, the hydrocarbon molecules present in the exhaust gas are cracked and undergo ionization. In the measuring cell, the flame

Fig. 3.61 Schematic diagram of an FID (flame ionization detector)

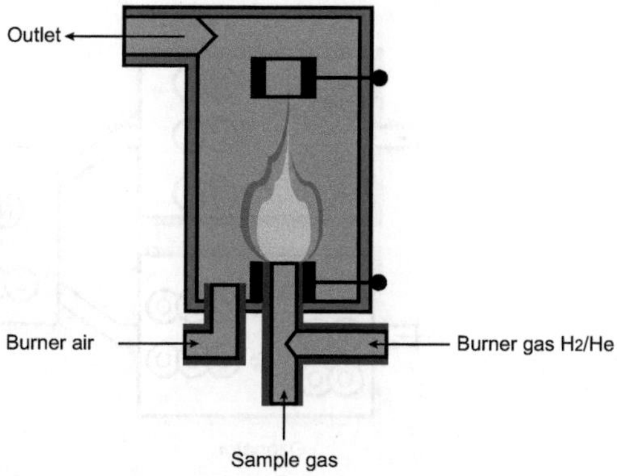

burns between a cathode and an anode, to which an electrical voltage is applied. The hydrocarbons produce ions which transport a small amount of current between the cathode and the anode. The induced flow of current is the measurement signal. The greater the flow of current, the higher the concentration of hydrocarbons will be in the exhaust gas. So there is a linear correlation between the concentration and the transferred current.

In an ideal case, all hydrocarbon molecules would be reduced to ionized parts, each containing a single carbon atom only. The ion current would be directly proportional to the number of carbon atoms bound to hydrogen atoms in the exhaust gas sample. All kinds of hydrocarbons could be measured this way. In reality, however, the cracking and ionization process is incomplete, but its effectiveness does remain constant.

Variation in the hydrocarbon molecules, however, is associated with minor differences in the effectiveness. This phenomenon is referred to as structural linearity or response factor Response factors indicate the difference between the value measured by an FID and the actual concentration of the different hydrocarbon compounds. Characteristically, the response factors for the hydrocarbons that are present in relevant quantities in the engine exhaust gas are between 0.9 and 1.1, based on the measurement of propane.

CLD: Chemiluminescence Detectors

Critical to exhaust gas measurement in engines is the sum of nitric oxide (NO) and nitrogen dioxide (NO_2). The sum of NO and NO_2 is referred to as NO_x. However, this summation concerns only the concentrations because NO and NO_2 have differing densities and the emitted NO_x mass would theoretically have to be calculated with the respective densities of NO and NO_2. As defined by the exhaust emissions legislation, NO and NO_2 are both calculated with the density of NO_2. This can be explained by the fact that NO_2 is far more dangerous to human beings than NO is. Apart from that, some of the NO from the exhaust gas converts to NO_2 once emitted to the atmosphere. NO_2 has a higher density than NO, which results in a calculated NO_x mass which is higher than what was actually emitted. However, from an environmental point of view, this makes sure you are on the safe side.

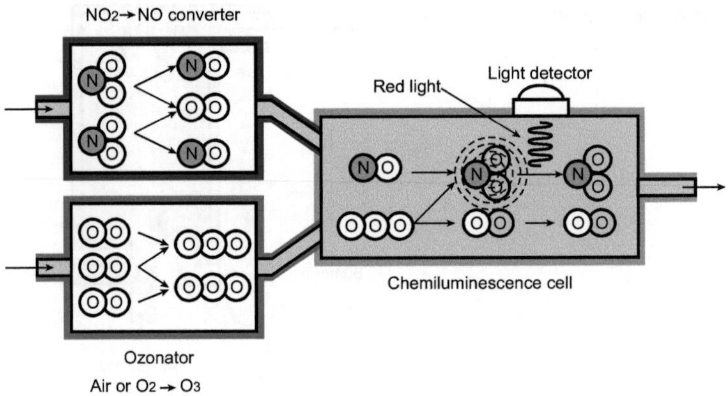

Fig. 3.62 Schematic diagram of a CLD (chemiluminescence detector)

In principle, a chemiluminescence detector (see Fig. 3.62) measures nothing other than the concentration of nitric oxide (NO). Chemiluminescence (sometimes chemoluminescence) is the emission of electromagnetic radiation (light) as a result of a chemical reaction. The exhaust gas is mixed with ozone (O_3) inside the CLD measuring cell. Reacting with the NO present in the exhaust gas, the highly reactive O_3 converts to NO_2 and O_2. About 10 % of these reactions result in the production of a NO_2 molecule that is in an energetically excited state (NO_2*). After a short while, the excited molecules return to their energetic base state. The excess energy is emitted as electromagnetic radiation (light). The intensity of the light is linearly proportional to the NO concentration in the measuring cell and is measured using photodiodes or photomultipliers.

$$\begin{array}{ll} NO + O_3 \rightarrow NO_2 + O_2 & \text{in approx. 90\% of the NO molecules in the sample} \\ NO + O_3 \rightarrow NO_2{*} + O_2 & \text{in approx. 10\% of the NO molecules in the sample} \\ NO_2{*} \rightarrow NO_2 + hv & \text{where h Planck's constant} \\ & \text{hv light quantum (photons)} \end{array} \quad (3.21)$$

Equation 3.21 CLD measurement reaction

In the analyzer itself, an ozone generator produces the O_3 needed for this reaction from oxygen O_2. This means that a CLD is only capable of measuring NO. To be able to measure NO_x ($NO_x = NO + NO_2$), all NO_2 molecules are transformed into NO upstream of the CLD detector. This is done in a heated NO_2/NO converter which accomplishes this conversion similar to a catalyst.

With NO_2 being partially soluble in water, water condensation must be prevented across the entire exhaust sample system all the way to the NO_2/NO converter. Heating the entire gas paths and the analyzer itself generally achieves this.

PMD: Paramagnetic Detector

A paramagnetic detector measures the oxygen content in the exhaust gas. Oxygen (O_2) is neither a harmful substance nor must it be measured as a result of the emissions legislation, however in engine development, it provides some essential information about the combustion.

Oxygen is one of the few gases with magnetic properties. The PMD makes use of this fact for measuring. Inside the measuring cell, the exhaust gas sample passes through a strong magnetic field. The magnetic properties cause the oxygen molecules to flow to the center of the magnetic field, where a quartz sphere without any magnetic properties is positioned. Such detectors are always designed symmetrically, having two magnetic fields and two quartz spheres. A rigid arm connects two spheres. Owing to their appearance, the two spheres are also referred to as a dumbbell. As they are mounted on a rotating axis, the oxygen molecules surging into the magnetic field and attempting to displace the dumbbell's spheres are able to rotate the dumbbell (Fig. 3.63).

The higher the oxygen concentration in the exhaust gas, the greater the oxygen molecules' attempt will be to displace the spheres. A mirror is mounted on the dumbbell's rotating axis and by means of a beam of light and a detector the dumbbell's rotation is measured. Either the deflection itself is the measurement signal, or there is a control unit that always keeps the dumbbell centered. In that case, the electrical current required to prevent the dumbbell from rotating is the measurement signal.

FTIR: Fourier Transform Infrared Spectrometer

Fourier transform infrared spectroscopy (FTIR) is a measuring method based on infrared light (see Fig. 3.64). Much like an NDIR, the measurement relies on infrared light that is absorbed through the individual components of the gas. Unlike an NDIR, however, FTIR

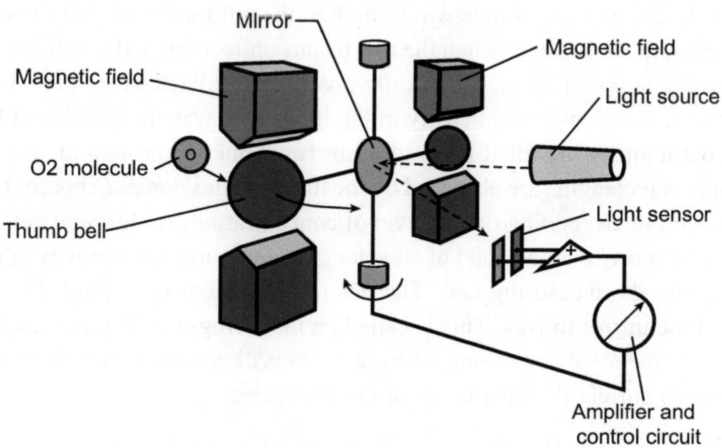

Fig. 3.63 Schematic diagram of a PMD (paramagnetic detector)

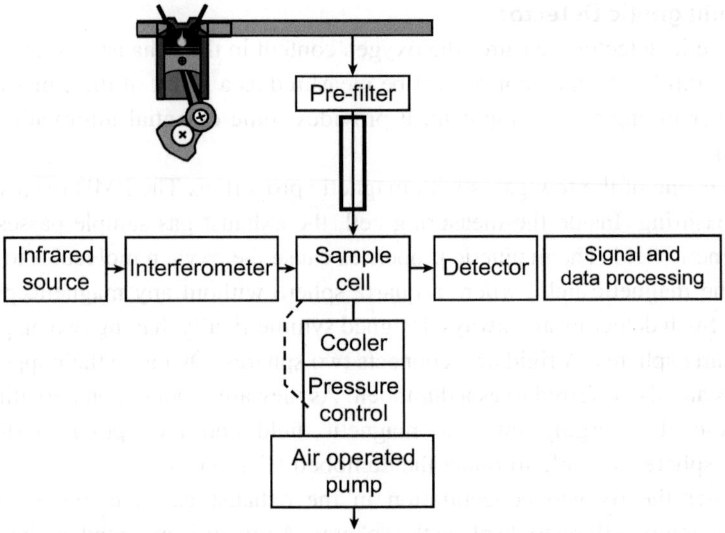

Fig. 3.64 Schematic diagram of an FTIR multi-component measuring system (AVL i60FT)

employs a broad band of infrared wavelengths, allowing the simultaneous measurement of a multitude of different exhaust gas components (approx. 30).

A Michelson interferometer continuously varies the intensity of the individual wavelengths from a broadband infrared source. For this purpose, a beam splitter divides the light source's infrared radiation into two beams. One of the beams hits a movable mirror, the other a stationary mirror and the mirrors reflect the light back to the beam splitter, where the two reflected beams rejoin to a single beam.

The moving mirror's continuous motion produces a difference in the two beams' path lengths and, as a result, interference effects when the two beams reunite. The interference effects are different for every single wavelength of the infrared band. If the two waves of a certain wavelength are in phase when the two beams unite, they will amplify each other; if they are exactly opposed to each other, they will cancel each other out. However, the resulting continuous change of intensity in the individual wavelengths is not known.

The so continuously modified infrared beam is conducted through the measuring cell and individual wavelengths are absorbed by the different gas components contained in the exhaust gas sample, depending on the level of concentration. An infrared detector, which is capable of capturing a broad band of wavelengths, measures the intensity of the infrared radiation leaving the measuring cell. The determined intensity is plotted over the time during which the mirror moves. This is called an interferogram. Typically, a background interferogram is recorded once a day when zero gas is directed through the measuring cell and used later to correct the measurement interferogram.

Fig. 3.65 Example of an FTIR measuring system AVL SESAM i60FT

A Fourier transform is then applied to the interferograms to mathematically calculate an infrared spectrum with intensity as a function of wavelength. The Fourier transform is a complex and computationally intensive mathematical calculation method that transforms the curved shape of an interferogram into a multitude of individual sine curves. Each of these sine curves has a different amplitude and frequency. The sine curve's frequency corresponds to the infrared wavelength (or frequency) and the signal's amplitude matches the signal strength downstream of the measuring cell (Fig. 3.65).

The infrared spectrum during the measurement and the background infrared spectrum are used to calculate the absorption spectrum and determine which gas components are contained in the exhaust gas sample (wavelength) and in what concentration (intensity) they are present. This is done using evaluation methods (see Fig. 3.66). To achieve the measuring accuracy required in engine development and exhaust emissions measurement, it is not sufficient to resort to the generally accepted evaluation methods. For our purpose, evaluation methods are used that were specifically designed to suit the properties of engine exhaust gas and that also have to be adapted to suit the used fuel.

Unlike single analyzers that are always built into an exhaust gas measuring system, the FTIR is an independent measuring system that already includes all the necessary components like sample gas conditioning system or control software.

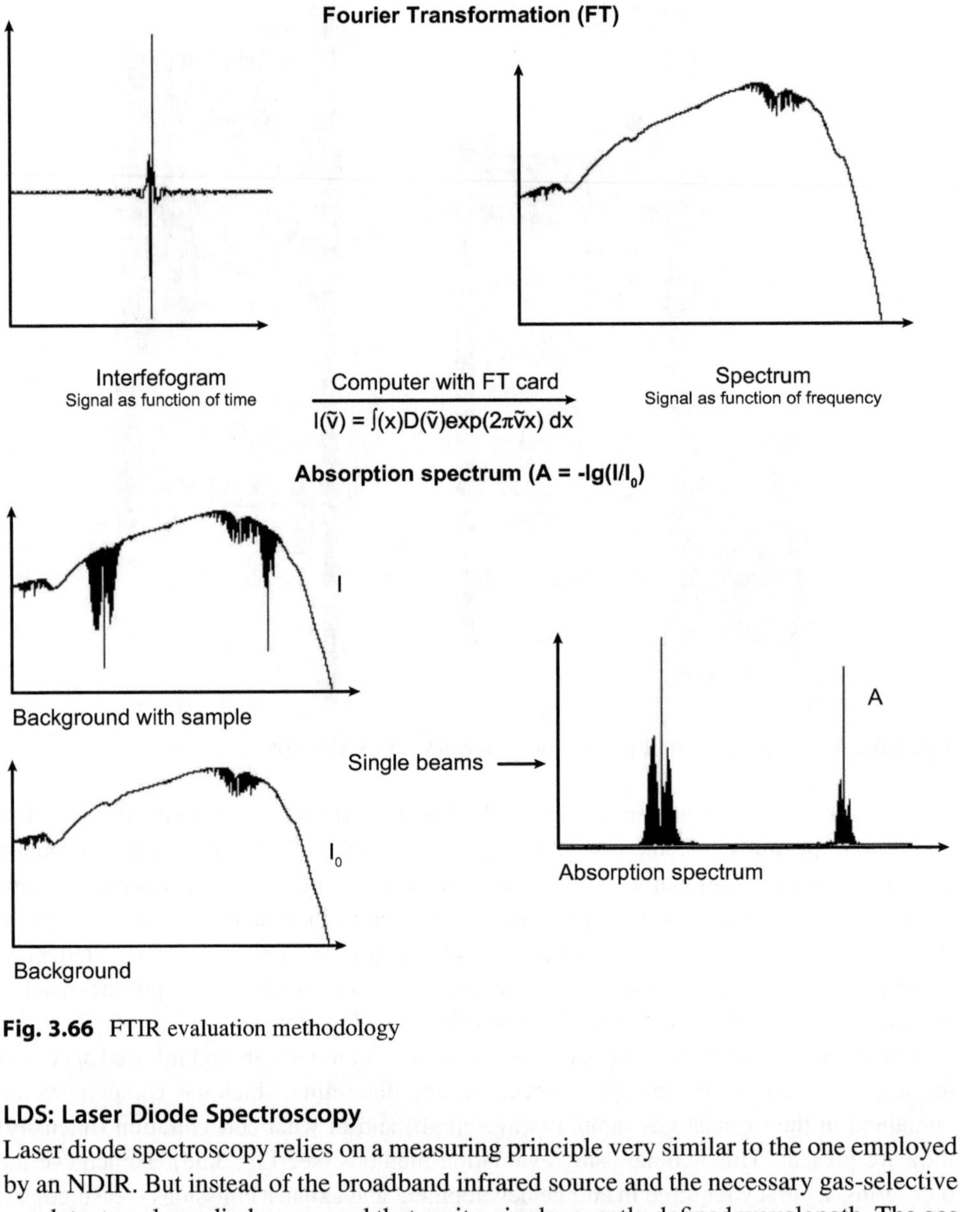

Fig. 3.66 FTIR evaluation methodology

LDS: Laser Diode Spectroscopy

Laser diode spectroscopy relies on a measuring principle very similar to the one employed by an NDIR. But instead of the broadband infrared source and the necessary gas-selective gas detectors, laser diodes are used that emit a single, exactly defined wavelength. The gas component to be measured in the exhaust gas sample then absorbs this wavelength. At present, the technology is mainly only used to measure ammonia (NH_3), but once further advancements are made in the field of laser diodes and the necessary wavelengths become available at acceptable prices, it will also be used for further gas components.

Further Measuring Methods

In addition to the aforementioned measuring techniques, there are a wide range of other methods, e.g. mass spectrometry (MS) or absorption of ultraviolet light (UV), that are used on occasion or for specific measuring tasks. Up to now, however, they have not found widespread use and we will therefore not be addressing them here any further.

3.3.13.5 Exhaust Emission Analyzing Systems

The individual gas analyzers are installed in a cabinet together with the required pumps, valves and other components, forming a complete exhaust emission analyzing system. An exhaust emission analyzing system characteristically contains single analyzers for the exhaust gas components CO_2, CO, HC, NO_x and O_2. Depending on the task to be accomplished, there might be further analyzers for measuring CH_4, NO or other gases. With all individual components being combined into an overall system, it is also possible to control it with a central computer, allowing effective integration with the entire test bed automation system (Fig. 3.67).

Besides having individual gas analyzers, an exhaust emission analyzing system is provided with a complex pneumatic system of lines, pumps, filters, coolers and valves. Beside the main task of extracting the exhaust gas sample from the engine's exhaust system, conditioning it and feeding it to each of the analyzers for measurement, there are a

Fig. 3.67 Exhaust emission analyzing system illustrated by the example of the AVL AMA i60 Emission Bench

variety of additional valves, tubes and components needed for operation, e.g. for supplying the operating or span gases.

Sample Gas Conditioning

A small amount of sample gas (approx. 10 l/min) is drawn from the engine's exhaust system and directed to the actual analyzers via various tubes, filters, pumps and valves. For the process to succeed, it is necessary to condition the sample gas first so that, firstly, it is suitable for the measurement analyzers and allows precise measurement and, secondly, to prevent any unintended changes from occurring in the composition of the exhaust gas on its way to the analyzer, e.g. caused by chemical transformations, deposits or condensation. This conditioning takes place in the sample conditioning system (Fig. 3.68).

Sample gas conditioning prevents the occurrence of changes in the composition of the exhaust gas sample on its way to the analyzers, e.g. to prevent undesired water condensation or deposits of certain hydrocarbon compounds as they may occur in diesel engine exhaust. Sample conditioning starts immediately at the exhaust gas sampling point and ends in the analyzer's measuring cell.

When measuring undiluted (raw) engine exhaust, observe the following details:

- The type and position of the sampling probe (SP in Fig. 3.69) should be chosen to obtain a representative average exhaust gas sample. If the installation situation is complicated, for example in manifolds or downstream of abrupt changes in the cross-section, there may be a marked heterogeneous distribution of exhaust gas concentrations over the cross-section of the exhaust pipe. Advantageous are sampling probes that have several holes spread over the entire cross-section of the exhaust pipe. When in doubt, several installation situations should be tried out by moving the sampling probe around while the engine is running before screwing it tight. If the readings change when moving the probe, it might be advisable to try out a different installation position or another probe design.

Fig. 3.68 Part of the sample gas conditioning system in an exhaust emission analyzing system

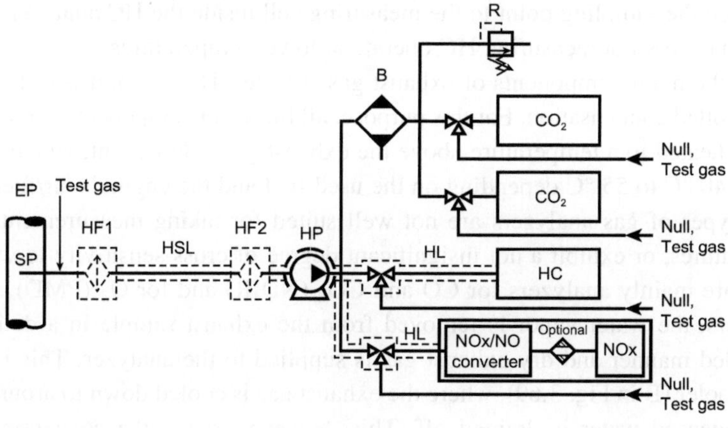

EP	Exhaust pipe	HP	Heated pump	CO	CO analyzer
SP	Sampling probe	R	Back pressure regulator	HC	THC analyzer
HF1	Heated pre-filter	B	Dryer/cooler	NO_x	NO_x analyzer
HSL	Heated sample line	HL	Heated line	NO_x/NO	NO_2-to-NO converter
HF2	Heated filter	CO_2	CO_2 analyzer		

Fig. 3.69 Schematic diagram of an exhaust emission analyzing system in accordance with ISO 16183 [7]

- The permissible pressure range at the sampling probe is specified for all exhaust emission analyzing systems and is typically in a range between −200 mbar to +200 mbar of relative (gauge) pressure. Sampling points upstream of an exhaust gas turbocharger or a heavily loaded diesel particulate filter may have significantly higher pressures. In such cases, additional pressure reducing regulators need to be used that reduce excess pressures to levels that suit the measuring system. On the other hand, measurements with long sample lines, or if the engine is in a high-altitude simulation chamber, an extra sample gas pump may be needed to overcome the pressure difference to the exhaust emission analyzing system.
- When measuring gaseous exhaust gas components, all solids (particulate matter) must be filtered out of the sample to avoid contamination in the system and the analyzers and valves, in particular. The filtering (HF1 and/or HF2 shown in Fig. 3.69) must be performed as close as possible to the sampling point to keep the complete system, including the exhaust gas lines, free of particulate matter.
- Hydrocarbons (HC) may condense at falling temperatures. In particular, some of the types of HC produced when diesel is burned are long-chain compounds that condense at temperatures even below 170 °C. Condensed hydrocarbons like these are, firstly, not available for measurement and, secondly, they not only would contaminate the system but also could re-vaporize later and so distort the measurement. In this respect, the term hydrocarbon (HC) hang-up is used. To prevent this from happening, all lines and components carrying exhaust gas are heated to a temperature of approx. 190 °C, all the

way from the sampling point to the measuring cell inside the HC analyzer. Of course, other analyzers not measuring HC operate at lower temperatures.

– One of the main components of exhaust gas is water. Here, too, it is critical to avoid uncontrolled condensation. For this purpose, all lines and components carrying exhaust gas are heated to a temperature above the exhaust gas's dew point, which is typically around 40 °C to 55 °C depending on the used fuel and the engine's air/fuel ratio.
– Some types of gas analyzers are not well suited for taking measurements at higher temperatures, or exhibit a not insignificant degree of cross-sensitivity to water vapor. These are mainly analyzers for CO and CO_2 (NDIR) and for O_2 (PMD). With these analyzers, the water vapor is removed from the exhaust sample in a deliberate and controlled manner and dry exhaust gas is supplied to the analyzer. This is done in a dryer/cooler (B in Fig. 3.69), where the exhaust gas is cooled down to around 4 °C and the condensed water is drained off. This, however, raises the concentration of the remaining exhaust gas components in the "dry" exhaust gas sample, as the fraction of water reduced the total exhaust gas amount. So the dry concentration measured by the analyzer needs correcting before the results can be calculated further. This correction of the measured concentrations is performed mathematically on the basis of the removed amount of water, which is not measured but calculated from the exhaust gas components CO, CO_2 and HC as well as the fuel composition.
– Further areas belonging to the sample gas conditioning system include switching units that allow switching an exhaust emission analyzing system between several sampling points, e.g. pre-catalyst and post-catalyst sampling points on an engine, or between different test beds. When measuring diluted exhaust gas, as is common with a CVS exhaust gas dilution system, the sample gas conditioning system is far less sophisticated because the exhaust gas is already diluted in the first place. The dilution already prevents water condensation, for example, and there are also no major pressure fluctuations at the sampling point. The following details must nevertheless be observed:
  • As the engine exhaust has already been diluted with air preventing water condensation to occur at room temperature, such exhaust gas measurement systems normally need no heating. Cooling systems are not required to remove the moisture from the exhaust gas sample either. That explains why the measured concentrations are "wet" concentrations and require no further corrections as would be the case with undiluted measurements. In the area of statutory exhaust emissions testing, water condensation achieved with a cooler, as is common with raw exhaust gas measurement, is not allowed. Nevertheless, analyzers have to be tested to establish a possible cross-sensitivity to water. A maximum acceptable level of cross-sensitivity is specified by the regulations. This provision concerns NDIR analyzers for CO and CO_2.
  • When measuring ultralow-emission engines, typically both the CVS and the exhaust emission analyzing system are heated moderately to approx. 30–40 °C. This raises the dew point and the exhaust gas does not need as much diluting to prevent the

water from condensing. The resulting lower level of dilution means that the already low exhaust gas concentrations in such engines are not diluted unnecessarily high—a fact that ultimately leads to improved measurement accuracy.

- But even in diluted exhaust gas, precautions must be taken to prevent the condensation of heavy hydrocarbons (HC) as are likely to occur in diesel engines. To achieve this, the complete measuring chain, including the HC analyzer, has to be heated to about 190 °C. Rather than being performed later using bags from the CVS system, the measurement is instead run continuously and straight from the diluted exhaust gas.

Functions of an Exhaust Emission Analyzing System

Exhaust emission analyzing systems are typically controlled by a built-in computer and integrated into the test bed automation system. While its main function is obviously to measure exhaust gas, there are a number of other features that these kinds of systems need to provide. Particularly in the context of the legally required exhaust gas tests, a vast number of extra quality control and check features have been added (Table 3.5).

3.3.13.6 Exhaust Gas Dilution Systems

Sometimes the exhaust gas is diluted before it is measured. From the results obtained the pollutant masses are determined (see under Sect. 3.3.13/Determining total pollutant mass emissions on page 190). Dilution is performed for three reasons:

- The formation of exhaust gas particulates is a process that takes some time and is also heavily dependent on the ambient conditions such as temperature or concentration level. Particulate matter found directly at the engine exhaust valve is different from what is sampled from the end of the tailpipe or from the kind present in the environment. To measure particulate matter properly in view of its environmental relevance, the exhaust gas is diluted in the measuring system to obtain a particulate formation very much like that present in the environment. Having passed through a sufficiently long mixing section, the actual particulate measurement takes place. Gaseous exhaust emission components typically no longer change after they have exited the tailpipe.
- The dilution of the exhaust gas also leads to a reduction of the relative water vapor content present in the exhaust emissions. An adequately high dilution prevents water from condensing in the measuring system, which could otherwise contaminate the system, thereby distorting the measuring results.
- Exhaust gas dilution greatly simplifies the measurement of total mass emissions of the individual pollutant components, particularly in transient engine operation. Accordingly, there is no need any more to measure the engine's exhaust gas volume flow, for example.

Table 3.5 Characteristic features and checks in an exhaust emission analyzing system

Function/Operating state	Description
Pause	In Pause state, the system switches to a power saving mode. The operating gases and all pumps switch off. The heated components are regulated to a low temperature. When switching from Pause state to Standby state, the system requires a short while to restart. But this period is much shorter than what it would be if the system had been switched off completely. This function is typically activated during the night.
Standby	In Standby state, the system is ready to measure. The temperatures in the heated components are regulated to the values required for measuring. Depending on whether the system was previously set to Pause or switched off, it may take a while for the system to become ready for operation. Standby cancels all the other running functions too.
Measure	The sample pumps switch on and the sample gas starts flowing to the analyzers, where it is tested.
Purge	The system and the analyzers are purged with zero gas to clean them from residual gases. The gas flows in the same direction as it would when sampling.
Backflush	Pre-filters and lines are purged with clean, pulsating compressed-air in the direction opposite the sampling direction.
Leak check	The system is checked for leaks, particularly in sections of the system upstream of pumps, because leaks here would draw in ambient air, leading to incorrect results.
Auto-range on/off	Most analyzers have multiple measuring ranges. When Auto-range is switched on, the analyzers automatically select the measuring range best suited.
Zero/span adjust	The system is calibrated, which takes place as a two-point adjustment with a zero gas and a span gas at the end of the measuring range. The term "calibration" is common, but it does not conform to definitions in other standards, where the term only means establishing a deviation without performing an adjustment too.
Activate zero gas	Zero gas starts flowing through the analyzers, but no adjustment is made to the analyzers.
Zero gas adjust	Zero gas flows through the analyzers and the measured values are adjusted to the zero point.
Activate span gas	Span gas starts flowing through the analyzers, but no adjustment is made to the analyzers.
Span gas adjust	Span gas flows through the analyzers and the measured values are adjusted with the span gas values.
Linearity check	Based on several (five or more) gas concentrations evenly distributed across the respective measuring range, the analyzers' linearity is checked.

(continued)

Table 3.5 (continued)

Function/Operating state	Description
Linearization	As in a linearity check, the feature locates deviations. The linearization function additionally calculates a new adjustment curve to ensure the analyzer's linearity. The system typically uses polynomials up to the fourth order.
ANR Check—accuracy, noise, repeatability	This check verifies the accuracy, noise and repeatability of the values measured by an analyzer.
Drift check	The system runs a check to establish how an analyzer's measured values change (drift) over an extended period of time.
Analyzer response time check	The analyzer response time is measured. Usually the T_{10}-to T_{-90} time is determined. If there is a sudden change in the concentration level, the system measures how much time it takes to reach 10 % and 90 % of the final value.
System response time check	The system response time is measured. Usually the T_0-to-T_{90} time is determined. If there is a sudden change in the concentration level at the sampling point, the system measures how much time passes by between the leap in concentration (T_0) and the time the system reaches 90 % of the final value.
NO_2/NO converter check	The efficiency of the NO_2/NO converter in the CLD analyzer is checked. The procedure is subject to criteria defined by legislation.
FID analyzer—CH_4 cutter efficiency check	The purpose of the CH_4 cutter of an FID analyzer is to destroy all hydrocarbons except methane. This test is run to check the efficiency of the cutter, which is an expendable part. The results of this test are usually needed for further result calculations.
CLD analyzer—quench check	The measured value of a CLD analyzer is reduced by quenching. This means that the presence of CO_2 and H_2O in the exhaust gas sample lowers the measured value. This test, which is subject to legally defined criteria, determines the magnitude of this negative effect.
CO analyzer – cross check	An NDIR analyzer for measuring CO exhibits cross-sensitivity to CO_2 and H_2O. This test, which is subject to legally defined criteria, determines the magnitude of this negative effect.
CO_2 analyzer—cross check	An NDIR analyzer for measuring CO2 exhibits cross-sensitivity to H2O. This test, which is subject to legally defined criteria, determines the magnitude of this negative effect.
FID analyzer response factor check	FID analyzers exhibit somewhat differing levels of sensitivity to various hydrocarbon compounds. The magnitude of such response factors is determined by means of methane, toluene, propylene and isopentane in relation to propane. The procedure is subject to legally defined criteria.
FID analyzer O_2 interference check	FID analyzers may exhibit sensitivity to O_2 if the individual gas flows to the combustion chamber are not optimally adjusted, as this could affect the flame temperature. This test determines the magnitude of this negative effect.

(continued)

Table 3.5 (continued)

Function/Operating state	Description
HC hang-up check	Condensed hydrocarbons in the sample gas lines can lead to incorrect readings because, on the one hand, they are unavailable for measurement and, on the other, they might re-vaporize in the heated lines later. This test, which is subject to legally defined criteria, determines the magnitude of this negative effect.
System check	In a system check, a span gas or gas mixture is applied to the system's entire sample gas path, similar to what would be done in an engine exhaust emission measurement. Having knowledge of the concentrations in the used gas allows the overall system to be checked for malfunctions.

The two reasons mentioned first (particulate formation and water condensation prevention) apply regardless of whether the engine's total exhaust gas is diluted (as with full-flow dilution), or whether only a small exhaust gas sample is being diluted (as is the case with partial-flow dilution).

However, the simplification of mass determination, mentioned above as third reason, is only true for the full-flow dilution method. With this method, the engine's entire exhaust gas passes into a dilution system. Such systems are called constant volume samplers (CVS). In terms of complexity and calculation, partial-flow dilution is very similar to raw exhaust sampling.

Full-Flow Dilution System CVS

The entire exhaust emissions of an engine pass into the CVS (constant volume sampler, see also page 190) and are instantly diluted with filtered ambient air in a special mixing section. The system keeps the total flow rate (exhaust gas and dilution air) constant, hence the name "constant volume sampler." This is usually accomplished using supercritical Venturi nozzles. An alternative option are pumps that have a constant ratio between speed and volume (PDP—positive displacement pumps) or pumps regulated via flow rate measurement.

With the CVS system's entire flow being constant, the exhaust gas is diluted to a smaller or greater extent, depending on the exhaust gas volumetric flow produced by the engine. An open inlet in the CVS draws in the difference between the exhaust gas flow and the CVS total flow as dilution air from the environment. This difference should be far greater than the engine exhaust flow. So if the exhaust flow from the engine is low, a greater amount of dilution air is required, while an engine running at full load will require less dilution air to maintain the total flow rate. This process results in a mass-proportional weighting of the exhaust gas concentration in the diluted exhaust gas. The concentration of diluted exhaust gas is therefore proportional to the product of undiluted exhaust gas concentration and the total exhaust volumetric flow of the engine.

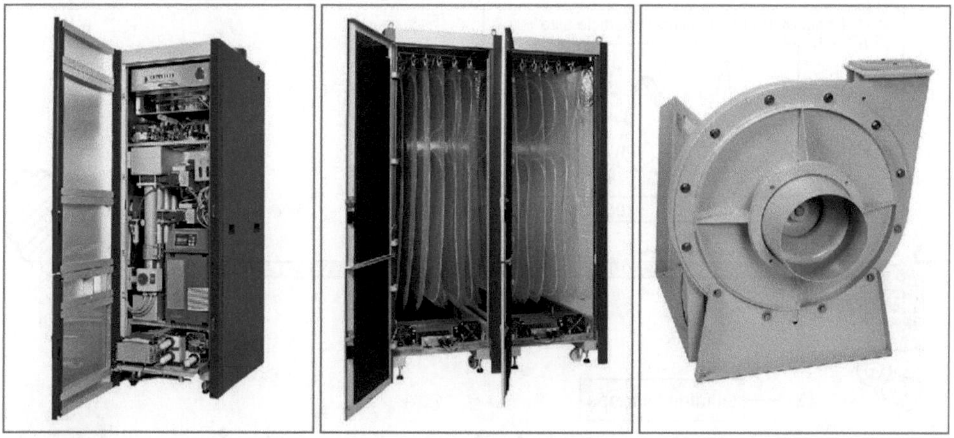

Fig. 3.70 Example of a CVS system: main cabinet with AVL CVS i60 components (*left*), two cabinets with analysis bags (*middle*) and CVS blower (*right*)

Determining the diluted exhaust emission concentrations and the diluted total volume flowing through the CVS during the measurement permits calculation of the total mass of pollutant emissions. Although the diluted concentrations undergo continuous change, it is not necessary to measure them continuously. The final result calculation only requires the average pollutant concentrations plotted over the diluted exhaust gas's sampling time. These are measured continuously and calculated mathematically or, as an alternative, a small sample amount is collected in an analysis bag over the length of the sampling time. The sampling is followed by an analysis of the concentrations contained in the bag, which were averaged in a pneumatic-mechanical process (Fig. 3.70).

To benefit from all the advantages of the CVS method (realistic particulate formation, prevention of water condensation and simplified mass determination), different CVS flow rates are needed, depending on engine size and test run. The following flow rates are typical for specific applications:

- Heavy-duty engines 120 . . . 180 m^3/min
- Passenger cars 8 . . . 30 m^3/min
- Motorcycles 1 . . . 5 m^3/min

Dilution Tunnel and Particulate Sampler of a CVS

As mentioned in the previous section, one of the reasons for performing exhaust gas dilution is to obtain realistic particulate formation as it would occur in the environment. For this purpose, a dilution tunnel and a particulate sampler are added to the CVS system. This also serves the measurement of particulate emissions. During engine combustion, the smallest of primary particulates are produced, which mainly consist of elemental carbon.

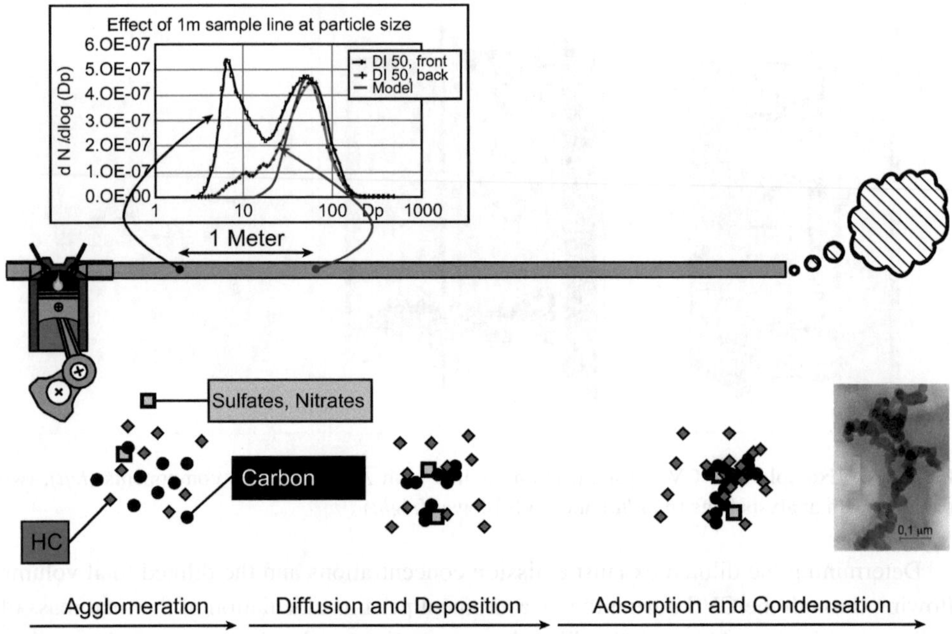

Fig. 3.71 Schematic diagram of particulate formation

The particulates already start to change in an engine's exhaust system. They impact each other and stick together (agglomeration), a process which reduces the number of particulates but makes them bigger. As the exhaust cools down, some of the vaporous exhaust gas components, such as high-boiling hydrocarbons, condense on the particulates. In addition, gaseous components may be absorbed by the carbon. This particulate formation process continues until the exhaust reaches the ambient air near the end of the tailpipe, where somewhat fewer but larger and more heterogeneous particulates form than those resulting from combustion in the engine. The diagram in Fig. 3.71 illustrates the change in particulate size distribution from many small particulates to fewer but larger particulates over one meter of exhaust pipe.

To realistically simulate this type of particulate formation in the measuring system, the exhaust gas is diluted in a CVS system. For this purpose, the CVS is extended to include a dilution tunnel. The dilution tunnel consists of a mixing zone, where a mixing module rapidly mixes the exhaust gas with dilution air, from which it passes through a pipe under turbulent flow conditions (Reynold number > 4000). The exhaust gas should remain sufficiently long inside the pipe to enable realistic particulate formation. Typically, a pipe length that is ten times the pipe diameter satisfies this requirement.

At the end of the dilution tunnel, a sample of the diluted exhaust gas is extracted. Under precisely controlled flow conditions, the sample is filtered through an analysis filter that extracts the particulates contained in the diluted exhaust gas. Prior to the measurement,

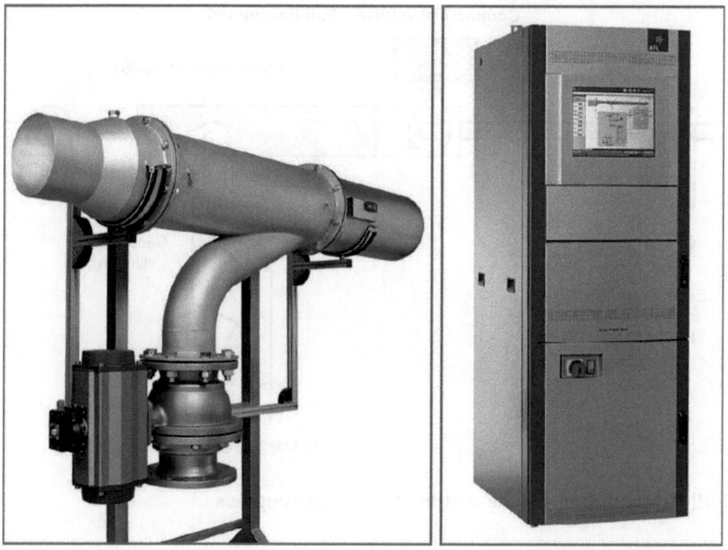

Fig. 3.72 Inlet and mixing zone of a dilution tunnel (*left*) and a particulate sampler (*right*) with an integrated particle counter

the analysis filter is weighed empty and after the measurement loaded. The increase in weight and the known flow rates serve as a basis to calculate the particulate mass. Given the involved weighing process, the method is referred to as gravimetric particulate matter (PM) measurement.

Modern exhaust emission control legislation no longer restricts its attention to particulate mass, but also focuses on the measurement and limitation of the number of emitted particulates. As is the case with particulate mass determination, it is again necessary to dilute the exhaust gas prior to measurement to obtain the particle number. To save space on the test bed, particle number counters (see Sect. 3.3.14) are often integrated into the cabinet of a particulate sampler (see Fig. 3.72).

3.3.14 Particulate Measurement and Exhaust Gas Opacity

3.3.14.1 Gravimetric Measurement of Particulate Mass

The limits for particulate emissions specified by statutory regulations are based on an integral measurement by gravimetric measurement of the particulate mass following dilution in a full-flow or partial-flow tunnel (see description starting on page 209).

The first definition of the measuring procedure was specified by the US EPA (United States Environmental Protection Agency) in [8] and later adopted worldwide (see also 70/220/EEC or TRIAS 60-2003). The exhaust gas is mixed with filtered air in accordance with the CVS principle. A partial flow of the diluted exhaust gas, having a

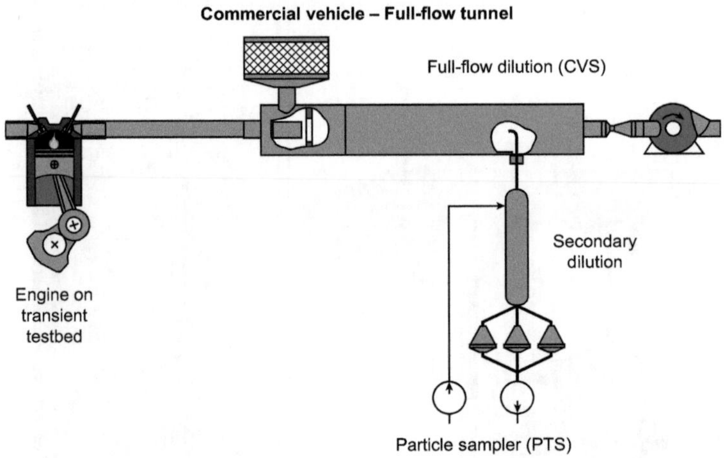

Fig. 3.73 Full-flow dilution for the certification of HD engines

temperature < 52 °C (or more precisely 47 °C ± 5 °C) is drawn through inert filters with > 99 % filtration efficiency.

The emission is calculated from the mass difference between the loaded and unloaded filter. Figure 3.73 shows a schematic diagram of a system with secondary dilution, a method commonly employed for heavy-duty vehicles. Vehicle emissions measurements for passenger cars on chassis dynamometers basically follow the same principle, but no secondary dilution is involved.

Particulates are composed of non-volatile soot components, such as abrasion particles and ash, adsorbed organic components, condensed and adsorbed sulfuric acid, etc. Condensed and adsorbed substances essentially do not form until they have reached the dilution tunnel; but, according to [9], even the soot concentration is not fully stable between the engine and the measuring filter either. It is therefore understandable that even minor changes to the design of the dilution and particulate sampling system already have an impact on the measured particulate mass.

To increase repeatability and reproducibility of the measuring method for decreasing particulate emissions, particularly soot emissions, the US EPA started to specify the dilution, particulate sampling and weighing system in further detail in 2007, see e.g. CFR 1065 [10].

In the European Union (in accordance with 2005/55/EC), and since 2011 also in the U. S., the use of partial-flow dilution tunnels that dilute a constant [9] part of the exhaust gas, as defined in the ISO 16183 standard, has also become acceptable for commercial vehicles. The advantages these systems offer in terms of space requirements and costs (see Fig. 3.74) are offset by the elaborate effort involved in regulating the mass flows. As described in [11], attention must be paid to several peripheral conditions in order to ensure that the measured emissions are essentially the same as those in a full-flow system.

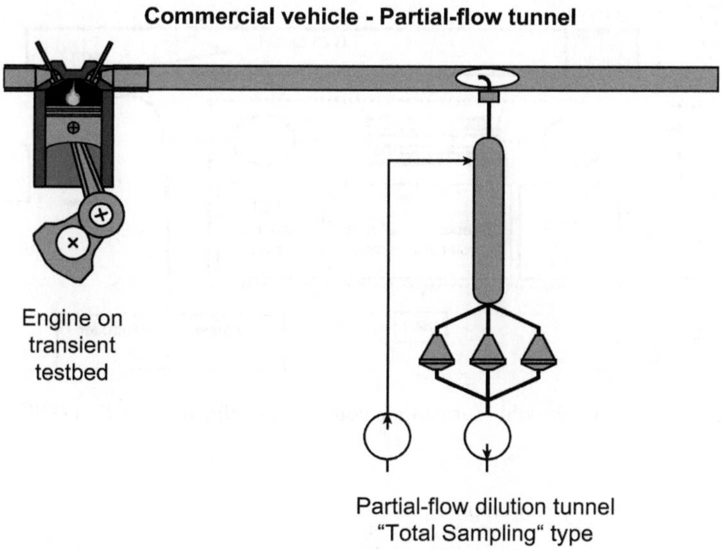

Fig. 3.74 Partial-flow dilution for the certification of HD engines

3.3.14.2 Determining the Particle Number in the Exhaust Gas

As the soot emissions in modern combustion engines can only be captured using extremely sensitive particulate measuring equipment, the PMP (Particle Measurement Program) expert group under the auspices of UNECE GRPE has investigated new measuring techniques for particle measurement. The group's recommendation for future homologation procedures includes a modified US-EPA particulate measurement procedure (US-2007) as well as the inclusion of particle number counting, as described in further detail in [12, 13].

The PMP group has defined a complex system to condition already diluted exhaust gas, which is outlined in Fig. 3.75:

- Step 1: Separation of coarse particles that stem from reintroduced wall deposits rather than directly from combustion.
- Step 2: High exhaust gas dilution (dilution factor ≥ 10) and subsequent heating to temperatures between 300 °C and 400 °C. This not only leads to a low particle number in the condensation nucleus counter (PNC), which is required in practical operation, but also to the fact that the volatile nanoparticles are transferred into the gas phase, leaving only nonvolatile particles (e.g. soot particles) left to be counted. This has two reasons: On the one hand, nonvolatile particles are toxicologically more relevant to human health, and on the other hand, it has proven to be extremely difficult to achieve reproducibility in measurements of the particle number concentration in volatile particle emissions. This is not a problem associated with the measurement itself—volatile particles can be counted just as well as solid ones. However, the way

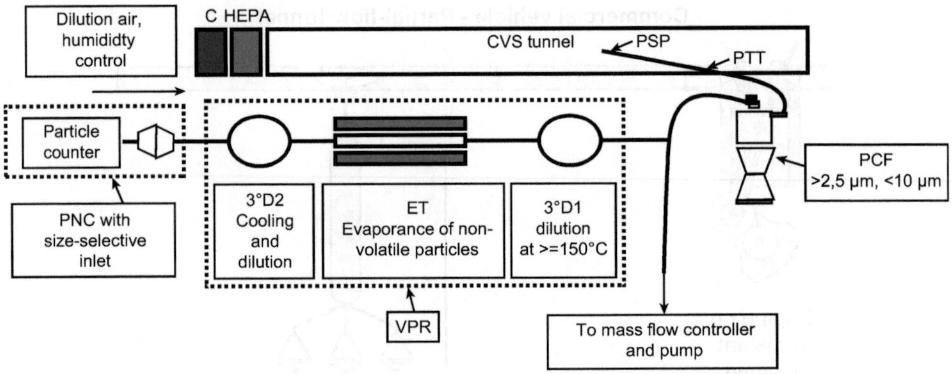

Fig. 3.75 Exhaust gas conditioning for particle counting as defined by PMP in UNECE Regulation No. 83

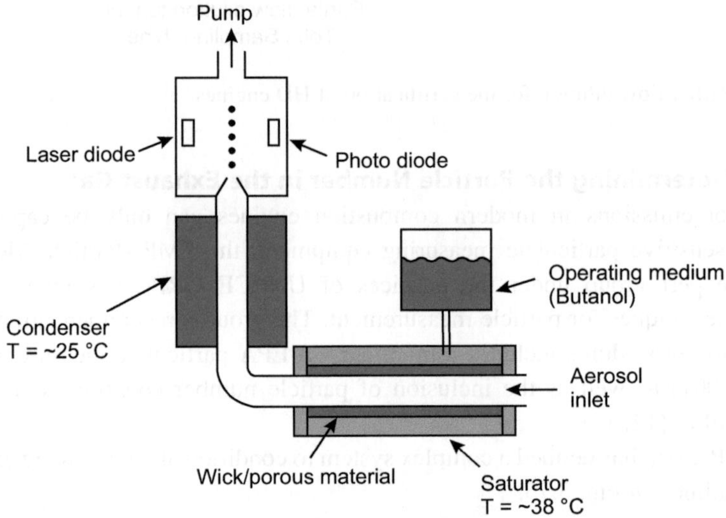

Fig. 3.76 Operating principle of a condensation nucleus counter (CPC) as applied in [14]

homogeneously condensed hydrocarbons and sulfates form downstream of particulate filters is extremely sensitive to the slightest modifications in the engine or exhaust gas conditioning system.

– Step 3: A further dilution upstream of the PNC cools down the exhaust gas and prevents any volatile components from re-condensing.

Condensation nucleus counters (PNCs), also known as condensation particle counters (CPCs), are very sensitive systems for particle number counting in the range of sub-μ (sub-microns) to a few nanometers. Figure 3.76 illustrates the operating principle of a

CPC. Heterogeneous condensation of supersaturated vapor (typically n-butanol) generates micro-particles from nanoparticles, which are subsequently counted by using a light scattering method.

3.3.14.3 Alternative Methods to Determine Particulate Properties in Exhaust Gas

The gravimetric method of determining particulate emissions has two major drawbacks when applied in engine or exhaust gas aftertreatment development activities: The method is, firstly, extremely time-consuming and, secondly, integrating. Engine development, however, very often requires rapid measurement execution and time-alignment of emissions to the associated dynamic driving modes. A series of simpler and more dynamic measuring procedures were therefore developed to meet the demands of present-day engine development. One of the disadvantages of these alternative measuring methods is that the measuring quantities generally deviate from the statutory particle requirements (particle number, particulate mass) and that the determined correlations apply to a limited extent only.

Of particular significance in this respect is the measurement of soot emissions. This is because soot is also a key indicator for determining the combustion quality. For this purpose, several methods were developed, which are almost exclusively based on soot's relatively strong ability to absorb radiation (in the near-infrared or visible range). Newer methods have shown very good time resolution (typically in the one-digit hertz range) or a very high level of sensitivity (typically in the range of a few μg per cubic meter).

Table 3.6 gives a summary of the most important alternative measuring methods while the schematic diagrams Figs. 3.77, 3.78, 3.79, 3.80, 3.81, 3.82 and paragraphs further below provide illustrations and brief descriptions of the methods. There are a variety of different designs and commercial suppliers for all of these measuring methods. For further, more detailed or summarized information on unconventional state-of-the-art diesel particulate measurement methods, please refer to the references.

Smoke Meter

The underlying principle of smoke metering is that filter paper is loaded with exhaust gas for a certain period of time. Then an optical method is used to measure the blackening of the filter paper. The resulting quantity is the filter blackening or filter smoke number (FSN).

Opacimeter

An opacimeter (or opacity meter) is a device that measures the attenuation of a beam of light shining through a cell of a defined length that is filled with exhaust gas. The result of this process is called light extinction. Other commonly applied terms for this quantity of measurement are obscuration or opacity. An opacimeter is an optical instrument that takes advantage of the heavily absorbing effect of soot in the visible light spectrum. It measures the light obscuration of exhaust gas.

For this purpose, either back pressure or a pump draws the exhaust gas into a measuring chamber that commonly has a measuring length of 430 mm. On one side there is a light

Table 3.6 Advantages and disadvantages of alternative particulate measurement methods

Method	Advantages	Disadvantages
Opacimeter according to [15]	– Required for some certification test procedures, e.g. ELR (European Load Response) – Robust, cost-efficient, established method for measuring exhaust gas opacity – Very good time resolution, 0.1 s – High sensitivity (0.1 % opacity, corresponds to roughly 300 μg/m³ soot) – With a special gas path design, the device is suitable for exhaust gas pressures up to 400 mbar; an option for higher pressures is in development – Acceptable correlation with soot concentration (mg/m³) can be established for engine families.	– Sampling systems require sample flow rates of up to 40 l/min – High sensitivity is only achievable with a sophisticated system design: large optical path length L, good thermal conditioning – Relatively strong cross sensitivity to NO_2
Smoke meter according to [15]	– Robust, cost-effective system – Established method – High sensitivity (0.002 FSN, equals approx. 20 μg/m³ soot) in long sampling periods – A special sampling option allows exhaust gas measurement upstream of diesel particulate filters – Good correlation to soot concentration, (mg/m³), minimum cross sensitivity to other exhaust gas components	– Integrating method—time resolution approx. 1 min
Photo-acoustic soot sensor according to [16, 17]	– High sensitivity—soot typically < 5 μg/m³ – Sensor signal is directly and linearly sensitive to soot concentration, minimal cross sensitivity – Good time resolution, ≈ 1 s – Applicable for diesel particulate filter tests – Moderately priced – High dynamic range (1:10,000)	– Requires exhaust gas dilution – Method of calibration not firmly established – Measurement upstream of the diesel particulate filter (DPF) requires exhaust gas conditioning – Maintenance is easy but necessary on a regular basis

(continued)

Table 3.6 (continued)

Method	Advantages	Disadvantages
Light scattering sensor	– Compact, cost-effective system – Measures the intensity of the light scattered by particulates – High sensitivity—particulates typically < 50 g/m^3 – In some cases, the signal has been found to empirically correlate with the particulate emissions of diesel engines. – Heavily dependent on the size of the particulates in the aerosol	– Cost-effective system – Good time resolution < 0.1 s – Moderate correlation with particulate mass
Laser induced incandescence (LII) according to [18]	– High sensitivity—soot typically < 5 μg/m^3 – Sensor signal is directly and linearly sensitive to soot concentration, minimal cross sensitivity – Good time resolution, ≤ 1 s – Applicable for diesel particulate filter tests	– Very expensive – Method of calibration is not established – High dynamic range is only achievable with optical attenuators (insertion of absorber filters)
Photoelectric aerosol sensor according to [19]	– Compact, cost-effective system – High sensitivity—soot typically < 1 μg/m^3 – In most cases an empirical correlation of the electric signal with soot emission of diesel engines can be established.	– Time resolution ≤ 10 s – Heavily influenced by substances with high photoemission (PAH)
Diffusion charging sensor according to [19, 20]	– Compact, cost-effective system – Measures the active particulate surface (Fuchs surface) – High sensitivity—particulates typically < 1 μg/m^3 – In some cases, the signal has been found to empirically correlate with the particulate emissions of diesel engines.	– Not proportional to particulate mass – Time resolution of a few seconds

Fig. 3.77 Measurement principle of the smoke meter according to [15]

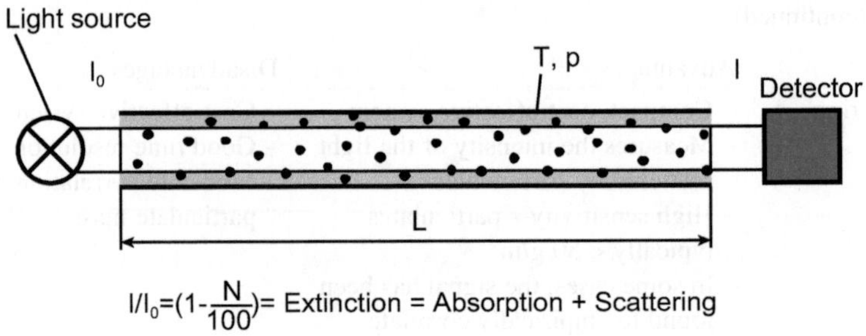

Fig. 3.78 Opacimeter measuring principle

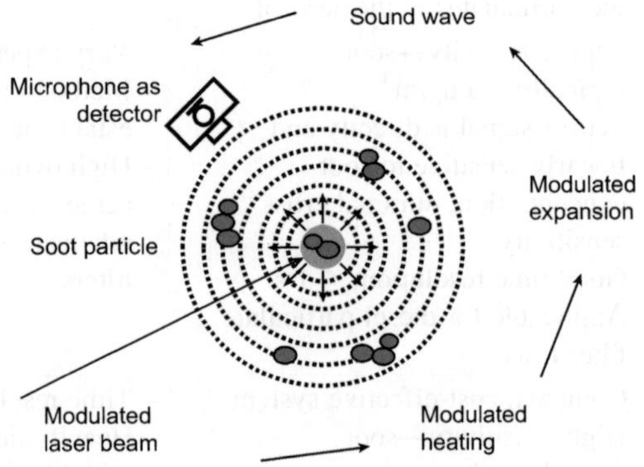

Fig. 3.79 Photoacoustic measuring principle according to [16]

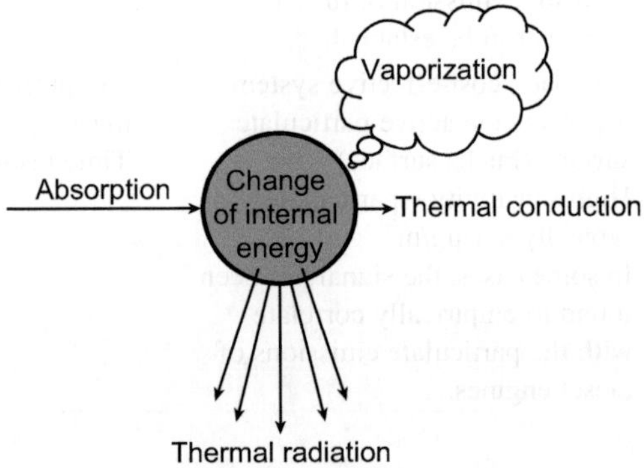

Fig. 3.80 Principle of laser induced incandescence (LII) according to [18]

source and on the other there is a detector that measures the attenuation of the light shining through the exhaust gas. This attenuation (light extinction) is caused by the light being absorbed and scattered by the particulates. The primary quantity measured by the opacimeter is opacity N (expressed in per cent), which indicates the relative decrease in light intensity:

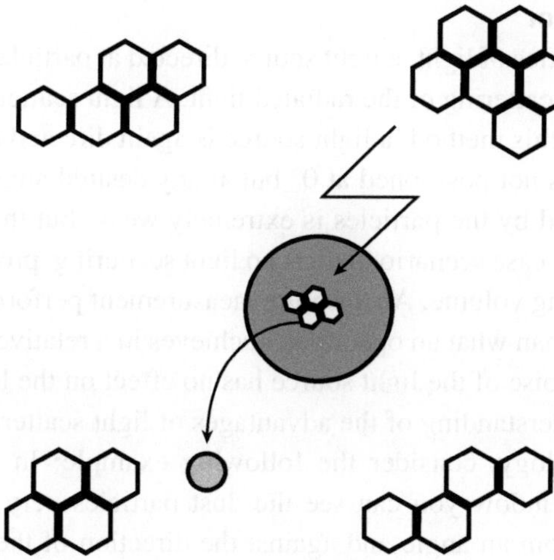

Fig. 3.81 Photoelectric measuring principle according to [19]

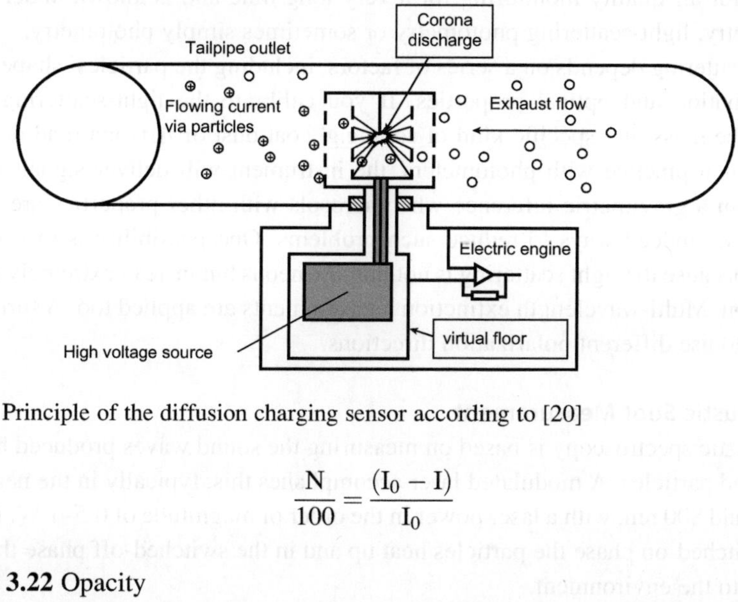

Fig. 3.82 Principle of the diffusion charging sensor according to [20]

$$\frac{N}{100} = \frac{(I_0 - I)}{I_0} \tag{3.22}$$

Equation 3.22 Opacity

I_0 is the intensity of the light source and I is the intensity measured by the opposite detector. It must be noted, however, that opacity depends on the measuring length, so it is common practice to use it to calculate a secondary measurement quantity independent of length, i.e. the absorption coefficient k in 1/m. The calculation follows the Lambert law: :

$$I = I_0 e^{(-kL_{eff})} \tag{3.23}$$

Equation 3.23 Lambert law

Light Scattering Sensor

Due to the wave properties of light, a light source directed at particles not only causes light obscuration but also a scattering of the radiated light. A light scattering detector measures the redirected light. In this method, a light source is again directed at the exhaust gas, but this time the detector is not positioned at 0° but at any desired angle to the incident light rays. The light scattered by the particles is extremely weak, but the arrangement has the advantage that in a best-case scenario there is no light scattering, provided that there are no particles in the scattering volume. An absolute measurement performed this way typically has a higher accuracy than what an opacimeter achieves in a relative measurement. This is because the intensity noise of the light source has no effect on the lower measuring limit.

To gain a better understanding of the advantages of light scattering measurement over the opacimeter technology, consider the following example: In a room with sunlight streaming in from a window you can see the dust particles very clearly by looking at the illuminated area from an angle and against the direction of the incident light. If you looked directly at the sun, you would never notice the slight light obscuration caused by the dust particles. Light scattering is a relatively old technology that has been employed as a method for air quality monitoring for a very long time and is known under the terms nephelometry, light-scattering photometry or sometimes simply photometry.

Light scattering depends on a series of factors, including the particles' shape, material, size distribution and optical properties. If you calibrate the light-scattering signal to calculate the mass of a specific kind of soot, e.g. coal dust or Arizona road dust (ARD), as is common practice with photometers, the instrument will deliver signals that differ widely from a gravimetric reference when aerosols with other properties are measured. But there are indeed ways to reduce such problems. One possibility is to use multiple detectors because the light scattering is not homogeneous but instead extremely dependent on direction. Multi-wavelength extinction measurements are applied too. A further option would be to use different polarization directions.

Photoacoustic Soot Measurement

Photoacoustic spectroscopy is based on measuring the sound waves produced by periodically heated particles. A modulated laser accomplishes this, typically in the near-infrared range around 800 nm, with a laser power in the order of magnitude of 0.5–1 W. During the laser's switched-on phase the particles heat up and in the switched-off phase they release their heat to the environment.

The process leads to the periodical expansion and contraction of the surrounding gas. The resulting pressure fluctuations produce a standing sound wave in the measuring chamber, which a microphone detects. The measuring chamber's dimensions must fit the laser's modulation frequency. In measurements on exhaust gas emitted by diesel engines, the choice of laser wave length in the near-infrared range has the effect that the laser energy is mainly absorbed by the soot; interference from other exhaust gas components is negligible. Alternative systems with wave lengths within the red range (e.g. 680 nm), on the other hand, require compensation for the effects of gaseous components.

Problems can occur if measuring cell windows are contaminated with soot. However the degree of contamination can easily be reduced by the choice of measuring cell geometry. The method features a very low noise level and is very sensitive to soot. The achievable sensitivity level is 5 μg/m^3 soot concentration with a minimal cross-sensitivity to other components. The time resolution is good (1–10 Hz), the dynamic range is very high (at around 1:10,000). However, measurements on older engines with higher emissions do require dilution. Though necessary on a regular basis, maintenance is found to be relatively easy.

Laser Induced Incandescence

Laser induced incandescence (LII) is a technique in which a high-power laser pulse heats up the particles to vaporization temperature. The resulting Planck radiation (black body radiation) provides information about the volume concentration and the particle size. Assuming average density, it is also possible to produce a signal that is proportional to the soot mass concentration.

Typical LII systems use an Nd:YAG laser to heat up the particles. The Nd:YAG laser is a solid-state laser that emits light at a wavelength of 1064 nm. In some cases, models are used in which the laser beam is frequency-doubled to 532 nm. The pulse length is typically 10 ns. The objective is to heat up the particles to the temperature of sublimation very rapidly within this short period of time without causing any significant loss in mass. Technical publications have specified that the laser requires a minimum energy density between 0.05 and 0.2 J/cm^2. After the laser pulse, the particles cool back down at a speed that depends on the ratio between surface and volume. The time-resolved acquisition of the emitted black body radiation and the analysis of the signal curve provide information not only on volume concentration but also on particle surface and size of the primary particulates. This is achieved by examining the maximum of emitted radiation.

This obviously requires instrumentation with nanosecond resolution. The technology's sensitivity toward soot is remarkably high—concentrations are detectable up to 5 μg/m^3. Its cross-sensitivity to other substances is minimal because liquid particles do not absorb the laser energy and any volatile components adsorbed onto soot particulates vaporize immediately. The time resolution is good and essentially depends on the laser's pulse frequency, which is typically within the range of 10–30 Hz.

Photoelectric Aerosol Sensor

A photoelectric aerosol sensor (PAS) measures the electric charge of particles that have received light radiation from a UV light source ($\lambda = 222$ nm). The method takes advantage of the photoelectric effect in which, depending on their size and surface material, particles are charged as a result of the radiation. The instrument exhibits a particularly high sensitivity to polycyclic aromatic hydrocarbon substances (PAHs). There is, however, an acceptable correlation to elemental carbon in the exhaust gas. A photoelectric aerosol sensor is a relatively cost-effective and compact device that achieves a high level of sensitivity of about 1 μg/m^3 for soot.

In diesel engines it is frequently possible to establish an empirical correlation to soot emissions, however the time resolution is rather moderate (0.1–1 Hz) and substances with high photoemissions, such as PAHs, have a heavy impact on the result.

Diffusion Charging Sensors

In diffusion charging sensors, a charging source, e.g. a corona discharger, produces small ions that attach to the particles by diffusion. The particle's charging efficiency is proportional to its active surface. The particles then pass through a filter where their charge is removed. This produces a measurable current flow, which is proportional to the aerosol's entire active surface.

To prevent volatile components from condensing in the device, dilution is recommended. Systems like these are compact and inexpensive to implement, achieve a high level of sensitivity in the range of 1 μg/m^3 particulate concentration. However, calibration to other known quantities has proved to be difficult. There is no correlation between the signal and the particulate mass; instead the signal has been found to be rather more proportional to a number between particulate diameter and surface ($d = 1.09–1.13$). Accordingly, the signals have been found to correlate with the emissions of diesel engines only in a few cases.

3.3.15 Swirl and Tumble

3.3.15.1 Purpose of the Measurement

Charge motion is a parameter relevant for spark-ignition and compression-ignition combustion engines that affects combustion significantly. We basically distinguish between swirl, a rotational charge motion around the cylinder axis, and tumble (also known as barrel swirl), a rotational charge motion perpendicular to the cylinder axis. In most cases, the two charge motions appear simultaneously, in a mixed form. In diesel engines, combustion methods have become well established that are more or less swirl-oriented, whereas in gasoline engines combustion methods are preferred with the tumble-type charge motion. The charge motion is generated and influenced not only by the shape and position of the intake ports, but also by the shape of the combustion chamber. There are special flow test beds (see Sect. 2.1.5) that capture both types of charge motion. For measuring these two parameters at a steady flow, both integrating and differential methods have gained acceptance.

3.3.15.2 Definition of Swirl and Tumble

The difference in the axis position of the two rotational swirl and tumble motions causes the swirl to be maintained during the piston stroke in the real engine as a result of the geometrical boundary conditions. It can even lead to an increase in the charge's rotational speed if the combustion chamber is shaped accordingly [21]. The tumble motion, on the

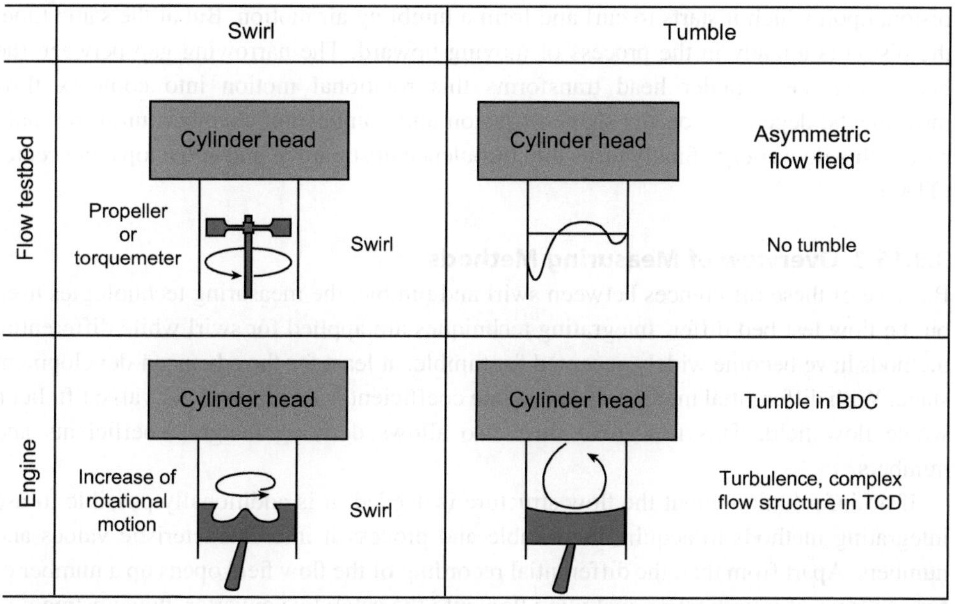

Fig. 3.83 Charge motions on the flow test rig versus the real engine

other hand, is virtually squashed between the piston and the bottom of the cylinder head and breaks up into complex flow structures and, eventually, into turbulence.

This significant difference has led to the fact that (where swirl is concerned) integrating methods supply favorable correlations between the combustion result and the swirl measurement. Integrating methods are given preference on steady-flow test beds. In the case of tumble, however, these correlations usually lead to a rather confusing picture.

Figure 3.83 illustrates the described processes in the form of a matrix. The columns refer to the charge motions swirl and tumble while the rows represent the charge motion in their respective environment, i.e. on the one hand, the steady-flow test bed and, on the other, the engine.

Swirl is present both on the steady-flow test bed and in the engine. On the flow test bed, a torque meter or a paddle wheel sensor measures swirl in the form of rotation coefficients for a number of valve lifts. Based on these rotation coefficients a swirl number is determined by integration over crank angle. Although the flow structures in the real engine are obviously different from those on a steady-flow test bed, good correlation is obtained between these results and those of combustion. Based on the angular-momentum conservation law and the geometric boundary conditions, the swirl is maintained as a dominating charge motion effect in both environments (engine and steady-flow test bed).

The situation is completely different where tumble is concerned (right-hand column in Fig. 3.83): On the steady-flow test bed no tumble is present at all. Instead a roughly asymmetrical forward flow occurs inside the cylinder, influenced by the shape of port and combustion chamber. In the engine, on the other hand, this asymmetric jet impinges the piston, upon which it starts to curl and form a tumbling air motion. But at the same time, the piston is already in the process of moving upward. The narrowing gap between the piston and the cylinder head transforms this rotational motion into complex flow movements, depending on the shape of piston and combustion chamber, until the main part of this flow energy finally turns into turbulence upstream of and at the top dead center (TDC).

3.3.15.3 Overview of Measuring Methods

Because of these differences between swirl and tumble, the measuring technologies used on the flow test bed differ. Integrating techniques are applied for swirl while differential methods have become widely accepted for tumble, at least for the advanced-development stage. With differential methods, not just one coefficient is determined per valve lift, but a whole flow field. This flow field then also allows deriving integral coefficients and numbers.

If no information about the flow structure is needed, it is additionally possible to use integrating methods to acquire the tumble and process it into characteristic values and numbers. Apart from that, the differential recording of the flow field opens up a number of further areas of investigation including flow into the catalytic converter, flow upstream of coolers (EGR, boost air), and so forth. In this regard, the parameters centricity and uniformity [22] have moved into the focus of interest.

Both in integrating and differential measuring methods, the measured physical quantities provide the basis for calculating the (flow, rotation and tumble) coefficients for a series of valve lifts. Based on a simple incompressible engine model, an integration is performed over crank angle (piston stroke), which then forms the basis to calculate the mean values, i.e. the characteristic numbers (flow, swirl and tumble numbers). To obtain comparable values for different types of engines, a standard valve lift curve (relating to the inner valve seat diameter) serves as a basis. To establish a correlation between the rotation and tumble coefficients and the mass flow, the measured mass flow is used as a basis to calculate a mean in-cylinder flow velocity. This flow velocity is equated with a mean piston speed, from which a fictitious engine speed is calculated. The speed of the air flowing into the cylinder and measured using e.g. a paddle wheel (see Fig. 3.84) can so be normalized to obtain a dimensionless rotation coefficient.

Another key parameter and development goal for port design alongside charge motion is the flow rate through the port. In the case of an intake port, development efforts will necessarily involve a trade-off between the two parameters charge motion and flow-rate, however only a single target parameter is required for the exhaust port: the maximum flow rate under the given boundary conditions.

To determine the flow coefficients, the mass flow is measured on the flow test bed for multiple valve lifts at a defined pressure difference across the port (see page 240) and put in relation to a theoretical mass flow for a reference cross-section. In line with the charge motion, this delivers a flow coefficient for each valve lift, and with a series of flow coefficients it is then possible to obtain a flow number by integration over the crank angle.

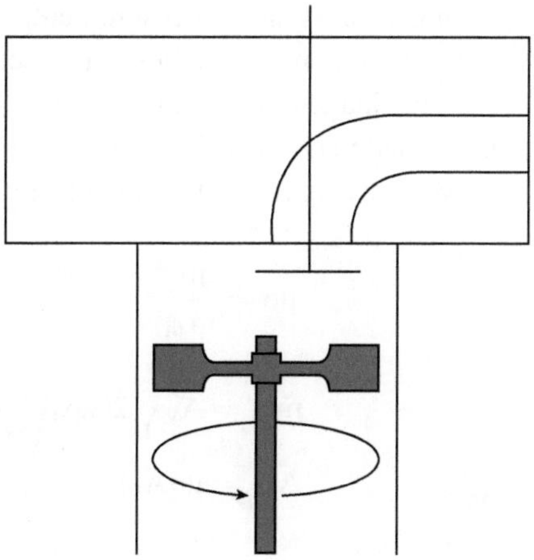

Fig. 3.84 Schematic diagram of a paddle wheel

To perform a comparative analysis of these coefficients, numbers and flow profiles, it is first necessary to normalize these quantities, a procedure which is essential for port development. A simple engine model with a referenced valve lift curve is used to calculate the mean values by integration over the crank angle. The effect of the stroke-to-bore ratio is eliminated when calculating the coefficients and reduced numbers for a stroke-to-bore ratio of 1 are determined.

Given the radical difference between flow processes on a steady-flow test bed and in real engines, engineers have long since been studying the transient flow processes in real engines—or at least in the combustion chambers of simplified research engines. Such research engines are simplified in that their combustion chamber is made optically accessible, in other words it is the combustion chamber's geometry that is simplified, an aspect that typically makes the engine as a whole rather more complicated. The optical accessibility is a critical factor since the measuring technology to be applied generally involves laser-optical methods. Consequently, the challenge the engineers face is not only to handle the optical accessibility but also to manage and interpret immense quantities of data. Relying on a combination of computational fluid dynamics (CFD) calculation and measurement on the steady-flow test bed and a subsequent analysis of the transient processes, exclusively using the CFD model, has proven to be a very helpful approach to the problem.

3.3.15.4 Integrating Measuring Methods for Flow and Swirl

Flow

To determine the flow rate it is necessary to measure the mass flow $\dot{m}$. There are a variety of different measuring systems that can be used for this purpose (see Sect. 3.3.9).

Using a mass flow measurement, a dimensionless flow coefficient ($\mu\sigma$) is derived, which is based on the smallest valve seat cross-section possible (calculated from the inner valve seat diameter d_v), which is defined as the ratio between the measured mass flow and a theoretical mass flow (i.e. a mass flow resulting without loss from the pressure difference Δp). Multiplying $\mu\sigma$ by the reference cross-section A_V yields the effective flow area A_{eff}.

$$\mu\sigma = \frac{\dot{m}}{\dot{m}_{th}}$$

$$\dot{m}_{th} = A_V\sqrt{2p_0\rho_0}\sqrt{\frac{\kappa}{\kappa-1}}\sqrt{\left(1-\frac{p_1}{p_0}\right)^{\frac{2}{\kappa}} - \left(1-\frac{p_1}{p_0}\right)^{\frac{\kappa+1}{\kappa}}} \quad (3.24)$$

$$A_{eff} = \mu\sigma A_V$$

ρ_0 Density of air
p_0 Pressure upstream of port
p_1 Pressure downstream of port
A_V Reference cross-section (inner valve seat cross-section)
κ Isentropic exponent

Equation 3.24 Flow coefficient

Multiplying by the area ratio β yields the bore-related flow coefficient $\mu\sigma\beta$, i.e. the flow capacity ($= \alpha_K$).

$$\beta = z\frac{{d_v}^2}{D^2} \quad (3.25)$$

d_v Inner valve seat diameter
D Cylinder bore diameter
z Number of valves

Equation 3.25 Area ratio

Using the calculated coefficients, the flow number $(\mu\sigma)_m$ is calculated, which is obtained by integration over the crank angle from the top dead center (TDC) to the bottom dead center (BDC), while taking into account a valve lift curve and the piston motion. For purposes of comparison, a fictitious "standard" valve lift curve is used. This curve is a purely theoretic, parabolic shaped curve that generally does not correspond to an engine's valve lift curve [23].

This means that the value $(\mu\sigma)_m$ represents a flow number calculated on the basis of the "standard" valve lift curve and weighted with the piston speed. The value is a criterion for the quality of the port. Obviously, other valve lift curves can also serve as a basis, provided this is taken into account when making comparisons.

$$(\mu\sigma)_m = \frac{1}{\sqrt{\frac{1}{\pi}\int_0^{\pi}\left(\frac{c(\alpha)}{c_m}\right)^3 \frac{1}{(\mu\sigma)^2}\,d\alpha}} \tag{3.26}$$

$c(\alpha)$ Piston speed
c_m Mean piston speed

Equation 3.26 Flow number

Multiplying by the area ratio β yields the bore-related flow number $(\mu\sigma)_m \cdot \beta$, i.e. the flow capacity (= mean α_K). This means that the value $(\mu\sigma)_m \cdot \beta$ gives indication of the quality of the port based on the bore cross-section.

Measuring Swirl Using a Paddle Wheel

The use of a paddle wheel allows the measurement of the speed of the air flowing into the cylinder and rotating around the cylinder axis (see Fig. 3.84) [24]. The shapes of the paddles on the paddle wheels used in one and the same laboratory are typically geometrically similar; the main dimensions are based on the bore diameter. Other laboratories may have a variety of different impeller drive installations and designs.

From the measured paddle speed a dimensionless number (rotation coefficient n_D/n) is calculated by dividing the paddle speed n_D by a fictitious engine speed. This engine speed is calculated from the mass flow measured on the flow test bed by equating the mean in-cylinder flow velocity (this is obtained from the measured mass flow) with a mean piston speed, which means it depends on the piston displacement.

$$n = \frac{30\dot{m}}{\rho\ s F_K}$$
$$\frac{n_D}{n} = \frac{n_D\,\rho\,V_h}{30\dot{m}} \tag{3.27}$$

ρ Density of air
s Stroke
F_k Bore cross-section
n_D Paddle speed
V_h Piston displacement
$\dot{m}$ Mass flow

Equation 3.27 Rotation coefficient, paddle wheel

Finally, the dimensionless swirl number $(n_D/n)_m$ is determined as a mean value, which is once again obtained by integration over the crank angle.

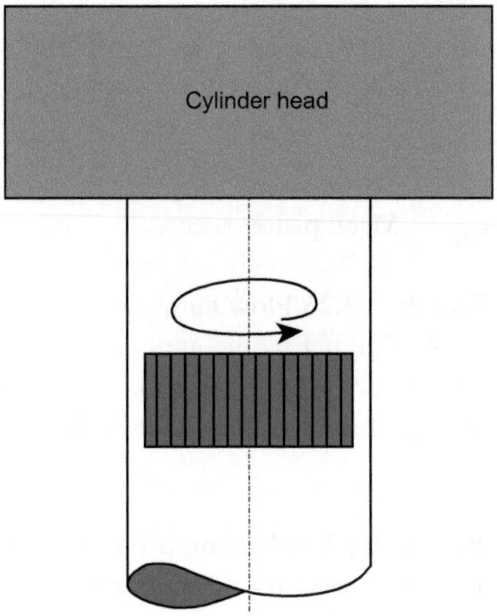

Fig. 3.85 Schematic diagram of a torque meter

$$\left(\frac{n_D}{n}\right)_m = \frac{1}{\pi}\int_0^{\pi} \frac{n_D}{n}\left(\frac{c(\alpha)}{c_m}\right)^2 d\alpha \tag{3.28}$$

Equation 3.28 Swirl number

Due to the fictitious engine speed, both the coefficient and the number depend on the stroke-to-bore ratio. The "reduced" coefficients and numbers, $(n_D/n)_{red}$, $(n_D/n)_{m,red}$, on the other hand, are based on a stroke-to-bore ratio of 1, enabling a comparison between the ports of different engines.

$$\left(\frac{n_D}{n}\right)_{red} = \frac{n_D}{n}\frac{D}{s} \text{ and } \left(\frac{n_D}{n}\right)_{m,red} = \left(\frac{n_D}{n}\right)_m \frac{D}{s} \tag{3.29}$$

Equation 3.29 Reduced rotation coefficient, reduced swirl number

Measuring Swirl Via Torque

A torque meter consists of a flow straightener (also termed flow rectifier) that removes the entire angular momentum from the air rotating around the cylinder axis, thereby exerting torque on the axis of the flow straightener [25, 26]. The occurring torque is measured (Fig. 3.85).

Compared with the paddle-wheel measuring technique, this measuring method has the advantage that it captures the entire flow field. Accordingly, flow structures are also measured that are located at the center of the cylinder, and even the axial flow distribution is taken into account. Unlike the rotary paddle whose bearing is merely checked periodically, the toque meter is an instrument that can be calibrated, which makes it far more suitable for comparisons with other labs than the paddle wheel. Furthermore, due to its higher level of sensitivity, the torque meter should be given preference over the paddle wheel for low-swirl systems (swirl number <1).

The torque measurement is again used to calculate a rotation coefficient n_D/n by assuming a solid body rotation and a constant axial flow profile and calculating the speed n_D of this solid body rotation and dividing it by a fictitious engine speed as is done in the measurement using a paddle wheel. The torque around the cylinder axis z results from the axial and tangential velocity components u and v, generally expressed as cylinder coordinates:

$$T_z = \rho \int_{r=0}^{R} \int_{\varnothing=0}^{2\pi} u(r,\varnothing)v\,(r,\varnothing)r^2 dr d\varnothing \qquad (3.30)$$

Equation 3.30 Torque around the cylinder axis

Assuming that u = constant and v = r ω (solid body rotation), it follows that

$$T_z = \rho \dot{V} \omega \frac{D^2}{8} \qquad (3.31)$$

Equation 3.31 Torque around the cylinder axis

ω represents the angular velocity of the rotating solid body. This results in a fictitious paddle speed n_D. As noted previously with regard to the paddle wheel, the normalization is performed with the fictitious engine speed n, resulting from the measured mass flow.

$$\frac{n_D}{n} = 2 T_z \frac{s}{\rho V_{st}{}^2} \qquad (3.32)$$

R Cylinder radius
D Cylinder diameter
s Piston stroke
V_{st} Volumetric flow
ρ Density of air
Ø Angle at the circumference

Equation 3.32 Normalized fictitious paddle speed

The swirl number $(n_D/n)_m$ and the reduced swirl number $(n_D/n)_{m,red}$ are calculated in the same way as described for the paddle wheel.

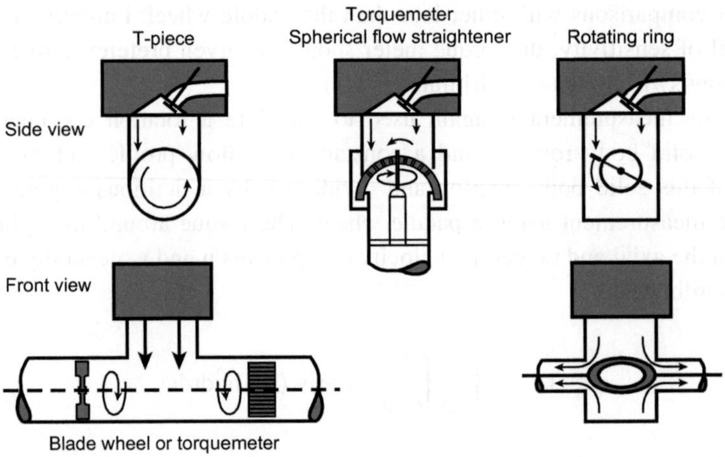

Fig. 3.86 T-piece [27], spherical torque meter [28], rotating ring [29]

3.3.15.5 Integrating Measuring Method for Determining Tumble

Overview of Methods

Understandably, combustion development engineers were interested in developing the same kind of integrating methods for the tumble charge motion as were already available for swirl. Figure 3.86 shows the three most commonly used methods in a schematic diagram.

- The **T-piece method** can be seen from two side views in Fig. 3.86: The illustration below shows a cylinder head fitted on top of a pipe T-piece with the already familiar integrating swirl measuring methods, paddle wheel sensor or torque meter, each in one section of the transverse pipe.

  The illustration above schematically indicates the process that leads to the rotating motion inside the transverse pipe. The asymmetric flow from the intake port hits the transverse pipe where the asymmetry generates a swirl flow. By twisting the cylinder head until the maximum measuring value is reached, it is additionally possible to capture the tumble angle.
- On the right side of Fig. 3.86, there is an illustration of the rotating ring method. The rotating ring corresponds to the rotary paddle in the swirl measuring technique for tumble acquisition. However, if the flow through the cylinder is simple and steady, this rotating ring is hardly likely to rotate at all given what was said in the introduction to this topic. Furthermore, its speed would depend very heavily on the distance to the combustion chamber, as it is very likely that there are local, rotating flow fields below the intake valves.

  That is the reason why the cylinder is closed by a dummy piston and the air is drawn off sideways in the area of the rotary paddle's axis. Again, the air jet asymmetrically impinging the dummy piston induces a rotational motion that is recorded by the speed of the rotating ring. The tumble angle is obtained by twisting the position of the rotating ring or cylinder head until a position is reached that produces maximum ring speed.

- In the middle of Fig. 3.86, the torque meter is shown that consists of a spherical flow straightener with radially oriented flow ports. Since this is the method that allows measuring both tumble and swirl, it will receive more detailed attention further below.

Measuring Tumble with Spherical Flow Straightener

Because of its radially arranged bores, the spherical flow straightener is capable of capturing not only an angular momentum around the z-axis but also a tilting moment around an axis in the x/y plane, allowing it to respond very elegantly to the flow field's asymmetry. As all three spatial moments are captured, only a single measurement is needed to obtain swirl, tumble and tumble angle.

As with swirl calculation, the use of the spherical flow straightener also requires calculating the dimensionless tumble coefficient n_T/n. For this purpose, the two measured moments on the x- and y-axis are needed. The moment on the z-axis is used to calculate the rotation coefficient (see page 228).

$$\frac{n_t}{n} = 2\frac{s}{\rho \dot{V}^2}\sqrt{{T_x}^2 + {T_y}^2} \tag{3.33}$$

Equation 3.33 Tumble coefficient

The tumble number and the reduced tumble number are calculated in the same way as the swirl number.

$$\left(\frac{n_T}{n}\right)_m = \frac{1}{\pi}\int_0^{\pi}\frac{n_T}{n}\left(\frac{c(\alpha)}{c_m}\right)^2 d\alpha \quad \left(\frac{n_T}{n}\right)_{m,red} = \left(\frac{n_T}{n}\right)_m \frac{D}{s} \tag{3.34}$$

Equation 3.34 Tumble number and reduced tumble number

3.3.15.6 Differential Measuring Method for Swirl and Tumble

Given the widely different boundary conditions for swirl and tumble (on the one hand on the steady-flow test bed and on the other in the engine), the integrating measuring methods for tumble are barely capable of giving much useful information for combustion development, which is why the investigation focuses on a differential recording of the axial flow field and its structure instead. From this structure, it is then possible to extract arbitrary integral parameters.

Considering what was said previously, it is now clear that the objective is to assess the asymmetry of the flow in order to determine the tumble. The more asymmetric the axial flow structure, the greater the tumble will be that forms in the presence of the moving piston inside the engine. Swirl calculation requires the tangential velocity components in a plane vertical to the cylinder axis.

Laser Doppler Anemometry (LDA), Also Known as Laser Doppler Velocimetry (LDV)
This point-by-point measuring technique, which is based on flow velocity, measures the axial flow structure in a plane at a distance from the bottom of the cylinder head which corresponds to half the diameter of bore (D) (see [21, 30–32]). For this purpose, light-scattering particles in the form of tiny droplets (seeding) are added to the intake air slightly upstream of the measuring object.

The measuring volume of an LDA system is directed through 200–300 points in the measuring plane, thereby capturing the axial flow field. Figure 3.87 shows a possible design featuring automatic traversing of the measuring volume. Below the measuring plane, there is another cylinder with bore diameter, establishing the connection to a stabilizing vessel.

The flow field w_{LDA} measured this way serves as the basis for calculating a tumble coefficient ω_{FK}/ω_{Mot} see Fig. 3.88.

From the w_{LDA} (measured axial velocities), a mean velocity $\overline{w}$ is calculated first. By reducing the flow field w_{LDA} by $\overline{w}$, the reduced, quasi-rotating flow field w_i follows under the assumption of a tumble axis which passes through the central point of the cylinder and stands perpendicular to the cylinder axis.

$$w_i = w_{LDA} - \overline{w} \tag{3.35}$$

Equation 3.35 Calculation of quasi-rotating flow field

This is used to calculate the angular velocity ω_i in each measuring point.

$$\omega_i = \frac{w_i}{r_i} \tag{3.36}$$

Equation 3.36 Calculation of angular velocity in each measuring point

Fig. 3.87 LDA setup on the steady-flow test bed

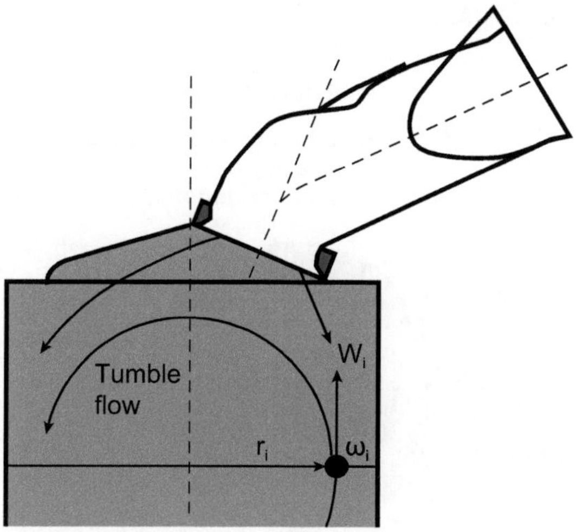

Fig. 3.88 Calculation of tumble coefficient

If each measuring point i is assigned the surface element f_i, the angular velocity of the flow field ω_{FK} is obtained by a weighted averaging using f_i and r_i over all i.

$$\omega_{FK} = \frac{\sum \omega_i {r_i}^2 f_i}{\sum {r_i}^2 f_i} \tag{3.37}$$

Equation 3.37 Calculation of angular velocity

Equating the mean axial flow velocity $\overline{w}$ with a fictitious piston speed c_m yields a fictitious engine speed n or engine angular velocity ω_{Mot}.

$$c_m = \frac{sn}{30} = \overline{w} \qquad \omega = \frac{2\pi n}{60} \qquad \omega_{Mot} = \frac{\pi \overline{w}}{s} \tag{3.38}$$

Equation 3.38 Calculation of a fictitious engine speed or engine angular velocity

As with the paddle wheel or torque meter, the tumble coefficient can now be calculated as ω_{FK}/ω_{Mot} or n_T/n.

$$\text{Tumble coefficient} = \frac{n_T}{n} = \frac{\omega_{FK}}{\omega_{Mot}} \quad \text{Tumble coefficient}_{red.} = \left(\frac{n_T}{n}\right)_{red} = \frac{\omega_{FK}}{\omega_{Mot}}\frac{D}{s} \tag{3.39}$$

Equation 3.39 Calculation of tumble coefficient

In addition, the tumble angle is indicated for the rotating axis with the maximum tumble coefficient. If tumble coefficients are determined for several valve lifts, it is

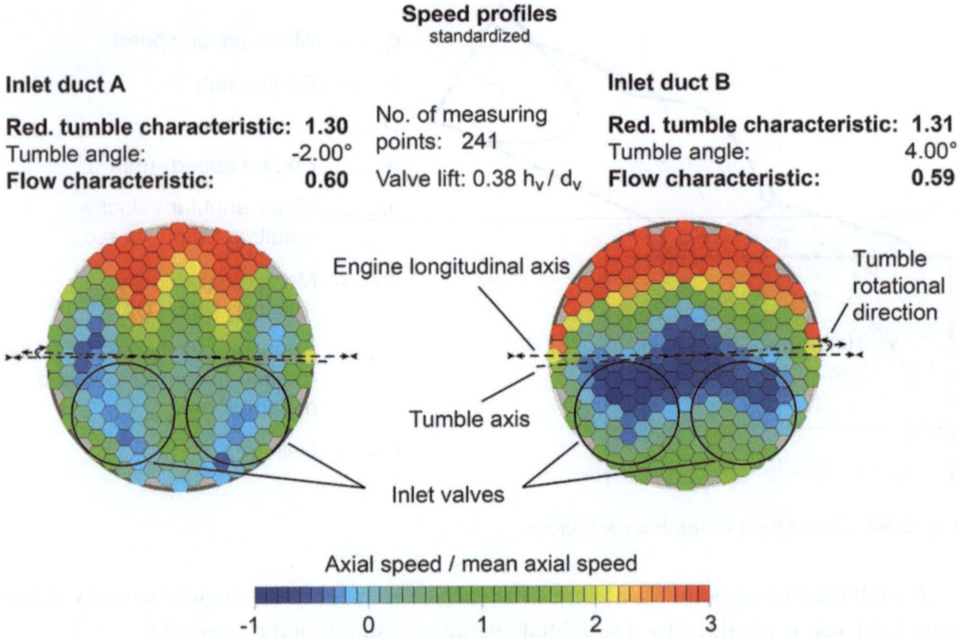

Fig. 3.89 Typical result—same tumble coefficients

possible to calculate the tumble number by performing an integration over the crank angle as shown previously (see Eq. 3.28).

$$\left(\frac{n_T}{n}\right)_m = \frac{1}{\pi}\int_0^{\pi} \frac{\omega_{FK}}{\omega_{Mot}} \left(\frac{c(\alpha)}{c_m}\right)^2 d\alpha \tag{3.40}$$

Equation 3.40 Calculation of tumble number

The advantage of the differential measuring technique is the way the flow structure is determined. Figure 3.89 shows a measuring result, which, although yielding very similar coefficients for tumble and flow, exhibits a markedly different flow structure. Despite being based on the same combustion chamber, the combustion results were very different too.

These LDA measurements are extremely time-consuming as each valve lift requires 200–300 speed measurements. In this case, particle imaging velocimetry (PIV) can be very helpful.

Particle Imaging Velocimetry (PIV)

A three-dimensional PIV system is used not only to measure axial velocity components (as with the previously described LDA system), but also velocity components in a

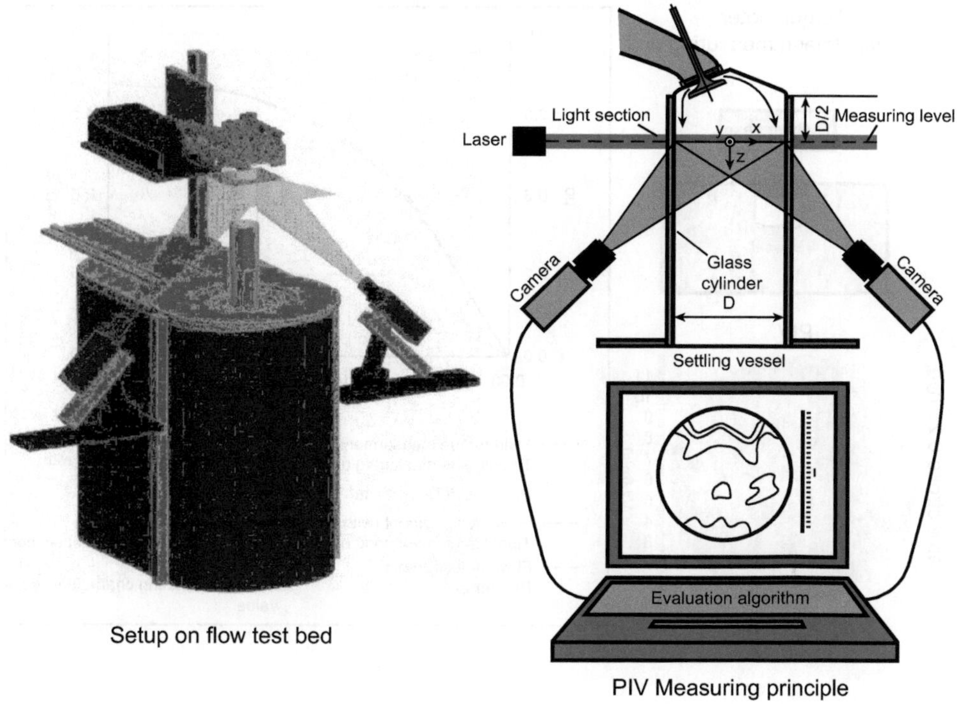

Fig. 3.90 PIV measuring principle and setup on the steady-flow test bed

tangential and radial direction. This makes it possible to measure not only tumble flow but also more complex mixed forms consisting of tumble and swirl [33–35].

A PIV system consists of two CCD (charge-coupled device) cameras and a lens, an Nd: YAG double pulsed laser, light sheet optics, a PTU (programmable timing unit) and a PC with the relevant analysis software. Figure 3.90 (on the left) shows a diagram of the measurement setup on the steady-flow test bed.

The measuring object—a cylinder head or flow box—is mounted on top of a glass cylinder with the same internal diameter as the engine bore. The two cameras are mounted symmetrically on each side of the cylinder and oriented at an inclined angle of about 90° towards each other in relation to the horizontal measuring plane under the bottom of the cylinder head. The double pulse laser and the optical system are placed in such a way that the light sheet is located on the measuring plane and completely illuminates the area to be investigated. Operating on the same principle as the LDA, light-scattering particles in the form of tiny droplets (seeding) are added to the intake air slightly upstream of the measuring object. The measurement principle outlined on the right in Fig. 3.90 makes it clear that PIV relies very much on the analysis of the visual information collected by both cameras. Specific correlation algorithms are used as a basis to calculate a single-plane

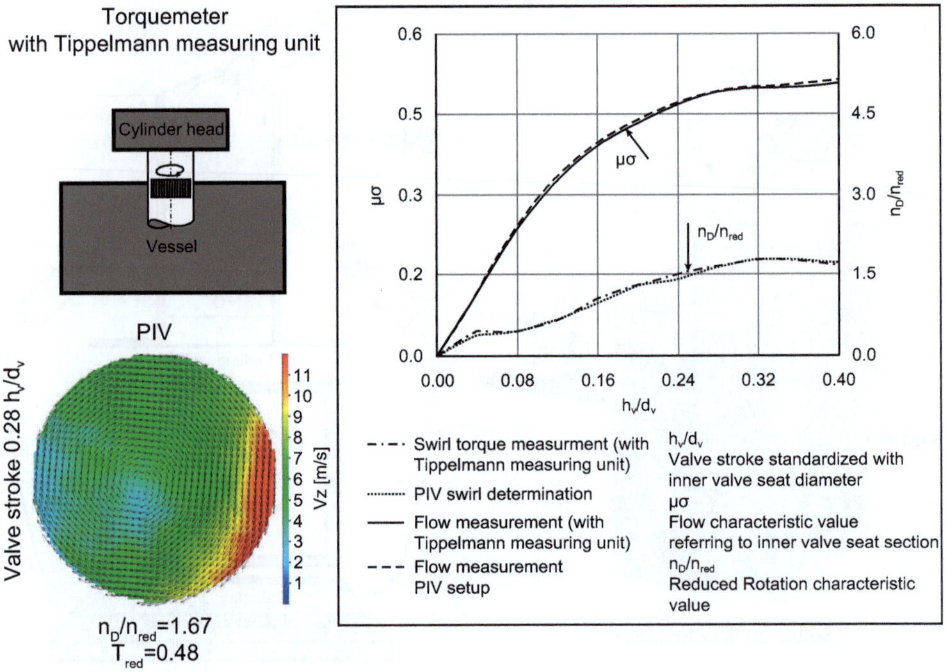

Fig. 3.91 Swirl—comparison PIV versus torque meter

flow profile that includes all three velocity components from the pictures taken by the two cameras.

Similar to the LDA technology, the axial velocity components can now be used to derive the tumble coefficients and a tumble number. Furthermore, together with the tangential velocity components it is possible to calculate a flow torque and, from that (as already demonstrated with the torque meter), a rotation coefficient for the respective valve lift. With several rotation coefficients, it is then possible to perform an integration over the crank angle to extract a swirl number. Figure 3.91 on the left shows the flow field generated by a swirl port as measured by PIV.

This port was also measured using a torque meter (see schematic diagram in the top left of Fig. 3.91). This results in the dot-and-dash lines shown in the diagram on the right in Fig. 3.91. The continuous line indicates the flow rate in the torque measurement. The excellent level of consistency between the differential measuring method PIV and the integrating torque measurement can be seen very clearly.

The colors show the distribution of velocities in the cylinder axial direction and it is worth noting that the asymmetry of the flow field also results in a quite considerable tumble coefficient. The arrows represent the tangential and radial components of the flow field. The rotation coefficient is calculated from the tangential and axial velocity components. This is performed for ten valve lifts. In Fig. 3.91 on the right, this result

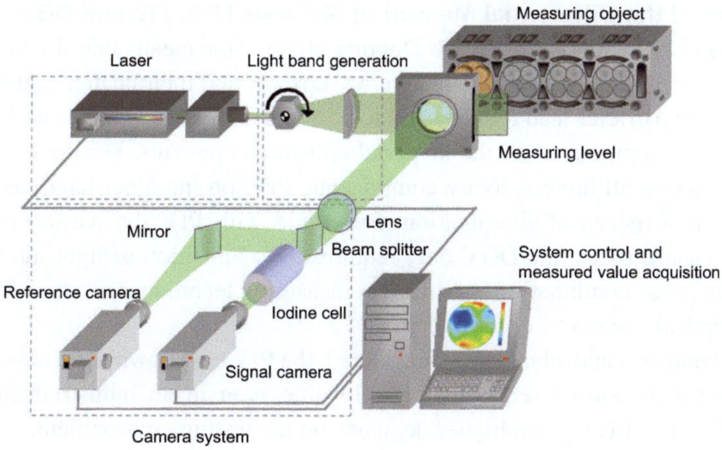

Fig. 3.92 DGV measuring principle

from the integral rotation coefficients is shown as a dotted line plotted against the normalized valve lift h_v/d_v. The flow coefficients (related to the inner valve seat cross-section) are shown as a dashed line.

Doppler Global Velocimetry Method (DGV)

DGV (also called planar Doppler velocimetry, PDV) is another laser-optical method that is used for flow measurement in all three spatial components of the flow field in a plane (measuring plane). The physical operating principle of DGV is depicted in Fig. 3.92 [36–39].

With the help of an optical system, a laser produces a band of light (laser light sheet), which illuminates a particle-seeded flow. Consistent with the Doppler effect principle, the light scattered by the particles exhibits a frequency shift. Through absorption in molecular iodine (iodine filled cell) the shift in frequency is converted into a shift in intensity that is detected by video cameras and converted into a velocity field. A velocity component is measured that is located in the direction of the angle bisector between the illumination direction and the observation direction. All three components of the velocity field of a measuring plane can therefore be obtained by combining different observation and illumination directions. Taking advantage of the physical operating principle, three illumination directions and one observation direction are combined for examining in-cylinder flow.

As with PIV technology, three illumination directions and one observation direction consequently yield all three velocity components of the illuminated flow field. Flow field analysis and the resulting calculation of coefficients and numbers are achieved in the same way as previously described in the context of LDA and PIV.

Comparison of the Differential Measuring Methods LDA, PIV and DGV

Both LDA and DGV are based on the Doppler effect. This means that the laser not only acts as a source of illumination but that the wave fronts and their change caused by being scattered by the particles lead to the signal to be evaluated. This demands high quality not only of the laser light but of all the involved optical components. For the LDA it is very difficult to capture all three velocity components. PIV, on the other hand, uses the laser exclusively as a means of illumination. But while with PIV the particles have to be brought into sharp focus, the DGV only requires the acquisition of light intensities. It is also possible to use combinations of different measuring techniques, as this offers benefits where the optical accessibility is concerned.

All disadvantages and advantages considered, the PIV has shown to be a comparatively straightforward measuring technology, deployable even in an industrial environment, whereas DGV and LDA place higher demands on the testing environment.

Comparison of Results of Different Measuring Methods

A comparison of the results of different laboratories and methods makes it clear that the produced characteristic values and numbers may vary greatly. There are three reasons for this:

- The geometry of the recording measuring system: Diverse impeller geometries or completely different measuring principles capture different areas and components of the flow field. This leads to different characteristic values for the charge motion.
- Different definitions are used for the characteristic values. If the same basic parameters and measuring methods are used, the characteristic values can be converted from one system into another.
- The averaging of characteristic values to characteristic numbers using either valve lift or crank angle uses different approaches. It is not possible to convert one characteristic number into another.

The following comparison analyzes the different measuring methods and compares the most common definitions. The correlations are generally fairly satisfactory.

Result of Swirl Measurement Using Torque: Paddle Wheel Method

While the vane anemometer only responds to the tangential flow velocity components, also depending on the shape of the vanes, the torque meter eliminates the flow's entire angular momentum, thereby including the axial flow distribution in the result (see formalism on page 228). But since the mechanism of swirl formation inside the cylinder is similar among different ports for different swirl levels, there are fairly good correlations between swirl numbers that were captured using the paddle wheel and those measured by the torque meter.

The more similar the swirl formation mechanism, the better the correlation will be. Figure 3.93, for example, shows the correlation between the swirl numbers for the

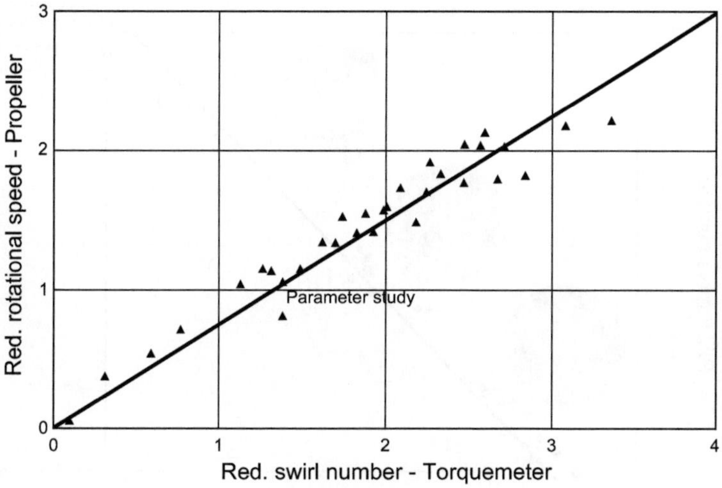

Fig. 3.93 Correlation of swirl number measured with paddle wheel and torque meter

intake ports of four-valve diesel engines measured with paddle wheel (y-axis) and those measured with the torque meter (x-axis).

Result of Tumble Measurement with LDA, PIV, T-Piece and Spherical Flow Straightener

As mentioned earlier (see page 233), there are several different tumble measurement technologies available to obtain tumble coefficients. The differential methods LDA and PIV lead to the same results within the required measurement accuracy. While all of the other methods produce values that vary quite widely, they show good correlation with each other due to the similarities in tumble formation. The relevant correlations shown in Figs. 3.94, 3.95 and 3.96 concern LDA and PIV, LDA and T-piece and, finally, PIV and spherical flow straightener.

Flow Coefficients from Different Laboratories

The following section addresses the differing definitions and, if available, any conversion relationships.

Rotation coefficient according to Tippelmann [26]:

$$K_z = T_z \frac{\rho D}{2\dot{m}^2} \tag{3.41}$$

Equation 3.41 Rotation coefficient according to Tippelmann

The Tippelmann method is based on torque measurement. The coefficient is converted to AVL laboratory rotation coefficients as follows:

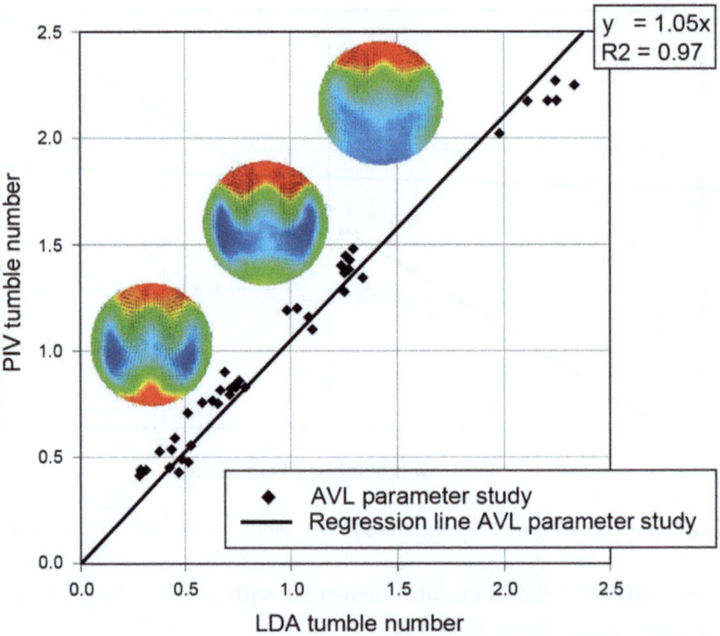

Fig. 3.94 Correlation of tumble number measured with PIV versus LDA

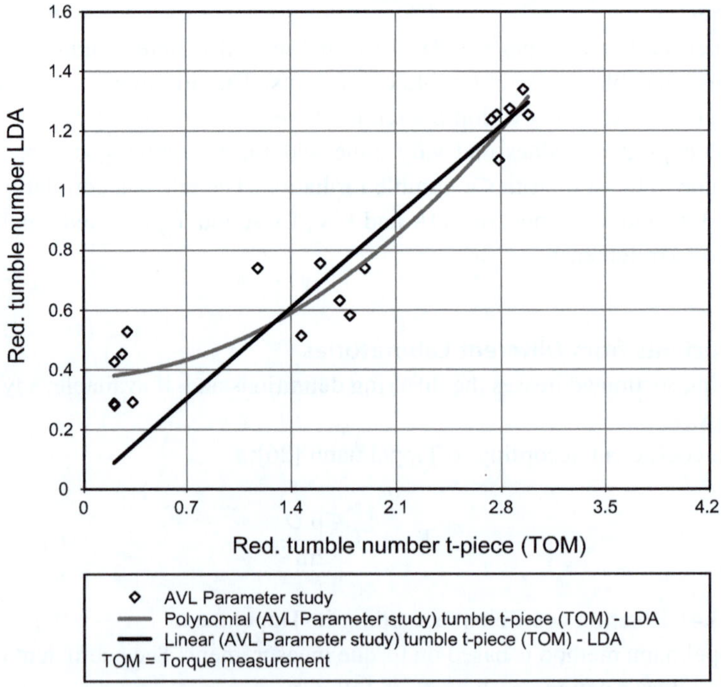

Fig. 3.95 Correlation of tumble number measured with LDA versus T-piece

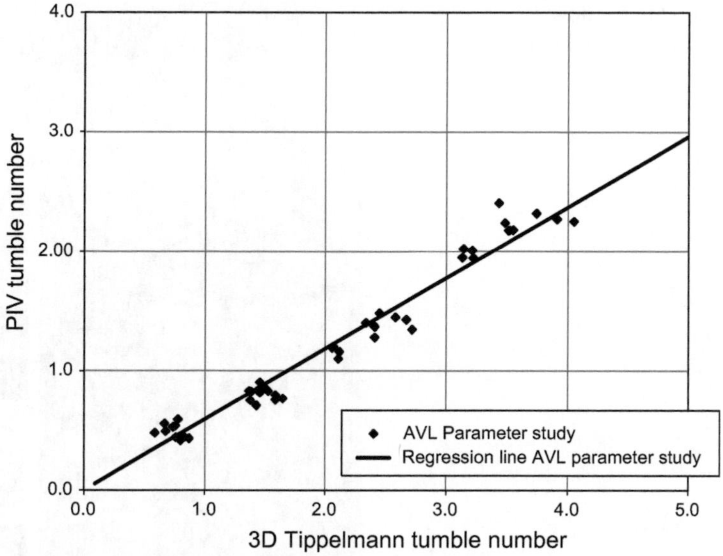

Fig. 3.96 Correlation of tumble number measured with PIV versus spherical flow straightener (3D Tippelmann)

$$K_z = \left(\frac{n_D}{n}\right)_{red} \frac{1}{4} = \left(\frac{n_D}{n}\right) \frac{D}{4\,S} \tag{3.42}$$

D Cylinder diameter
s Piston stroke

Equation 3.42 Rotation coefficient, AVL Tippelmann

The swirl number according to Tippelmann is represented as follows;

$$\overline{K} = \frac{\sum K_z \dot{m}\,\Delta t}{\sum \dot{m}\,\Delta t} \qquad \overline{K} \neq \left(\frac{n_d}{n}\right)_m \tag{3.43}$$

Equation 3.43 Swirl number, AVL Tippelmann

Further characteristics according to Riccardo [40] are explained in more detail in other publications. These characteristic values and numbers can be equated with or converted to AVL definitions to a limited extent only.

3.3.15.7 Measuring Transient Flow Processes

Flow measurements essentially also include measurements in transient flow fields. In this respect, there has always been a great variety of ways in which turbulence has been

Fig. 3.97 Test setup for flow measurement in motored engine operation [37]

investigated. In internal combustion engines, particularly where charge motion is concerned, such investigations are of great interest, as the boundary conditions for the real engine and those for the generally applied measuring methods on the steady-flow test bed could not be more different. Engineers perform measurements on the real engine in an attempt to gain a better understanding of charge motion and its development over time as a mean and a turbulent flow field, not only to learn how to better control the combustion process but also to validate the flow calculations (CFD, computational fluid dynamics), the majority of which have been conducted alongside other measurements over the past 35 years.

For these measurements on the real engine, the already mentioned optical methods are in use (see also pages 233, 237 and 226). Gaining optical access to the engine is a major challenge. If a flow field is to be captured to the fullest extent possible, the engine's geometry needs to be adapted accordingly (see Fig. 3.97). On the other hand, if optical access is achieved by optical endoscopic techniques with minimum space requirements, the flow field is generally only measurable in a section of the combustion chamber (see Fig. 3.98).

If such flow field curves are recorded over time and across multiple engine cycles, the question of turbulent and cyclical fluctuations arises, and how to separate these from the mean transient velocity curves. Different approaches are applied:

- Averaging over time slots of constant width
- Best-fit curve or spline through the measured points

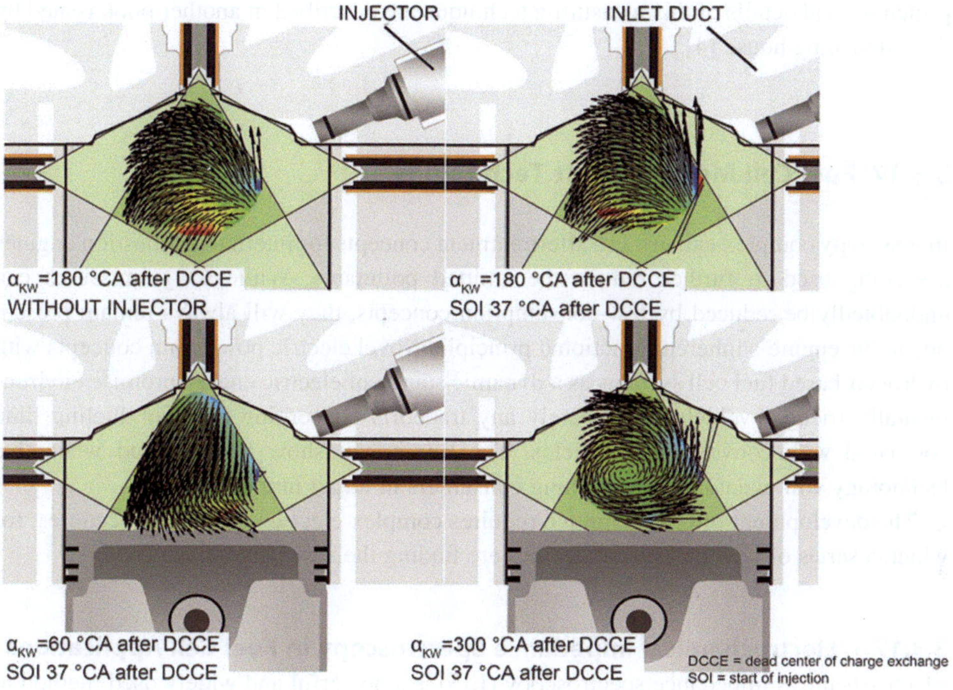

Fig. 3.98 DGV comparison of flow fields in motored engine operation [36]

- Conditional measurement: Additional parameters (cylinder pressure, flame propagation, etc.) are measured. Averaging is performed only over such cycles in which the additionally acquired parameters are within certain limits.
- Frequency analysis of curves and extraction of higher frequency components by defining a cut-off frequency

All of these methods are somewhat arbitrary. The conditional measurement seems to be the likeliest legitimate method for sorting individual cycles by what they have in common and using such groups to derive mean flow values and turbulence quantities by averaging over the grouped cycles.

3.3.16 Indicating Measurement Technology

Indicating is an important tool to develop optimal combustion in gasoline and diesel engines. High-pressure indicating measures the pressure inside the combustion chamber, usually using piezoelectric pressure transducers. An analysis of the captured pressure curve allows a meaningful assessment of the internal engine processes. The basic principles and details of this measuring technique are described in another book issued by this publishing house [4].

3.3.17 Fuel Cell Measurement Technology

Increasingly complex exhaust gas aftertreatment concepts for internal combustion engines are being used to further reduce the emitted pollutants. While CO_2 emissions can undoubtedly be reduced by low-consumption concepts, they will always remain present due to the engine's inherent functional principle. Novel electric powertrain concepts with hydrogen-based fuel cell systems as a dynamic source of electric energy promise environmentally friendly vehicles with barely any tradeoffs concerning range or fueling time compared with conventional vehicles. The future will show whether and when this technology will eventually be reaching consumers in larger numbers.

The development of this technique requires complex electro-chemical procedures, for which a series of new measuring devices are finding their way into R&D centers.

3.3.17.1 Electrochemical Impedance Spectroscopy in Fuel Cell Applications

Electrochemical impedance spectroscopy (EIS) is a powerful and widely used method to analyze electrochemical systems. Among other things, EIS provides a non-invasive and non-destructive way to study galvanic elements such as fuel cells or batteries. EIS measures the frequency-based impedance of electrochemical systems. The analysis and interpretation of the impedance spectrum enables one to identify specific electrochemical processes inside systems.

When analyzing fuel cells, this technique delivers valuable insights into the investigated system. Fuel cells produce electric energy from chemical energy through a redox reaction, for example, between hydrogen and oxygen.

$$\begin{array}{ll} \text{Anode reaction} & 2H_2 \rightarrow 4H^+ + 4e^- \\ \text{Cathode reaction} & O_2 + 4H^+ + 4e^- \rightarrow 2H_2O \\ \text{Overall reaction} & H_2 + O \rightleftharpoons H_2O \end{array} \tag{3.44}$$

Equation 3.44 Chemical reaction in a fuel cell

There are various types of fuel cells. Most common in the automotive sector is the low-temperature proton-exchange membrane fuel cell (PEM). In this type of fuel cell, hydrogen and oxygen are separated by a gastight membrane, which is selectively conductive for positive hydrogen ions. The membrane is an ionomer with a perfluorinated backbone terminated with sulfonic acid groups (SO_3H). A commonly used membrane is NAFION® (registered trademark of DuPont). This particular membrane is a

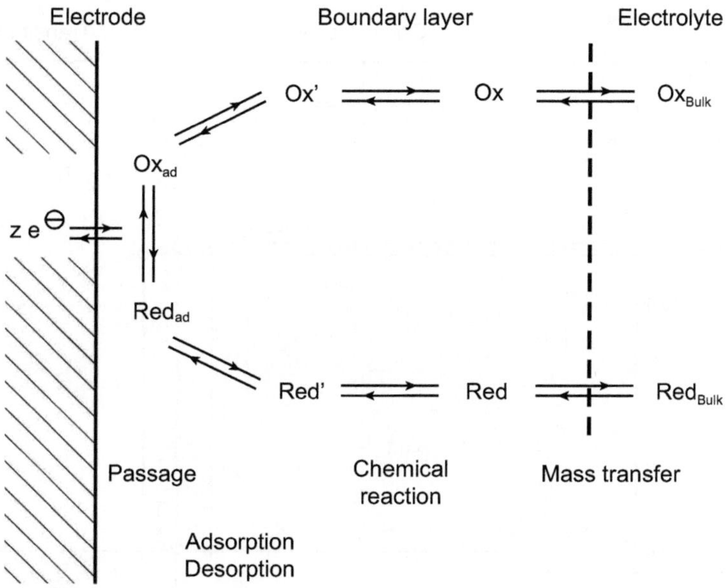

Fig. 3.99 Processes at the electrode/electrolyte interphase [41]

tetrafluoroethylene-perfluoro-3,6-dioxa-4-methyl-7-octensulfonic acid copolymer. To make it conductive to protons, the membrane is hydrated.

First hydrogen is fed to the anode where it adsorbs, dissociates and oxidizes. In the membrane the hydrogen ion reacts with water to form oxonium (H_3O^+) which binds to the sulfonic acid groups, thereby overcoming the barrier (the membrane). Via the electrode, the hydrogen electron follows an external path to do electrical work. At the cathode, the electron reacts with the hydrogen ion and is subsequently reduced to water after reacting with the (air) oxygen. The schematic diagram in Fig. 3.103 shows the example of a fuel cell system. Most fuel cells have a sandwich-like structure.

The redox reaction requires further reactions and transport processes beforehand. On the electrode surface these processes are as follows: First, the gaseous reactants (hydrogen and oxygen) move across the gas channels and the gas diffusion layer to the reaction zone. This happens by diffusive and convective transport processes. Next, surface reactions, such as adsorption, desorption and crystallization, take place, as well as homogeneous and heterogeneous catalytic reactions and electron transfer reactions (see [42]). Figure 3.99 shows the different chemical and electrical processes and their location in relation to the electrode surface.

The individual processes are kinetically inhibited, which means that the reactions are slow to proceed. Different processes have different time constants and this allows us to differentiate between the individual processes. We will take a closer look at this circumstance further below.

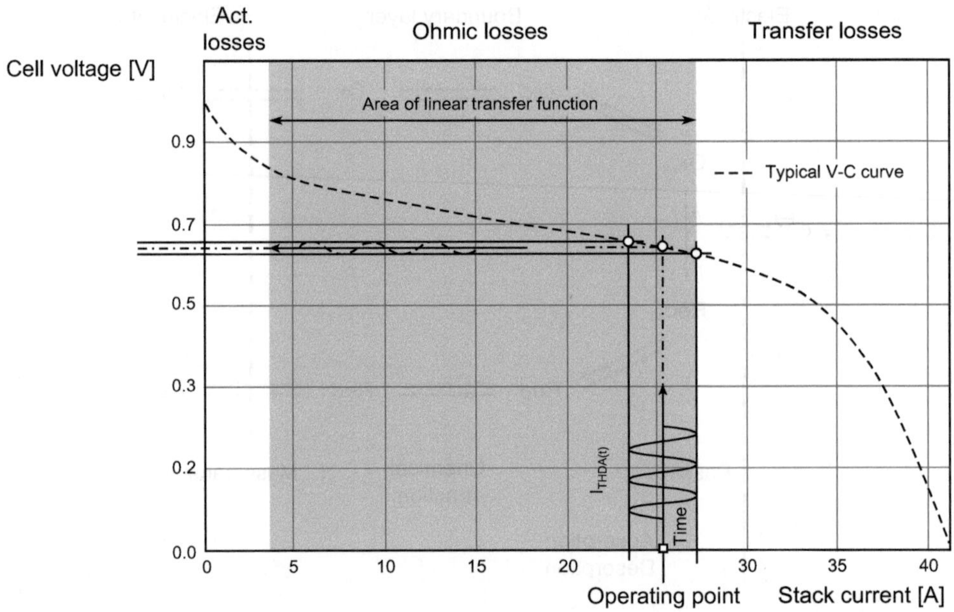

Fig. 3.100 Polarization curve of a fuel cell with dominant types of losses and schematic diagram of the EIS

Function Principle of Electrochemical Impedance Spectroscopy

The underlying principle of EIS is the external excitation of a system to analyze inherent system properties. An excitation signal, most commonly a sine-shaped voltage or current signal with sufficiently small amplitude, is applied to the unit under test. This perturbation throws the system out of balance. If the excitation signal is a voltage, the response is seen in the current response. If the excitation signal is a current, then the response is visible by the voltage response or by a change in the electromagnetic field. Here we can see a phase shift and a positive or negative gain in the amplitude, which is measured on the system's current-voltage characteristic (V-C curve or polarization curve), see Fig. 3.100.

Normally, electrochemical systems have a non-linear current-voltage characteristic. Despite this, the linear response theory allows a description of the non-linear characteristic curves using linear differential equations. This is possible by impressing exclusively signals with sufficiently small amplitudes. However, the external excitation signal has to be large enough to ensure a sufficiently high signal-to-noise ratio (SNR). This enables subsequent data analysis. After the perturbation, the system regains its balance in the form of a relaxation process.

The following equation describes the excitation signal:

$$\mathrm{i}_{(\mathrm{t})} = \mathrm{i}\,\sin(\omega\mathrm{t} + \theta) \tag{3.45}$$

Equation 3.45 Excitation signal

The system response results in:

$$u_{(t)} = u \sin(\omega t + \varphi) \tag{3.46}$$

Equation 3.46 System response

In these equations, i and u are the amplitudes of current and voltage, $\omega = 2\pi f$ is the circuit frequency with frequency f, and θ and φ are the phase angles.

Figure 3.100 shows s schematic diagram of the EIS; the exciting current signal appears as a continuous oscillation. At the operating point of the polarization curve, the signal is redirected, amplified or weakened and phase-shifted. The system response is then measured at the voltage output.

The voltage signal can be described mathematically using two terms:

1. an in-phase component that oscillates with the signal and
2. a $\pi/2$ phase-shifted component

The former determines the real part of the impedance; it is a measure for the energy dissipating in the system. The second component determines the imaginary part of the impedance: it is a measure for the electrical energy that is stored in the system. Both are expressed as a quotient of the voltage amplitude and the current amplitude assigned to this branch.

The impedance is described as:

$$Z_{(\omega)} = \frac{u_{(\omega,t)}}{i_{(\omega,t)}} = \frac{u}{i} e^{-j(\theta-\varphi)} = Z'_{(\omega)} + jZ''_{(\omega)}$$

Form of representation and interpretation

To evaluate the data, various diagrams are in use. The most widely used are the Bode diagram and the Nyquist diagram. We will now take a closer look at the Nyquist diagram. For more details on other forms of representation, please refer to the relevant publications.

Nyquist Diagram Figure 3.101 shows a Nyquist diagram of a fuel cell with its different overvoltages, divided into the four typical areas— ohmic losses, anode activation losses, cathode activation losses and transfer losses. The figure also shows an equivalent circuit diagram of a fuel cell. The elements in the equivalent circuit diagram are assigned to the different regions in the Nyquist diagram. For more information about the equivalent circuit diagram, please refer to the next section.

As described above, the different processes in a fuel cell have differing time constants. This makes them identifiable in the Nyquist diagram. While electrochemical processes are fast and barely have any latency, membrane humidification processes and mass transfer processes are relatively slow and may require several seconds or even minutes.

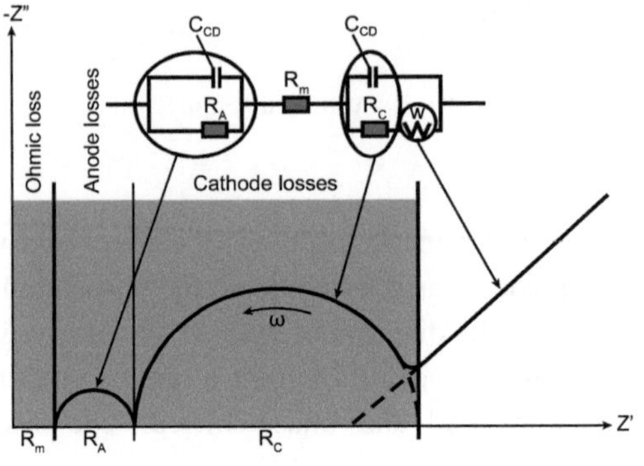

Fig. 3.101 Nyquist diagram of a fuel cell [43]

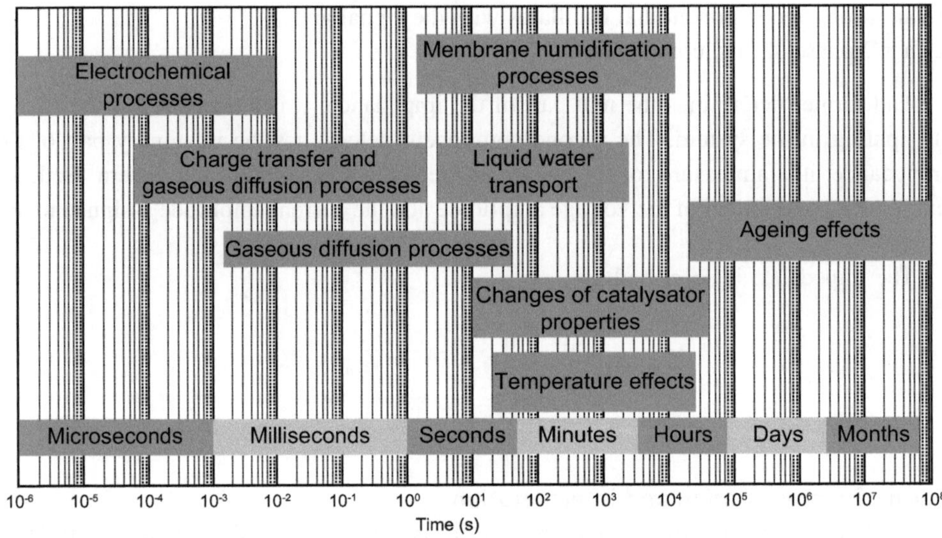

Fig. 3.102 Time constants for different processes in a fuel cell [44]

Figure 3.102 shows the different time constants for each of the processes. It becomes evident that electrochemical processes are among the fastest reactions (in a range of microseconds) followed by charge transfer and gas diffusion processes (in a range of milliseconds). Liquid water transport and membrane humidification processes take significantly longer, with durations that may be anything between a few seconds and several hours. The situation is similar if there are changes to the catalyst properties, or temperature effects. Aging effects are by far the slowest processes.

Equivalent Circuit Diagram Electrical equivalent circuit diagrams allow the description of non-linear system behavior with the help of well-known electrical standard components. Figure 3.103 shows the example of a fuel cell's electric equivalent circuit

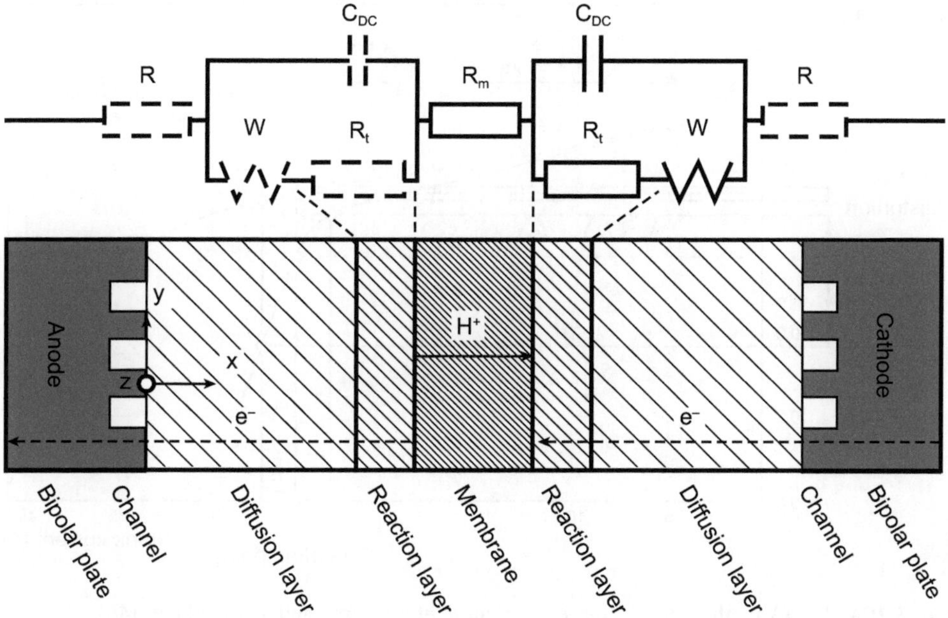

Fig. 3.103 Electric equivalent circuit diagram of a low-temperature polymer electrolyte membrane fuel cell [45]

diagram. The components shown as interrupted lines are the ones that are usually neglected. The components represented as continuous lines are the ones that are used most in order to describe a fuel cell.

Total Harmonic Distortion Analysis THDA

Based on a combination of hardware and software, THDA is a measuring and analysis method that is used as an online diagnosis tool for fuel cell stacks. The biggest advantage over conventional EIS measurements is that THDA allows measuring the fuel cell stack during dynamic load cycles and not just at steady-state operating points. The name *Total Harmonic Distortion Analysis* (THDA) refers to the harmonic distortion of a voltage response in a fuel cell stack induced by an impressed alternating current.

The underlying principle of the THDA is similar to that of the EIS. A signal composed of one or more sinusoidal current signals with different frequencies (in a range between 2 Hz and 1 kHz) is impressed on the system during normal operation. The stack's resulting alternating voltage response is measured and analyzed.

Provided that the operating point remains within the fuel cell stack's linear range of the polarization curve (current-voltage characteristic), which is usually the case in normal, optimal operating ranges, the system's alternating voltage response consists of the same frequency spectrum as the originally impressed current signal. If the operating state turns critical, for example as a result of water droplets collected in the gas channels, the

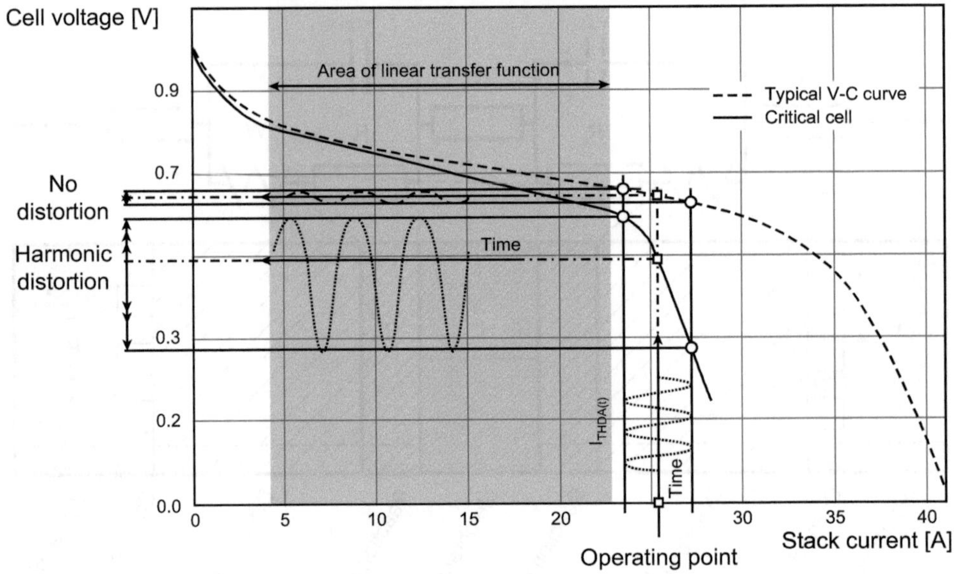

Fig. 3.104 THDA methodology, polarization curve of a healthy and a critical cell [46]

polarization curve will change, causing the operating point to depart from the characteristic curve's originally linear region into the non-linear region. This non-linearity induces harmonic distortions in the voltage response, resulting from the impressed current signal. Such distortions lead to the occurrence of harmonics (integer multiples of the original frequencies of the impressed signal). The total harmonic distortion (THD) describes the ratio of the sum of all harmonic components to the fundamental frequency [46].

Figure 3.104 shows the current-voltage characteristic of a single healthy cell (typical U-I curve) operating in the desired operating range (approx. 26 A) at the time of measurement. This operating range is generally located in the linear region of the transfer function, where ohmic losses prevail. The continuous line indicates the current-voltage characteristic of a "defective/critical" cell. However, at the time of measurement, the same operating point (approx. 26 A) is not within the non-linear region of the polarization curve. This area describes the losses resulting from mass transfer limitations, which may have various causes. Due to the collection of liquid water in the cell, for example, it may no longer be possible to ensure the reaction zone's sufficient supply of reactants (O_2 or H_2).

An efficient form of the discrete Fourier transform allows the detection of distortions in the voltage response caused by the non-linear transfer function. An analysis (THD calculation) of the distortions follows and helps to identify the cause of a critical operating state. Such undesired operating condition may be induced by liquid water formation in the cell, dehydration of the membrane or undersupply of reaction components. It is a great

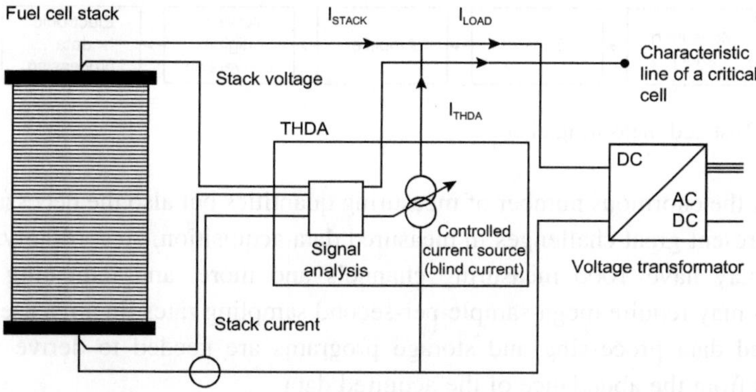

Fig. 3.105 Measuring setup for a fuel cell stack illustrated by the example of AVL THDA [46]

advantage if these disturbances can be detected in time, as they not only lead to output losses but also to accelerated aging of the fuel cell.

Figure 3.105 shows a measuring setup for a fuel cell stack with a THDA measuring device. The THDA measuring device produces an alternating current signal (I_{THDA}) which is impressed on the stack current (DC) (I_{STACK}). Voltage sensors at the stack's anode and cathode measure the total stack voltage. The device also processes the alternating component of the measured voltage signal. The measured data are transferred via an interface to a PC for further calculation and analysis.

Future advancements in the THDA methodology will include:

- Multi-channel measurement (two-dimensional resolution of individual cells)
- Fuel cell aging prediction
- Correlation with 3D-CFD simulation data, e.g. AVL FIRE
- On-board diagnostics unit for use in the automotive industry or for stationary use (μCHP)

3.4 Errors and Accuracy of Measurement

3.4.1 Measuring Chain

Test beds satisfy a very wide range of measuring tasks. What all these tasks have in common is the necessity of speed, temperature, torque, current, voltage or pressure measurement. Special applications are equipped with fuel consumption, air mass, particulate, exhaust gas component, blowby or indicating measurement systems, as well as impedance spectroscopy or similar instruments, as have been described in previous sections of this book.

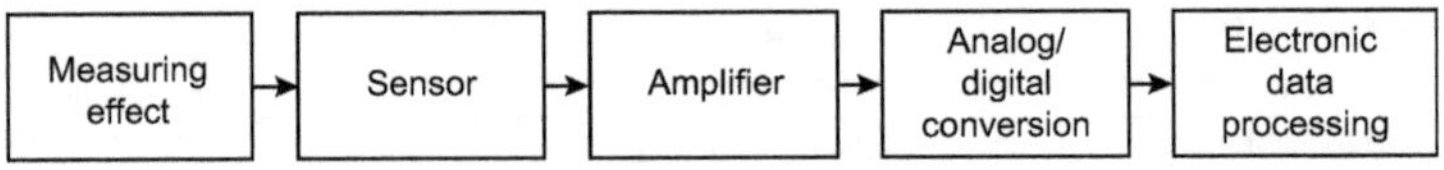

Fig. 3.106 Test bed measuring chain

Not only the enormous number of measuring quantities but also the necessary acquisition rates present great challenges in measured data acquisition. Accordingly, individual test beds may have 1000 measuring channels and more, and indicating measuring instruments may require mega-sample-per-second sampling rates. In any case, extremely sophisticated data processing and storage programs are needed to derive meaningful evaluations from the abundance of the acquired data.

On a test bed, all of the physical measuring quantities are measured electrically in order to allow automatic electronic data processing within what is called a measuring chain (see Fig. 3.106).

Each link of the chain causes measuring errors: :

- The measuring effect to acquire the physical signal generally has a zero error, a gain error, a linearity error and a hysteresis. All of these errors are subject to thermal effects.
- Sensors have a zero error, a gain error and a linearity error. Hysteresis errors, as is common with pressure sensors, for example, fall into the mechanical engineering category. All of these errors are subject to thermal and dynamic effects and aging.
- Analog-to-digital converters have a zero error, a gain error, a linearity error and a quantization error. The first three errors are subject to thermal and aging effects.
- Electronic data processing produces a reconstruction error when calculating values between the sampling instances and a synchronicity error when computing values measured on different channels. In a static (i.e. steady) state, both of these errors are small and other measuring errors in the measuring chain dominate.

3.4.2 Effect of the Sensor Installation Location

The effect of the measuring site is generally greater than the measuring chains' overall error. The choice of installation location is the responsibility of the user and has to be made very carefully.

A good example of this is the installation of pressure transmitters that measure the pressure of liquid media. Problems with their installation occur on a regular basis. The relationship for Δp shows that if the medium is water and the pressure transmitter has a measuring range of one bar at 0.1 % accuracy specification, a difference in installation height of 1 cm will be in the same order of magnitude as the measuring error:

$$\Delta p = \rho \cdot g \cdot \Delta h$$

where:

ρ Density
Δp Pressure change
g Gravitational acceleration
Δh Change in height

Example:
1 bar = 100,000 Pa
0.1 % bar = 100 Pa

$$\Delta h = \frac{\Delta p}{\rho g} = \frac{100 \frac{kgm}{s^2 m^2}}{1000 \frac{kg}{m^3} 10 \frac{m}{s^2}} = 0.01 \text{ m} = 1 \text{ cm} \tag{3.47}$$

Equation 3.47 Effect of installation location on accuracy of measurement

This means that if you fail to properly take into account the installation height by as little as 1 cm, the measuring error in the pressure measurement may under certain circumstances double.

3.4.3 Measurement Uncertainties

A description of measurement uncertainties can be found in the "Guide to the Expression of Uncertainty in Measurement" (in short: GUM) and in ISO 17025. The publication deals with the expression of measurement uncertainties caused by measuring and calibration device cascades. The cascades start at the national calibration centers. The publication also describes statistical procedures for establishing measurement uncertainty.

For voltage and resistance, the national calibration standards are based on quantum physical effects and can claim a level of accuracy that is almost absolute. Every step between the national calibration standard and the measuring devices increases measurement uncertainty in accordance with the specification of the used measuring devices and the ambient conditions at the calibration centers. The standard describes how to take into account the different kinds of errors and scattering measured values in order to create specifications for devices and their calibration. However, this specification only applies to stationary or quasi-stationary parameters in a steady thermal condition.

Dynamic measurements on the test bed with changing ambient conditions require the assumption of significantly higher measurement uncertainties. That is the reason why the device specifications for devices intended for dynamic measurement are based on a very conservative approach. Compliance with the standard has particular advantages for production and servicing, however it is not possible to reproduce fully identical results on the test bed. In addition to technical requirements, the standard also demands organization structures that guarantee the maintenance of defined procedures on a continuous basis.

3.4.4 Interpolation Errors

A closer look at a dynamic measuring case suggests that a linear interpolation between neighboring measured values in electronic data processing can cause an error in the percentage range. A (sin x)/x interpolation that takes into account several measured values can serve as a remedy. The error is then reduced below the per-thousand range, depending on the number of measured values taken into consideration. The Nyquist-Shannon sampling theorem suggests that by using a (sin x)/x interpolation one can approximate the values between the sampling points as precisely as desired if an adequate number of measuring points are taken into account. Figure 3.107 shows a triangular signal that is sampled six times and each sample multiplied by a (sin x)/x function. The resulting signal is shown above it as the sum of all (sin x)/x functions.

Figure 3.108 shows the summation function without zero shift. We can see that the summation function extends beyond the two topmost sampling values. This also shows that the sampling rate has to be at least twice as high as the highest frequency component.

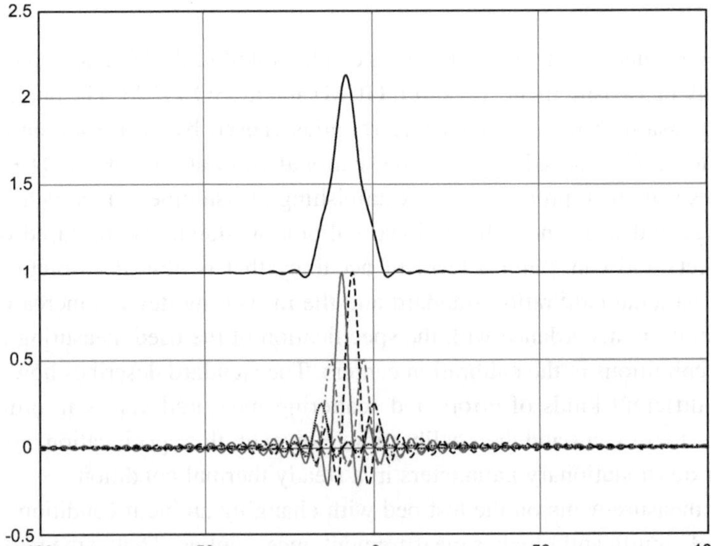

Fig. 3.107 (sin x)/x interpolation of a triangular signal

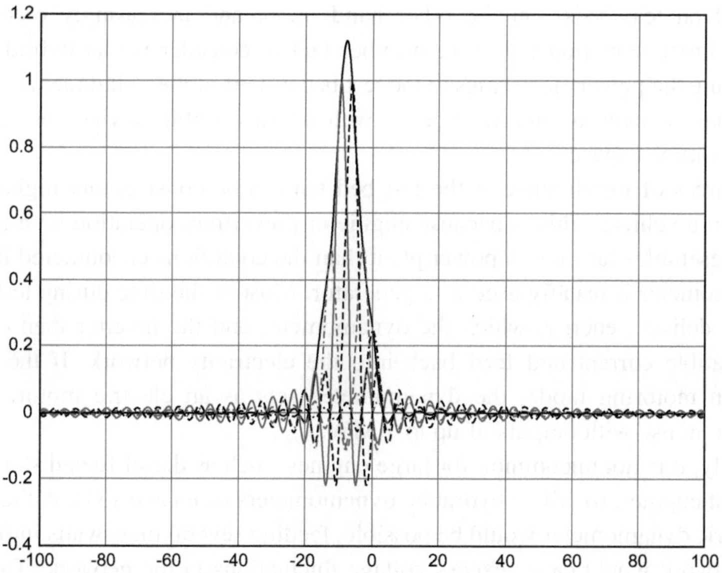

Fig. 3.108 (sin x)/x interpolation of a triangular signal without zero shift

3.4.5 Calibration and Adjustment

In German-speaking countries, calibration means to determine the measuring errors in a device, whereas adjustment means to remove such measuring errors. In English-speaking countries, both processes are called calibration. Calibration compliant with standards means determining the errors in a device that requires calibration by using dedicated measuring devices (subject to test equipment control) and calibrators in thermal steady-state conditions at suitable temperatures and without any interference. Furthermore, the evaluation procedure must conform to the standard too. As it is practically impossible to meet the aforementioned boundary conditions, calibration on the test bed requires sufficiently expanded error limits.

During test bed operation, people frequently use these terms incorrectly, so it is always advisable to check in what way the term is being used. For an example of calibration methods, refer to Sect. 3.3.6 "Calibrating torque measuring chains" on page 170.

3.4.6 Electromagnetic Compatibility (EMC)

3.4.6.1 Electromagnetic Compatibility on the Test Bed

Electromagnetic compatibility (EMC) on the test bed is gaining in significance. The reason for this is that, on the one hand, electrical inverters for drive systems or those used as electrical power sources are becoming ever more powerful and more and more widespread on test beds. On the other hand, more and increasingly precise electric measuring instrumentation is in use. Another fact to consider is that hybrid powertrain systems bring the power electronics to the test bed as part of the unit under test itself. As a result, strong sources of interference as well as susceptible devices are placed in a relatively confined space.

The sources of interference at the test bed tend to be considerably higher than they would be in a vehicle. This is because engine or powertrain operation on a test bed, for example, resembles far more a power plant than the conditions encountered in a vehicle. The dynamometer is usually used as a generator. Most of the time during test operation, the engine delivers energy, which the dynamometer and the inverter then convert into grid-compatible current and feed back into the electricity network. If the test bed is operated in motoring mode, the dynamometer acts as an electric motor. Nowadays, systems are in use with outputs of up to 1 MW.

Currently, it is not uncommon for large engines, such as diesel-fueled ship engines or power plant engines, to utilize hydraulic dynamometers (see page 134). Although the use of an electric dynamometer would be possible, feeding several megawatts into a medium-voltage network would cause excess voltage fluctuations in the network. The increased number of solar and wind energy plants are forcing grid operators to make the grids more tolerant toward power fluctuations, also on the producer side. In the future, this will open up the opportunity to sell energy generated during engine test runs even in large engine tests by feeding it back into the grid and not just using cooling systems to carry the thermal energy away.

Unit-under-test installation must also consider the differences between the geometric dimensions of a test bed and those of a vehicle. The cable lengths and their EMC treatment are different from what an installation in a vehicle would require. A third difference is the presence of high-precision and high-availability measuring equipment that needs to operate unaffected by the high interference potential caused by the transformation of energy. These factors result in high demands on the test bed planning, wiring, positioning and maintenance of the installed devices.

3.4.6.2 Coupling Mechanisms

To examine the way interference sources and sinks interact with each other, it is first necessary to take a brief look at the different coupling paths.

Direct Galvanic Coupling

Direct galvanic coupling occurs when the interference source and sink have one or more conductors in common, on which currents are distributed arbitrarily. For instance, two personal computers connected by an RS232 interface are galvanically coupled because they are both connected to ground by the protective conductor, establishing a further connection via the RS232 ground. In addition, there is also a connection via the protective conductor. This allows some of the protective conductor current to flow across the RS232 cable's ground connection, and consequently through both PCs. From a certain limit onward, this leads to interference or even damage.

The fact that protective conductor systems always carry various currents can lead to problems, particularly in industrial systems that use grounded neutral conductors. For example, when a 1-MW inverter was being switched on, potential shifts of 160 Vpp (volts peak-to-peak) were measured. These led to currents of 35 App (amps peak-to-peak) in an interface that usually carries no more than several milliamps of current. This circumstance very likely led to sporadic problems in a connected test-bed automation system. Electrical isolation and improvements to the grounding system helped to eliminate the problems.

Capacitive Coupling

Capacitive coupling is common where cables are laid very close to each other over long distances. Whether there is interference depends on the susceptibility of the input and the shape of signal in the interfering cable. The higher the voltage level and the steeper the voltage edge, the higher the capacitive effect will be. A measuring line routed next to a cable that connects inverter and electric motor is particularly unfavorable. The effect diminishes with increasing distance. Another possibility to protect against capacitive coupling is to use a shielded measuring line with the shield only being connected to ground on one side.

Inductive Coupling

Inductive coupling occurs in conductor loops through which another conductor's magnetic field passes, for example the field of an electric machine or a transformer. To keep a conductor's magnetic field small, the distance between the forward and return conductors should be as narrow as possible. As a result, the magnetic fields cancel each other out in the far-field region of the conductors. In electric drives supplied by single conductors, this compensation is sometimes very small, e.g. because the cable-bending radiuses have to be very small or the electric motors are movable. This kind of arrangement has a high potential for electromagnetic interference.

The presence of high-frequency common mode current also has very high potential for interference. This type of current flows back to the inverter via the ground system. Depending on the design conditions, the distance between the forward and return conductors may be very large. In that case, the current field remains uncompensated and has a high potential for interference. To reduce the influence, it is necessary to shield the measuring cable. The shield must be provided with low-inductance grounding on both ends. Low-inductance grounding must also be established between the generators. As a result, a shield current forms, whose field acts against the exciting, i.e. in our case interfering, field. Never use pigtail connections to establish the shield connection as this would confront the compensating shield current with a high level of impedance. In pigtail connections, the measuring cable braiding is twisted to a wire and connected to the device's ground. The better approach is to establish ground connections using wide, flexible straps.

3.4.6.3 Remedial Actions on the Test Bed

An analysis of the combination of inverter, electric motor and connection cables reveals the following chain of effects: The inverter transistors switch at maximum speed to keep the electrical power as low as possible during switching. This produces sharp switching edges that result in high-frequency interference currents. Using pulse pause modulation in the switching transistors, the inverters produce almost sinusoidal phase currents. This means that the inverters switch at a much higher frequency than the frequency of the phase currents. Only the fundamental component of the phase currents can do electrical work at the electric machine's shaft. All harmonic components of the current and voltage curves are potential sources of interference. This means that switching edges of around 400 V with very high dV/dt and repetition rates of several dozen kHz are present at the inverter's output. These are coupled to the electric motor's stator via an electrical pathway consisting of cable inductances, leakage inductance of electric motor, cable capacitances and coupling capacitances of the windings.

With every switching procedure, the coupling capacitance of the windings to the stator causes high-frequency currents to be injected into the grounding system, which flow back to the inverter via the widely meshed ground conductor system. These currents remain uncompensated in the conductor between the inverter and the electric motor and have a correspondingly strong magnetic field with high interference potential. The purpose of EMC measures is to direct these currents back to the inverter in a way that is short and able to offset the magnetic field in the far-field region.

Ideally, a shielded power cable is used to accomplish this task. Its shield is contacted using high-frequency capable EMC screw connections with minimum inductance values. The shield acts as a current return conductor, which—similar to a coaxial cable—allows the high-frequency currents to flow back to the source of interference, thereby minimizing interfering ground conductor currents.

Another source of interference is the shaft. The rotor is coupled to the windings by capacitive coupling and may also feed currents into the grounding system. A countermeasure is to conduct these currents away using grounded brushes.

Another category of interference sources are solenoid valves and contactors that have no installed surge absorbing diode or capacitor. While the system is switched off, high negative switching edges are produced that may be transferred to other cables by capacitive coupling. These sources of interference do not have a high repetition rate, but due to the fact that the forward lines are laid with other cables in cable ducts they do have a high interference potential.

3.5 Bus Systems

3.5.1 Overview

The demand for high reliability and precise timing in data transmission, even in the immediate vicinity of electromagnetic sources of interference, at low node and line costs, led to the development of a special kind of communication networks: bus systems.

Today bus systems are used in industrial production systems and building automation technology as well as in vehicles. In automation systems, bus systems mainly serve to connect a substantial number of sensors and actuators to a higher-level control system. As their central focus is on communication between what is referred to as the field layer and the process control layer, such bus systems are also called fieldbuses. In vehicles, on the other hand, bus systems are used to ensure communication between complex, intelligent control units for tasks such as engine management, transmission control, vehicle management, driver information, driver assistance and so forth, within so-called in-vehicle networks (IVNs).

In both applications, data exchange is either event triggered or cyclic. Typical examples for cyclic data exchange are sensor values or control signals for actuators. Error-related or diagnostic messages as well as driver interactions such as gear shifting are examples of event-triggered data exchange. On the test bed, interfaces must be operated for fieldbus systems and in-vehicle networks. A typical representative of applications suitable both for test-bed and in-vehicle use is the CAN bus (CAN = controller area network). The following section gives a brief introduction to the CAN bus and typical fieldbus systems, such as PROFIBUS and industrial Ethernet, as well as typical vehicle buses such as FlexRay and LIN.

Network protocols are defined using the Open System Interconnection (OSI) model's layer architecture [47]. In fieldbus systems only the layers one, two and possibly seven are implemented (see Table 3.7).

Fieldbus nodes generally consist of a transceiver (voltage level adaptation), a bus controller in Layer 2 and a microcontroller or FPGA (field-programmable gate array)

Table 3.7 Common characteristics of typical fieldbus protocols based on the OSI model

OSI Layer	Function in Fieldbuses
Layer 7 (application layer)	Application services for automation tasks such as cyclic data transfer, finite state machines, alarms and events.
Layers 3–6	Rarely used in fieldbus applications and therefore heavily simplified or done without completely.
Layer 2 (data link layer)	Ensuring data integrity using check sums and frame check, access mechanisms to the bus medium.
Layer 1 (physical layer)	Electric representation of the signal with measures to limit the impact of inductive and capacitive interference in cost-effective bus media, such as difference-signal transfer or interference detection.

for the higher layers in the fieldbus protocol. The fieldbus nodes interface with the node application software. For reasons of efficiency, fieldbuses are designed as serial, asynchronous networks that are usually connected in a bus (alternatively referred to as line) topology. In rarer cases, bus systems are also designed as star, tree or ring topologies. Different bus systems can also be connected via gateways.

Participating nodes access the common bus medium either randomly, by central control (master/slave or token principle) or according to a defined time-slice scheme. If the system is accessed randomly, it may occur that two or more nodes start transmitting simultaneously. As this leads to a collision, the participants have to detect such conflicts and resolve them. Central control by a master or by a fixed assignment of time slots, on the other hand, has the advantage of offering deterministic time relationships (real-time capability). However, the effort required to manage the network is higher and problems are encountered when transmitting unplanned (asynchronous) events.

A deterministic transfer based on the master/slave principle takes place in PROFIBUS. Access control based on message priority as implemented in CAN is a compromise between statically planned bus access and the transfer of asynchronous events. Random access, such as e.g. Ethernet (IEEE 802.11), leads to collisions on the bus medium. As such collisions have to be resolved, making the method non-deterministic and increasing the time and transfer energy requirements, the method is of minor significance in the field of automation technology. Bus systems have therefore established themselves under the term "Industrial Ethernet" which are specifically designed to address the needs of automation technology.

The data transmission protocol (communications protocol) is a frame format that establishes the rules for the exchange of information between bus participants. As shown in Figure 3.109, it consists of the frame information (header and trailer section) and the actual payload (body section). The frame information contains all the other information defined in the protocol such as start flags, address, transfer modes, error handling, etc. As it is common for the transferred data objects to be structured very finely (e.g. only a few bits of payload per message), optimal utilization of the bus bandwidth requires high data efficiency. This results in the following equation:

$$\text{Data efficiency} = \frac{\text{payload}}{\text{payload} + \text{frame data}} \cdot 100\% \tag{3.48}$$

Equation 3.48 Data efficiency

Table 3.8 shows an example of data efficiency in different bus systems. 8 bits of payload (PL) are transmitted.

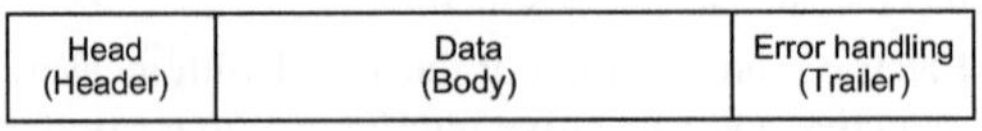

Fig. 3.109 General structure of a data transmission protocol

Table 3.8 Data efficiency in different bus systems

Data Bus	Data Efficiency for PL = 8 Bits	Note
Basic CAN	15 %	11-bit identifier; bit stuffing ignored; incl. EOF
PROFIBUS DP	7 %	SD2 telegram, 11 bits per character
TCP/IP Ethernet	1 %	

3.5.2 CAN

CAN (Controller Area Network) has been the standard in-vehicle communication solution worldwide for more than a decade. Developing CAN, engineers managed to combine the specifications for an optimum mixture of flexible extensibility and practical applicability with reliability, security and simplicity. Driven by the very high quantities produced for the automotive industry, CAN bus controllers have now become very low-priced and can be found fully integrated in a large number of microcontroller families. Accordingly, the bus has become increasingly widespread in automation technology, as an alternative to conventional fieldbuses such as PROFIBUS or INTERBUS.

CAN is operated with data rates of up to 1 Mbit/s. Usually, twisted pair wires in a shielded cable are used as bus line. The maximally allowed bus length is inversely proportional to the data rate. This is because the propagation time for one bit is relevant for the bus arbitration system used in CAN which is described below.

3.5.2.1 Access Methods

Unlike conventional fieldbuses, CAN allows real multi-master operation. This means that there is no central node to control bus access by the participants (nodes). As soon as the bus is free, a node can start sending a message at any time. If multiple nodes start communicating this way at the same time, the non-destructive arbitration mechanism sets in and resolves the pending collision by employing an arbitration method, according to which messages are transmitted via the bus based on a fixed priority scheduling. For this purpose, the first part of the message, the message ID, not only serves to identify the message but also to indicate its priority.

Nodes that start transmitting at the same time compare each bit in the message ID to identify the priority of their messages. The bit value "0" is transmitted as electrically dominant; "1" is recessive. By reading back the level which the bus assumes, every node can detect whether its own message was suppressed by a higher-priority (and consequently dominant) message. In this case, the node immediately quits transmitting and switches back to receiving mode. After the transmission/reception of the last bit in the message ID, at the latest, the message with the highest priority wins the arbitration over all simultaneously starting messages and the node finishes transmitting it without delay.

This method is implemented in the two lower layers of the CAN protocol, consequently in the CAN controller's hardware. The nodes that lost the bus arbitration as described above retransmit any simultaneously queuing lower-priority messages after the higher-priority message is finished. The iterative application of this method results in an automatic serialization of messages, scheduled according to their priority. This requires that the messages from different nodes differ at least in one bit of the ID. If there are not enough time slots on the bus to enable the transfer of lower-priority messages, the transmitting node reports an error state ("starvation") after "losing" a certain number of arbitration attempts. This state occurs at bursts of high-priority messages or if the bus load is excessive, which has to be avoided by a suitable communication design.

CAN identifies messages by their address and therefore initially needs no node addresses. Messages are always sent as broadcast, which means they are received by all nodes on the bus. This enables high flexibility in the system design because it is not relevant which node is sending a message. As a result, functions can be shifted between nodes (e.g. control units) without having to make any changes to the communication. Also, nodes can be added at any time, as long as there is sufficient bus bandwidth and no message IDs are used that are already taken. This enables modularity and improves serviceability in automotive systems and automation plants.

3.5.2.2 Error Detection and Error Handling

In-vehicle functional safety standards require reliable error detection and error handling. At the same time, engineers have to cope with high cost pressure. By combining several methods, CAN provides a high transmission error detection rate (the so-called Hamming distance is 6). Error handling takes place in the CAN controller hardware, which unburdens the node processors. CAN guarantees correct message transmission, i.e. the automatic retransmission continues until at least one node confirms the messages as being formally correct (ACK bit) and not a single node reports an error (ERROR frame). Nevertheless, this method only guarantees that at least one node has correctly received the message—not necessarily that it has been received by the node for which it was semantically intended. CAN guarantees an all-or-nothing principle. This means that a message has to be either recognized as correct by all receiving nodes or it is discarded by all if only a single node reports an error. This ensures a consistent process image in the nodes.

To a certain extent, the error handling defined in CAN also enables selective detection and response to defective nodes. If a node continuously transmits erroneous messages, e.g. as a result of hardware trouble, the related CAN controller will detect this and, as a last resort, exclude the faulty node from the bus (this is called the "babbling idiot problem").

3.5.2.3 Format of a CAN Message

Figure 3.110 shows the structure of a standard CAN message. The CAN protocol supports two different message formats for data transmission:

- The standard (or base) frame format supports a length of 11 bits for the identifier (arbitration field).
- The extended frame format supports a length of 29 bits for the identifier.

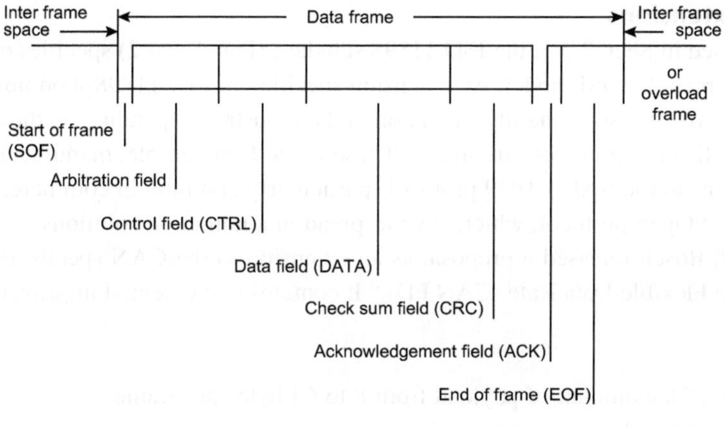

Fig. 3.110 Format of a standard CAN message

Table 3.9 Exchange formats to describe a CAN network

Description	Defined by	Format
A2L	ASAM	ASCII text file
DBC	Vector GmbH	ASCII text file
FIBEX	ASAM	XML

Provided that suitable controller hardware is installed, both message types can be transmitted on the same bus. In practice, all nodes are usually configured to the same format type. Another form of message is the RTR bit (i.e. remote transmission request bit). However this mechanism is rarely used in automotive applications. A special form of message is the error frame, which a node transmits the moment it detects a transmission error. An error frame consists of six dominant bits, so it wins the competition for bus access against all other messages, and by specifically violating the bit stuffing rule which allows no more than five consecutive bits of the same polarity, it triggers error detection in all receiving nodes.

Sometimes, what are referred to as multiplexer messages are defined at the application layer. This allows multiple use of one CAN message with a single identifier. One part of the payload (usually the first 8 bits, called "selector") is used to identify the specific assignment of the remaining payload, normally by means of an index in an assignment matrix.

Table 3.9 shows formats have become established for describing network communication based on CAN messages:

Among other things, the description contains a mapping of data (start bit and bit length within the 8 bytes of a CAN message payload), data format (for example, whether signed or unsigned, integer or floating point) and scaling (frequently a parameter of a fractional rational function). Using these descriptions helps to transfer message definitions from one application to another.

3.5.2.4 Extensions

As mentioned in Sect. 3.5.1, the ISO 11898 standard (Part 1 and 2) specifies only the two lowest layers in the OSI model. An extension specified in ISO 11898-4 on time-triggered CAN (TTCAN) has so far hardly caught on in the industry. Depending on the application, various application protocols are in use. These range from simple, manufacturer-specific specifications to the SAE J-1939 protocol (particularly common in commercial vehicles) and the CANOpen protocol, which is widespread in automation solutions.

In 2012, Bosch released a proposal as an extension to the CAN specification entitled "CAN with Flexible Data Rate (CAN FD)." It contains two essential improvements to the protocol:

- Increase of the amount of payload from 8 to 64 bytes per frame
- Increase of the data rate up to 8 Mbit/s

Messages in CAN FD are backward compatible with ISO 11898 CAN. Setting a previously reserved status bit in the header switches a message to FD mode. Previously unused codes of the DLC (data length code) are used to identify payload packages with more than 8 bytes. Step by step, this allows achieving a payload length of up to 64 bytes per message. After bus arbitration, the system switches to a higher bit rate for the rest of the message.

3.5.3 PROFIBUS

PROFIBUS (Process Field Bus) is the most widely used fieldbus in European system and automation technology today. As shown in Table 3.10, PROFIBUS is a "family" of protocols, specifically designed for automation technology tasks.

The following section focuses on PROFIBUS DP. Alongside the basic variation, the extensions V1 and V2 were defined in the application layer (Table 3.11).

RS485 is by far the most widespread implementation of the physical layer for PROFIBUS DP. Also in use are the two alternatives optical fiber and IEC 1158-2, which is based on impressed currents, but we will not deal with these any further in this publication.

Table 3.10 The PROFIBUS family

PROFIBUS DP	DP (Decentralized Peripherals) is the most widely used variation.
PROFIBUS FMS	FMS (Fieldbus Message Specification) has the broadest functionality.
PROFIBUS FDL	FDL (Fieldbus Data Link) is used for simple connection on the OSI Layer 2.
PROFIBUS PA	PA (Process Automation) is common in the chemical and petrochemical industry and supports intrinsically safe transmission in potentially explosive systems using IEC 1158-2 transmission technology.

Table 3.11 PROFIBUS DP extensions

PROFIBUS DP V1	Provides acyclic communication, e.g. for nonrecurring transmission of larger quantities of data (such as parameter sets) as well as the multicast mode (a data source for several data sinks).
PROFIBUS DP V2	Enables direct slave-to-slave communication (previously set up by the master) as well as time synchronization between network nodes.

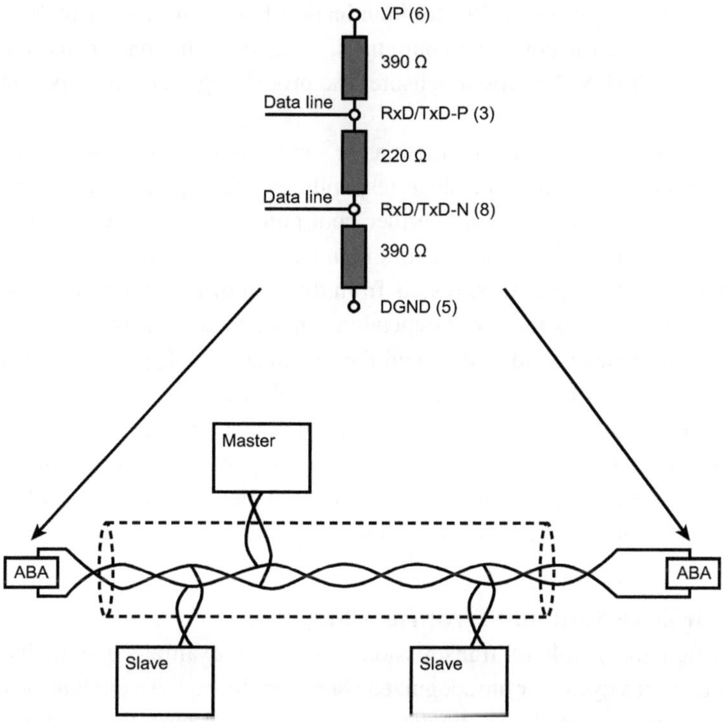

Fig. 3.111 PROFIBUS DP network of three nodes based on RS485

Besides absorbing line reflections, active bus termination serves to set a defined open-circuit potential on the line (potential of the line when none of the participants are active) and therefore requires electric power supply. A shielded twisted pair wire serves as bus medium. The allowed total length of a bus segment is inversely proportional to the transmission rate. At 12 Mbit/s this is no longer than 100 m (Fig. 3.111).

Data rates within the range between 9.6 kbit/s and a maximum of 12 Mbit/s are defined as gross data rates (see [48]). The underlying RS485 interface allows up to 32 nodes per bus segment; an extension is possible by connecting multiple segments using repeaters. You can address a total of 126 nodes in a PROFIBUS DP network. Similar to missing or poorly dimensioned bus termination, the use of stub lines may lead to signal reflections and, consequently, to transmission errors, and should therefore be avoided.

3.5.3.1 Access Methods

In PROFIBUS DP, the bus master dictates communication and also controls network management and the exact timing. In a typical network configuration, the central automation system assumes this master role; all of the other controllers and peripheral systems are the related slave nodes. Although PROFIBUS DP basically allows multiple masters in the same network, only one master can be active at a time. This active role passes cyclically between the master nodes (token passing), enabling continued deterministic behavior (real-time capability). The communication based on this master/slave principle with token passing first needs to be configured. After that, the master that happens to be active in the PROFIBUS DP network ensures the processing. Periodic repetition results in the bus cycle.

At the beginning of a new cycle, the master sends the first of its assigned slave nodes command and output data and the slave responds by sending status and input data. The master repeats this sequence for all further configured slaves. At the end of the cycle, messages can be transferred to the network management; after that the next cycle begins. The lower bound of the cycle time results from the sum of transfer and response times to and from the slaves and is therefore dependent on their number, the data quantities, the processing times in slaves and master and the bus data rate. Typical bus cycle times for PROFIBUS DP are in a range between 2 ms and 100 ms.

Multiple PROFIBUS DP networks are interlinked using DP-DP couplers. Such a coupler is a node with two bus connections, enabling its integration into two separate PROFIBUS networks while acting as slave in each one. Within the node the coupler allows data exchange between the two sides.

3.5.3.2 Error Detection and Error Handling

Errors are often the result of transmission errors, for example due to incomplete or excessively long bus cycles or unrecognized slaves on the bus. An evaluation of the status information in the master allows diagnosing such cases; a more detailed analysis, however, requires full monitoring of the data traffic on the bus. For this purpose, various diagnosis tools are available. Controlling the signals in the physical layer, for example by using a suitable storage oscilloscope, provides an even more detailed analysis. This reveals such causes as faulty bus termination, too long stub lines, incomplete cable shielding or effects from ground loops.

3.5.3.3 Extensions

The PROFIBUS & PROFINET International (PI) Community released PROFINET which includes methods and protocols to connect classic PROFIBUS systems to networks based on Ethernet. Variations also support real-time applications (PROFINET RT = Real-Time and IRT = Isochronous Real-Time). For communication tasks in systems with elevated requirements on functional safety, the protocol PROFIsafe was developed. This supports both PROFIBUS and PROFINET systems and enables applications up to SIL 3.

3.5.4 Industrial Ethernet

The wide distribution of Ethernet in PCs and servers at the end of the nineties led to an intensified discussion on whether to use this standard on the field layer too and replace the existing fieldbuses. The cost benefits of using components from the mass market, the high bandwidth compared with fieldbus systems, the consistency when integrating higher-level planning and control systems as well as the use of freely available diagnosis and configuration tools are seen as key advantages. The drawbacks that have to be mentioned too include the access method that is basically not real-time capable and also not collision-free, the particularly harsh conditions in industrial plants and the possibility of unauthorized access to the lowest layers.

To avoid undesired collision-prone transmission behavior in industrial plants, there are the options of selecting the bandwidth (compared to the net transmission rate) big enough to ensure that a collision is unlikely (keyword "Fast Ethernet") or using a special transmission variant based on standard Ethernet (keyword "Industrial Ethernet").

3.5.4.1 EtherCat

EtherCAT belongs to the group of real-time Ethernet systems that are most widely used in test bed automation technology. As no more than two messages are sent per cycle regardless of the number of nodes, this results in high protocol efficiency which, combined with a high gross data rate (100 Mbit/s) and the underlying Ethernet technology, leads to cycle frequencies of up to 30 kHz.

An EtherCAT network consists of a master node and one or more slave nodes that are connected in sequence. From a logical point of view, the slaves have the structure of a distributed shift register. The master sends data which the first slave synchronizes and then sends to the second one, and so forth. This means that the sent process image must take into account the slaves' spatial and sequential position, as changes in their position are impossible without making any changes to the master configuration.

EtherCAT slaves require two unidirectional Ethernet ports (in, out) with special EtherCAT hardware to implement Layer 2 of the OSI reference model. The EtherCAT master, on the other hand, is generally operated with standard Ethernet hardware, using special software stacks. If the master supports redundant operation via two network interfaces, extending the network from the last slave back to the master produces a closed-ring topology. As a result, if a network segment malfunctions, the EtherCAT system splits into two separate networks and operation continues.

Since EtherCAT is based on the physical layer of IEE 802.3 standard Ethernet, it is possible to use standard tools (for example WireShark) extended by protocol decoder plug-ins for network diagnosis. EtherCAT supports well-known CANopen configuration standards (such as Object Directory, state machines, SDO and PDO principle), which simplify network management and facilitate migration from existing systems.

Table 3.12 Conformance classes

CC-A	Standard scope of functions (cyclic and acyclic data exchange, device diagnostics, device identification, alarms).
CC-B	The functional scope of CC-A plus network diagnostics and it supports redundancy networking.
CC-C	All the functions of CC-A plus hard real-time by bandwidth allocation for 250 μs cycle time and faster (reservation and guarantee of a certain bandwidth and latency time).

3.5.4.2 PROFINET

PROFINET complements the PROFIBUS environment by adding Ethernet technologies and a variety of IP protocol families. In a PROFINET physical layer, Ethernet replaces the classic RS485 interface. PROFINET exclusively supports 100-Mbit/s connections running on twisted-pair copper cables (100BASE-TX) or optical fiber cables (100BASE-FX). With the exception of PROFINET IO CC-C (see Table 3.12), communication utilizes standard Ethernet hardware on which the protocols UDP and TCP are used if communication is not time-critical, and VLAN or LLDP for highly efficient, low-latency communication.

PROFINET CBA (component-based automation) defines a communication task as a system composed of cooperating objects. To describe the interfaces and communication relations, PCDs (PROFINET component descriptions) are used that are implemented in run-time via TCP/IP messages. These systems are used for non-real-time tasks with cycle times in the range of 100 ms.

PROFINET IO, on the other hand, focuses on networks, devices and connections. The data to be transferred and the necessary protocols are defined very similarly to classic fieldbuses. Depending on the applications' time requirements, all devices must satisfy one of the following three conformance classes:

For CC-A and CC-B, standard Ethernet hardware is sufficient. Redundant operation, however, requires the implementation of MRP (media redundancy protocol). For CC-C special Ethernet controllers are used in the devices that are able to serve >100 nodes with cycle times of $<=1$ ms in hard real-time with a maximum jitter of <1 μs. The additional hardware acts as intelligent switch, which also allows mixed operation with non-real-time TCP/IP data exchange on the same network.

3.5.4.3 Ethernet Powerlink

A Powerlink network consists of a master—the managing node (MN)—and several slaves—the controlled nodes (CN). In a Powerlink network with one managing node, up to 239 controlled nodes are operated. The communication is organized according to a mixed time-slot and polling procedure. This method is also called Slot Communication Network Management (SCNM).

A Powerlink communication cycle consists of three phases. First, the managing node transmits the Start of Cycle (SoC) frame to synchronize all devices. Then the cyclic isochronous exchange of data takes place between the managing node and the controlled nodes, using a polling procedure. The managing node transmits a polling request (PRq) to a controlled node, which then responds by sending its data in a polling response (PRs). Each controlled node has its own defined time slot in the cycle. This prevents any collisions from happening on the bus.

Since all messages (with the exception of the polling request) are transmitted using multicast addressing, all participants can listen to all messages. This particularly enables cross traffic between the controlled nodes, i.e. direct communication without requiring the assistance of the managing node. The last phase of the cycle is the asynchronous phase (Start of Assync, SoA), which is where non-time-critical data are transferred. This is frequently used for status or diagnostic messages and for transmitting data saved on the node. This asynchronous phase can also be used to integrate participants that are not real-time capable, via standard Ethernet.

Both the physical layer (PHY) and the media access control (MAC) layer correspond to Fast Ethernet standard. So an Ethernet Powerlink network is provided with all physical transmission media that comply with the 100BASE-X standard. In the data link layer, an additional Powerlink data link layer is added. The majority of the Powerlink functionality described in this section is implemented in this layer; among other things the managing node and controlled node functionality definitions, the processing of isochronous and asynchronous communication or the specification of an Ethernet POWERLINK frame.

The specifications that go beyond the 802.3u Ethernet standard, in other words those from the network layer upward, are divided into two categories:

- The hard real-time communication is implemented using the PDO concept (process data objects), which we are already familiar with from CANopen, and the already described mixed time-slot and polling procedure. Here, Layers 3 and 4 of the ISO/OSI reference model are bypassed.
- Non-time-critical data exchange uses the parallel implementation of the protocols IP, TCP, UDP, HTTP and FTP. Combined with the application layer and the concept of service data objects, the data are communicated in asynchronous time slots, e.g. with the help of UDP/IP.

Since Ethernet Powerlink can be implemented as a pure software solution, it is possible to provide the managed node and the controlled nodes with a suitable software stack. There are several solutions on the market. For the slaves, we still recommend resorting to the commercially available FPGAs because, firstly, they supply the entire Ethernet Powerlink protocol stack and, secondly, they act as a hub, allowing very easy network extension. The master requires a certain level of computing power because the software solution has to organize the entire time-synchronous data exchange using the time-slot and polling procedure. In addition, the master has to identify whether the incoming data is a standard Ethernet frame coming from a non-real-time-capable terminal device.

3.5.5 Further Vehicle Buses

3.5.5.1 FlexRay

FlexRay is a time-controlled, redundant bus system that was specifically designed for automotive applications. The FlexRay Consortium defined a standard that establishes a maximum data transfer rate of 10 Mbit/s in redundant operation or up to 20 Mbit/s if both channels are used independently.

Time-controlled protocols have been in use in a variety of applications for some time. The exact definition of transmission slots guarantees time-determinism and synchronism, preventing the occurrence of transmission collisions. But this comes at the cost of reduced flexibility and is not easy to combine with subsequent extensions during the processing schedule, event-triggered transmission or flexible use of the available bandwidth reserves. FlexRay utilizes its full potential if communication and processing activities are scheduled very carefully in the connected control units. This achieves highly deterministic communication and redundant signal flow with minimum latency periods. The necessary tool chains create the resulting bus configuration as FIBEX files and normally also the source code for schedule control in the control units.

3.5.5.2 LIN Bus

For connecting a variety of simple peripherals to vehicle networks, CAN is often complemented by a LIN (Local Interconnect Network) bus. LIN is a simple serial-communication system based on a single-wire bus (signal against ground) that supports maximum data rates of 19.2 kbit/s. Unlike CAN, LIN implements the master/slave principle, so no mechanism is needed to resolve collisions.

A LIN message consists of a byte identifier and up to 8 bytes of data (payload). The master initiates each communication and additionally acts as a gateway to the higher-level CAN bus.

3.5.5.3 MOST Bus

The MOST (Media Oriented Systems Transport) bus was developed for the networking of multimedia and infotainment systems in motor vehicles. Transmitting audio and video data requires high data rates and the synchronization of data transmission between the bus participants. Apart from that, the transmission of such data in a motor vehicle is associated with high electromagnetic compatibility requirements (EMC) [1].

The MOST Cooperation standardizes and maintains the specifications. In its current version, the MOST bus offers a data rate of 24.8 Mbit/s (MOST25 standard). For higher data quantities of 50 Mbit/s (MOST50) or even 150 Mbit/s (MOST150) the relevant concepts are already available [49].

Participants in a MOST network are arranged in a ring topology. To satisfy the high data rates and the robustness versus electromagnetic interference, plastic (or polymer) optical fiber (POF) is generally used as a transmission medium. Data transmission is organized in data frames. One of the participating devices assumes the role of the timing master which generates these frames at a fixed data rate and passes them on to the subsequent ring participants. The data frame contains administrative information, an area for control commands (control channel), an area for synchronous data and an area for asynchronous data. In MOST25, the data frame has a total length of 512 bits.

The MOST Cooperation has also defined a higher-level protocol for the application layer, which even allows controlling various collision detection (CD) systems with a uniform set of commands.

3.6 PC Interfaces

The personal computer (PC) led to the standardization of the x86 architecture, also in the area of test bed systems. This architecture is also used for control and automation tasks. The freely accessible system bus (ISA, later on PCI and PCIe) allows extending the functionality in a flexible and modular fashion, using plug-in cards. For example, cards with fieldbus controllers provide the option of integrating external devices via CAN or PROFIBUS (see Sects. 3.5.2 and 3.5.3). In addition to the featured open extensibility, standard PC communication connections have become established that are often a firm part of a PC's circuit board (e.g. Ethernet, RS232, USB). The paragraphs below give a short overview of the most widespread PC interfaces.

3.6.1 RS232

RS232 is a bit-serial, asynchronous data interface that is established as a point-to-point connection between two digital systems. Transmission is broken down in data words, often in ASCII code, containing a start bit followed by a number of data, stop and parity bits, depending on the configuration. "9600/8n1," for example, means transmission at 9600 bits/second, with every data word consisting of 8 data bits (plus the implicitly provided start bit), no parity bit and 1 stop bit, i.e. in total 10 bits per data word. Transmission take place using the non-return-to-zero (NRZ) code, with a voltage level ranging between −15 V and −3 V for logic one and between +3 V and +15 V for logic zero.

Some devices interpret certain words as a control sequence. These may serve to control the data flow (example: XON/XOFF; specifically with 3-wire connection cables) or trigger defined actions in terminal devices such as "cursor up" or "clear screen." Such sequences are defined based on earlier terminal types (VT100, VT220, etc.).

RS232 interface cables consist of at least three connections for each data signal per direction and one common ground. With two equal devices (DTE = data terminal equipment), the two data signal lines connect one device's transmit wire to another device's receive wire and vice versa; this is achieved by using a crossover cable (null modem cable). When using a nine-pin D-sub connector, this corresponds to a connection between Pin 2 of one connection and Pin 3 of the other connection; ground is connected through via

Pin 5. In a connection between DTE and DCE (data communication equipment, e.g. a modem) the signal lines must not be crosslinked ("1:1 cable"). Furthermore, depending on the terminal device, there will be additional control lines (RTS = Request to Send, CTS = Clear to Send, DSR = Data Set Ready, DTR = Data Terminal Ready, RI = Ring Indicator and DCD = Data Carrier Detect). In the application layer, it is possible to use protocols such as the AK protocol or XMODEM.

3.6.2 RS422 and RS485

Similar to RS232, an RS422 interface represents an asynchronous serial interface for point-to-point communication, allowing up to ten receivers to be connected in parallel. Unlike RS232, a transition to symmetric signal transmission extends the line length and increases the allowed data rate to up to 10 Mbit/s.

Symmetrical signals represent one bit as the difference between two signal voltages. This efficiently suppresses noise that affects both lines to equal extent (common mode). Twisting the pairs of wires improves this symmetry even further. An RS422 cable consists of at least five wires: two twisted pair wires and a ground wire. Since RS422 establishes the signal level but not the pin assignment or the protocol, it is not possible to establish a general standard for it. Optionally, additional wires are used to control communication. If data rates are higher, we recommend terminating the line at each end in the cable's characteristic impedance (typically 100–200 Ω).

Similar to RS422 networks, RS485 networks are based on differential signal transmission, but additionally support the equal connection of multiple parallel nodes within a segment of the bus structure ("multipoint systems"). Usually there is no more than one twisted pair wire for data transmission that is used bi-directionally (half-duplex) and has to be terminated with a resistor at both ends. Up to 32 nodes share the line in parallel connection. It is important to ensure that no more than one node actively impacts the voltages on the data line ("transmits"); all the other nodes have to act passively ("read"). Wherever possible, stub lines should be avoided in the bus topology. If stub lines cannot be avoided, they must not exceed a defined maximum length (usually in the one meter range) and they should never be terminated with a resistor.

3.6.3 Ethernet, TCP/IP and UDP

Ethernet is probably the most widely used wired data network technology available for use between all kinds of computer systems. This is not least due to the systematic definition and standardization of the hardware (transceiver, level, cable, connectors and their assignment) and protocol for Layers 1 and 2 in the ISO-OSI reference model (see Overview). There are even well established solutions for the higher layers, as the examples in Table 3.13 illustrate.

Table 3.13 Ethernet and standard protocols based on the ISO-OSI reference model

Number	OSI Layer	Example
7	Application	HTTP, OPC-UA, SMTP, etc.
6	Presentation	
5	Session	
4	Transport	UDP, TCP
3	Network	IPv4, IPv6
2	Data link	Ethernet
1	Physical	

The following paragraphs give an initial overview of Ethernet; for more information, please refer to [50–52] or [53]. By the mid-1980s, Ethernet already was the standard technology for local computer networks (LAN). Later on, it also caught on in industrial environments, especially by the time real-time Ethernet was introduced in the form of EtherCAT (see Sect. 3.5.4). Ethernet provides serial, asynchronous communication. As it implements a bus system that allows all nodes random access (multi-master principle), there may be data package collisions on the transmission medium when several participants attempt to transmit at the same time. Standard Ethernet resolves this situation by applying the carrier sense multiple access/collision detection (CSMA/CD) mechanism. The system detects the collision; nodes stop transmitting and start retransmitting after a waiting time of random length that is different for each node.

The advantage of this method is the network's ability to organize itself, because no central coordination is needed to resolve the collision. However, high communication loads may lead to congestion effects and, consequently, a decline in the effective data throughput. Also, the lack of determinism makes such networks unsuitable for real-time tasks. Standardized transmission rates are 10 Mbit/s, 100 Mbit/s, 1000 Mbit/s and above. The bus topology's wiring was initially implemented using coaxial cables (10BASE5, 10BASE2), which led to correspondingly high collision domains (a large number of nodes that can potentially lead to a collision when attempting to transmit simultaneously).

Most contemporary systems use the switched-Ethernet method. With this method, twisted pair wiring (for example 100BASE-TX or CAT5) is used to connect several terminal devices to a central switch, which divides the entire network into point-to-point connections. The cable includes one twisted pair wire for each direction, i.e. four single wires, enabling full-duplex data traffic without collisions. Before forwarding the messages, the switch stores them until the line to the target device is no longer busy. "Power over Ethernet" (PoE, IEEE 802.3af-2003) allows using the wires of the network cable to pass on not only data but also electrical power to supply terminal devices. Depending on the appliance class, the power is limited to several watts, but enables simple design and wiring of compact terminal equipment such as Ethernet-based sensors or VoIP (voice over IP) devices. The network layer (Layer 3) supports the Internet Protocol (IP) in Version 4 or 6. This is where, among other things, the address resolution takes place (with IPv4 4 bytes and with IPv6 16 bytes). The protocol for the transport layer

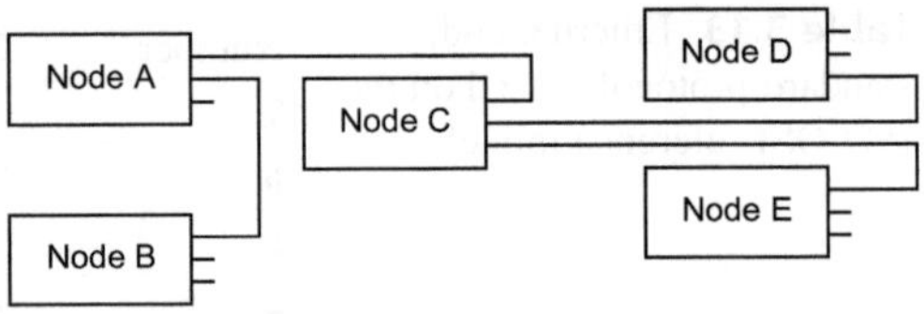

Fig. 3.112 Tree topology

is TCP (Transmission Control Protocol), a connection-oriented protocol, and UDP (User Datagram Protocol), a connectionless but somewhat more efficient protocol.

3.6.4 USB

Under the leadership of INTEL, the Universal Serial Bus (USB) was introduced from 1996 onward to replace and standardize a variety of widely used interfaces between PCs and peripherals such as serial COM (RS232), printer, keyboard and mouse interfaces. As shown in Fig. 3.112, USB supports a tree structure and "hot plugging" (adding components without rebooting the system). In a USB network, point-to-point connections are established between the ports of the participating nodes. Simple peripherals usually have one port while more complex systems have two or more.

USB is a serial asynchronous 4-wire bus, in which two wires are used for bidirectional symmetric data transmission (half-duplex system) and the other two wires are available for power supply (5 V). The line length between two ports must be no longer than 5 m. A USB network is administered by a master node (USB host controller). In a network, the PC usually assumes this function. The protocol supports identification and dynamic registration of devices, as well as automatic configuration of the corresponding software (drivers) in the PC. Up to 127 nodes and data rates in the range of 12 Mbit/s (USB Version 1.1), 480 Mbit/s (USB 2.0) and around 4 Gbit/s (USB 3.0) are supported.

3.6.5 IEEE1394

IEEE 1394, also known under the product name FireWire or iLink, is a high-speed data bus, which, similar to USB, is based on a tree topology and allows making changes while the system is running (hot plugging). Unlike USB, IEEE 1394 supports the allocation (reservation) of certain time slots for cyclic communication of the nodes, allowing deterministic, i.e. real-time, data traffic. The communication is briefly interrupted by bus re-initialization only if changes are made to the topology. It is also possible to set up direct peer-to-peer data traffic between the participating modes without having to route the data over the central host controller, as is the case with USB. The role of the network coordinator can be assigned at every re-initialization to any node that has the necessary capabilities. This means there is no dependence on a single host controller, as is the case with USB.

The IEEE 1394a standard, introduced in 1995, supports 400 Mbit/s; later versions support 800 Mbit/s (1394b) and lastly 3.2 Gbit/s (1394–2008). The addressing procedure is based on 64-bit addresses in a virtual memory distributed across all network nodes. Because of the mentioned features, IEEE 1394 is used for high-performance measuring systems with real-time data acquisition and for applications with video data transmission or disk clusters. The real-time capability and determinism led to the great amount of measurement data acquisition applications.

3.6.6 VXI, VISA, PXI

As an alternative to standard PC architecture with ISA, PCI or PCIe system bus, some applications also use VXI and PXI systems. VXI is based on the VME bus standard (VME = Versa Module Eurocard), which was introduced roughly at the same time as the PC and was initially developed for Motorola 68000 CPUs. VXI is a system bus for backplane systems in 19-in. rack technology that supports one or multiple computer cards and a variety of peripherals such as data acquisition (DAQ) cards. VXI (VME eXtension for Instrumentation Bus) extends the VME standard by the addition of features for interrupt handling, resource management, extension to multiple 19-in. racks, etc. For detailed information on VXI, please refer to the VXIbus Consortium [54].

VISA (Virtual Instrument Software Architecture) defines interfaces for measuring systems that are ideally based on backplane architecture (preferably VXI) or are coupled via GPIB. GPIB stands for General Purpose Interface Bus, a parallel bus, developed back in the 1960s, for connecting measuring devices. Later it was standardized as IEEE 488. Using VISA drivers, supplied by the manufacturer, measuring devices can so be integrated into VISA-based measuring system software.

The PXI standard is a further advancement of PCI definitions for measurement and automation applications.

References

1. K. Reif, *Bosch Autoelektrik und Autoelektronik* (Vieweg+Teubner Verlag|Springer Fachmedien, Wiesbaden, 2011)
2. E. Hering, G. Schönfelder, *Sensoren in Wissenschaft und Technik* (Vieweg+Teubner Verlag| Springer Fachmedien, Wiesbaden, 2012)
3. Wikipedia, Stromwandler, 2014. [Online]. http://de.wikipedia.org/wiki/Stromwandler. Accessed 28 Apr 2014
4. G. Merker, C. Schwarz, R. Teichmann, Verbrennungsdiagnostik, in *Grundlagen Verbrennungsmotoren* (Vieweg + Teubner|Springer Fachmedien, Wiesbaden, 2012), pp. 309–378
5. H. Huppertz, Zündung—Sekundärschaubild, 2014. [Online]. http://www.kfz-tech.de/ZuendSekundaer.htm. Accessed 28 Apr 2014

6. R. Maly, Spark Ignition: its physics and effects on the internal combustion engine, in *Fuel Economy in Road Vehicles Powered by Spark Ignition Engines*, ed. by J.C. Hilliard, G.S. Springer (Springer, New York, 1984), pp. 91–148
7. International Organization for Standardization, *ISO 16183—Heavy Duty Engines—Measurement of Gaseous Emissions from Raw Exhaust Gas and of Particulate Emissions Using Partial Flow Dilution Systems under Transient Test Conditions* (Beuth, Berlin, 2002)
8. Environmental Protection Agency, *Control of Air Pollution from New Motor Vehicles—Certification and Test Procedures* (Environmental Protection Agency, Washington, DC, 2002)
9. K. Engeljehringer, W. Schindler, The organic insoluble diesel exhaust particulates- Differences between diluted and undiluted measurements. J. Aerosol Sci. **20**(8), 1377–1380 (1989)
10. Environmental Protection Agency, *Engine Testing Procedures* (Environmental Protection Agency, Washington, DC, 2005)
11. N. Kreft, G. Marek, W. Schindler, W. Silvis, Diesel particulate measurement with partial flow sampling systems: a new probe and tunnel design that correlates with full flow tunnels. SAE Technical Paper 2002-01-0054, Detroit, 2002
12. UNECE, Regulation No. 83—Uniform provisions concerning the approval of vehicles with regard to the emission of pollutants according to engine fuel requirements, 2011. [Online]. http://www.unece.org/fileadmin/DAM/trans/main/wp29/wp29regs/r083r4e.pdf. Accessed 28 Apr 2014
13. J. Andersson, B. Giechaskiel, R. Munoz-Bueno, E. Sandbach, P. Dilara, *Particle Measurement Programme (PMP): Light-Duty Inter-Laboratory Correlation Exercise (ILCE-LD)—EUR 22775 EN* (Office for Official Publications of the European Communities, Luxemburg, 2007)
14. Grimm Aerosol Technik GmbH & Co. KG, GRIMM—The master in AEROSOL measurement, 2014. [Online]. http://www.grimm-aerosol.com. Accessed 28 Apr 2014
15. AVL List GmbH, AVL—Passion and results, 2014. [Online]. http://www.avl.com. Accessed 5 May 2014
16. W. Schindler, C. Haisch, H. Beck, E. Jacob, D. Rothe, A photoacoustic sensor system for time resolved quantification of diesel soot emissions. SAE Technical Paper 2004-01-0968, 2004
17. Z. Bozóki, L. Krämer, R. Niessner, Characterisation of a mobile photoacoustic sensor for atmospheric black carbon monitoring. Anal. Sci. **17**, 563 (2001)
18. C. Heimgärtner, A. Hemm, A. Leipertz, S. Schraml, S. Will, Application of a new soot sensor for exhaust emission control based on Time Resolved Laser Induced Incandescence (TIRE-LII). SAE Technical Paper 2000-01-2864, 2004
19. Matter Aerosol AG, Matter Aerosol—A Testo Company, 2014. [Online]. http://www.matter-aerosol.ch/. Accessed 5 May 2014
20. D. Ltd., Dekati—Excellence in particle measurements, 2014. [Online]. http://www.dekati.fi. Accessed 5 May 2014
21. R. Glanz, Differentielle Erfassung von Tumble-Strömungsfeldern. MTZ **61**(1), 40–45 (2000)
22. C. Hoßfeld, R. Kaiser, Schadensprävention am motornahen Katalysator durch Einsatz der CFD-Simulation. MTZ **64**(4), 278–285 (2004)
23. G. Thien, U. Trenner, *Anpassung der Ventilhub-Ersatzkurve für mittlere Strömungskennwerte von Ventilkanälen* (AVL List GmbH, Graz, 1982). FS-Mitteilung Nr. 47
24. G. Thien, Entwicklungsarbeiten an Ventilkanälen von Viertakt-Dieselmotoren. *ÖIZ* **8**(9) (1965)
25. G. Cser, Investigation of intake phenomena in 4-stroke DI diesel engines. *Acta Technica Scientiarium Hungaricae* **69** (1970)
26. G. Tippelmann, A new method of investigation of swirl ports. SAE Technical Paper 770404, 1977

27. N. Jackson, J. Stokes, M. Sadler, M. Heikal, L. Pommier, Correlation of the combustion characteristics of spark ignition engines with the in-cylinder flow field characterised using PIV in a water analogy rig. SAE Technical Paper 971637, 1997
28. G. Tippelmann, Räumlicher Drallmesser für Drall- und Tumble-Messung. MTZ **58**(6), 327 (1997)
29. D. Endres, H. Neussler, M. Breuer, Einrichtung zur Untersuchung der Strömungsbewegung in zylindrischen Bauteilen. DE Patent 4133277, 1993
30. F. Durst, A. Melling, J. Whitelaw, *Principles and Practice of Laser-Doppler-Anemometry* (Academic Press, London, 1976)
31. R. Glanz, F. Hödl, Verfahren zur Erfassung einer Tumble-Strömung in einem Zylinderraum einer Brennkraftmaschine. DE Patent 10063045 C2, 18 Dec 2000
32. Z. Zhang, *LDA Application Methods—Laser Doppler Anemometry for Fluid Dynamics* (Springer, Berlin, 2010)
33. J. Kompenhans, M. Raffel, S. Wereley, C. Willert, *Particle Image Velocimetry—A Practical Guide* (Springer, Berlin, 2007)
34. R. Glanz, E. Petschenig, H. Sorger, Differentielles Erfassen von Strömungsfeldern in der Motorentwicklung. MTZ **71**(5), 332–338 (2010)
35. R. Glanz, E. Petschenig, Verfahren zum Erfassen einer Wirbelströmung im Zylinderraum. Industrial Property Right AT Patent. Patent 507733 B1, 2010
36. J.F. Meyers, Doppler global velocimetry—The next generation? AIAA Paper 92-3897, Nashville, 1992
37. T. Seidl, H. Steuker, Doppler Global Velocimetry—Messung der Zylinderinnenströmung mit ottomotorischer Direkteinspritzung. MTZ **69**(4), 328–333 (2008)
38. I. Röhle, *Doppler Global Velocimetry* (DLR—National Aeronautics and Space Research Center of the Federal Republic of Germany, Cologne, 1999)
39. O. Dingel, T. Seidl, H. Steuker, Strömungsmessung während der Kompression mit Doppler Global Velocimetry. MTZ **65**(5), 368–375 (2006)
40. M. Bates, M. Heikal, A knowledge-based model for multi-valve diesel engine inlet port design. SAE Technical Paper 2002-01-1747, 2002
41. S. Keller, *Presentation from Fraunhofer ISE*, Freiburg, 2012
42. P. Kurzweil, Thermodynamik und Kinetik der Brennstoffzelle, in *Brennstoffzellentechnik* (Springer Fachmedien, Wiesbaden, 2013), pp. 17–52
43. P. Kurzweil, *Brennstoffzellentechnik: Grundlagen, Komponenten, Systeme, Anwendungen* (Springer Fachmedien, Wiesbaden, 2013)
44. H. Li, H. Wang, X.-Z. Yuan, *PEM Fuel Cell Diagnostic Tools* (CRC Press, Boca Raton, FL, 2012)
45. B. Davat, W. Friede, S. Rael, Mathematical model and characterization of the transient behavior of a PEM fuel cell. IEEE Trans. Power Electron. **19**(5), 1234–1241 (2004)
46. W. Baumgartner, V. Hacker, V. Peinecke, P. Prenninger, E. Ramschak, T. Schaffer, New approach for detection of different critical stack operating conditions for large volume fuel cell applications. ECS Trans. **1**(8), 367–375 (2006)
47. International Organization for Standardization, *ISO/IEC 7498-1:1994—Information Technology—Open Systems Interconnection—Basic Reference Model: The Basic Model* (Beuth, Berlin, 1994)
48. M. Popp, *PROFIBUS-DP/DPV 1—Grundlagen, Tips und Tricks für Anwender* (Hüthig, Heidelberg, 2000)
49. M. Cooperation, MOST Cooperation—Home, 2014. [Online]. http://www.mostcooperation.com/home/index.html. Accessed 5 May 2014

50. IEEE, IEEE 802.3 Ethernet Working Group, 2014. [Online]. http://www.ieee802.org/3/. Accessed 5 May 2014
51. W. Stallings, *Data and Computer Communications* (Prentice Hall, Upper Saddle River, NJ, 2011)
52. A. Tanenbaum, *Computer Networks* (Prentice Hall, Upper Saddle River, NJ, 2003)
53. C. Spurgeon, *Ethernet—The Definitive Guide* (O'Reilly, Sebastopol, 2000)
54. V. Consortium, VXI—Home, 2013. [Online]. http://www.vxibus.org. Accessed 5 May 2014

Software Perspective: Test Bed

4

After the extensive discussion of the two lower (hardware) layers of the generic test bed architecture, as can be seen in Fig. 3.1, the following chapter deals with the automation layer. This layer comprises the different software functions that are needed to automatically execute a variety of testing tasks. There are also other functions that ensure the unit under tests' safety and that of the test bed environment. Important are also functions that simulate components that are not physically present on the test bed but still affect the actual behavior of the physical components under test. In one form or another, these functions are present in all automation systems.

The first section of this chapter concentrates on the general software structure and introduces several interface standards. The use of standards allows users to add new software components to an automation system without changing the already available system and tackle new testing challenges in a fast and cost-efficient fashion. A description of individual key software functions follows.

4.1 Software Architecture and Interface Standards

4.1.1 Software Architecture

The automation layer requires a modular structure to be able to handle rapidly changing testing tasks automatically and efficiently. Figure 4.1 shows a schematic diagram of the necessary software architecture. It consists of the following layers (from bottom to top):

– Test objects that require appropriate modeling. Modeling languages such as SYSML are widely used for this purpose. Most automation systems allow addressing the test

M. Paulweber, K. Lebert, *Powertrain Instrumentation and Test Systems*, Powertrain,
DOI 10.1007/978-3-319-32135-6_4

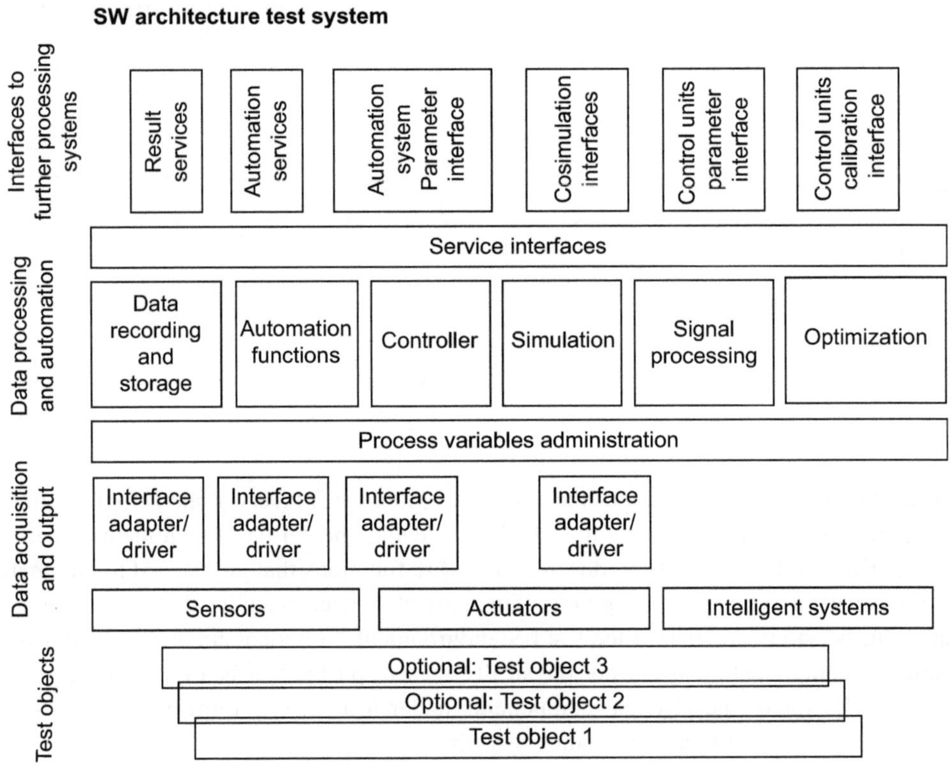

Fig. 4.1 Software architecture

objects using names that are valid across the system. Functionality is often modeled using languages as, for example, Simulink®, Dymola® or Cruise-M®.

- Sensor and actuator objects, which are also described in data structures and are dynamically extensible and adaptable to new testing tasks with the help of editors. It is necessary to address these objects separately.
- Driver layer: Most automation systems allow you to add further device drivers later, which establish the connection to sensors, actuators or intelligent measurement or control systems.
- A process variable management layer provides the acquired data for further calculations or automation functions. In most automation systems this is a layer where new variables can be dynamically created or deleted. Names or unique identification keys (IDs) are used to access these variables.
- The process variables are processed further in the automation function layer. Links to the variables are established via the unique names or IDs and described either graphically or in the form of scripts, formulas or simulation models. Typical automation modules are formula calculation modules, controller modules, simulation

modules, monitoring modules, evaluation modules, storage modules, optimization functions and the core of the test automation systems, the test run executor. The majority of test bed automation systems allow the description of new test procedures in text form (scripts are usually written in smart editors with automatic text completion features, etc.) or graphical form. Widespread script languages are Visual Basic (Microsoft) or Python; commonly used graphical tools are the standard IEC 1131 (PLC technology), the widely used simulation language Simulink or similar.
- Service oriented architectures (SOAs) are commonly used to improve the way automation tasks are embedded in the development procedures. SOAs allow external software programs to address all publicly accessible functions in an automation system. The functions (services) provided by the test bed systems vary. Figure 4.1 shows typical interfaces.

To make it easier for the automation systems to integrate measuring devices, control units or other development tools, there now exist several organizations which specialize in defining and standardizing the necessary interfaces. As mentioned in the list above, most of the interfaces are based on service-oriented architecture. These types of standardized interfaces facilitate test system extension and modification considerably, making such processes much faster and more cost effective. They also satisfy the key requirements addressed at the beginning of this section for easy adaptability to new testing tasks. The following section provides brief descriptions of the most frequently used standards.

4.1.2 Interface Standards

Standardized interfaces are of key significance for modern test systems. In the automotive industry, a standardization organization formed at the end of the 1990s focusing on interfaces for measuring and testing systems in the automotive industry. ASAM (Association for Standardization of Automation and Measuring Systems) now has more than 150 members worldwide. Among them are all the major car manufacturers and suppliers. Figure 4.2 shows the interface standards defined by ASAM [1] between the different components of an automatic test system.

Apart from ASAM, there are other organizations that deal with the standardization of interfaces between software development tools which also include automatic test systems. In Europe, organizations worth mentioning in this respect are ARTEMIS-IA and EICOSE which define interfaces between development tools for embedded software in the context of European Union projects. The standards were created in several related EU projects (CESAR, MBAT, iFEST and CRYSTAL) and are based on the communication architecture OSLC [4], which was standardized by OASIS [5].

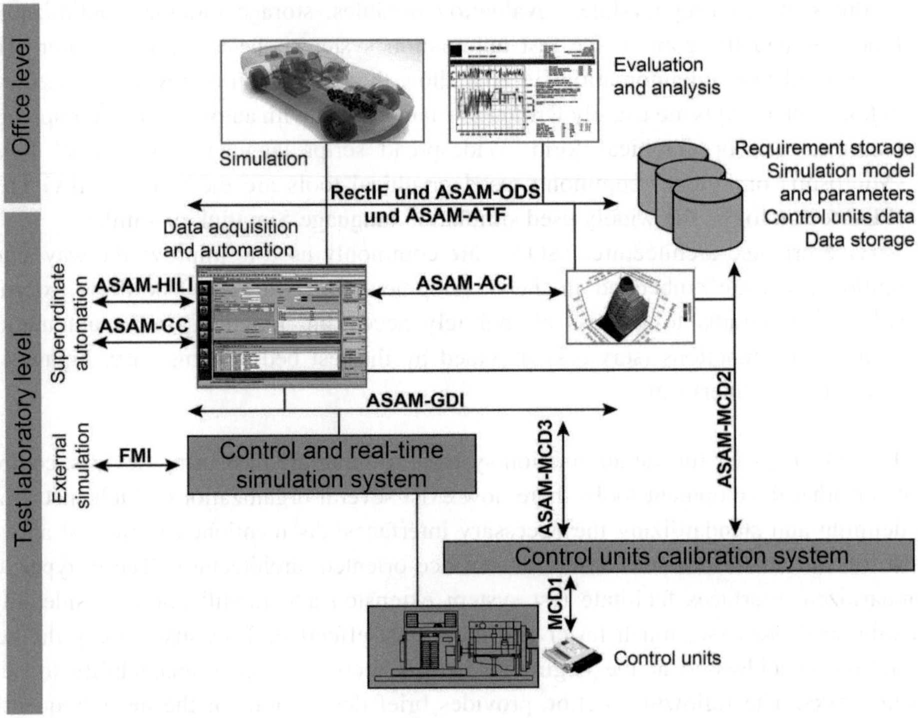

Fig. 4.2 Standardized interfaces

Another important standard formed in the area of simulation. This one allows integrating complex environment simulations into automatic test procedures: The FMI (functional mock-up interface) standard, which is defined and managed by a Modelica Association Project (MAP) belonging to the MODELICA Association [6], allows linking a variety of simulation models, as is required for calculating environment conditions for test objects.

The following sections address the standardized interfaces shown in Fig. 4.2 (for further details, please refer to [1]). ASAM has divided the standards into three groups:

- AE: Automotive electronics
- CAT: Computer aided testing
- COMMON: Common standards used in AE and CAT

ASAM uses common description methods for the technology definitions in the standards:

- Format description: defines syntax and semantics of a file format for the purpose of data exchange.

- API: defines the interfaces and functional behavior of executable routines for the purpose of calling service routines or exchanging data between computer programs.
- Protocol definition: defines syntax, semantics and synchronization of bus communication for the purpose of establishing a communication link between two computer systems.
- Technology reference: specifies a technology-dependent interpretation of a technology-independent part of the standard, typically via mapping rules or program code.
- Application area companion: defines an extension of a base standard for a specific application area or a specific type of device.
- Transport layer specification: defines how a generic protocol definition is to be interpreted using a concrete physical layer.

4.1.2.1 ASAM AE

ASAM AE standards (Automotive Electronics) are primarily applied during the design and implementation phase of electronic control unit (ECU) software development (see Table 4.1). They focus on:

- design of functional and interface specifications for software components
- carrying out measurement, calibration and diagnostics tasks on an ECU
- automation of HiL tests
- description of development artifacts
- exchange of development requirements

4.1.2.2 ASAM CAT

ASAM CAT Standards (Computer Aided Testing) are primarily applied during the verification & validation phases of ECU software development (right side of the V-model), and during automated calibration and system testing on engine and vehicle test beds (see Table 4.2). They focus on:

- automated calibration of ECU parameters
- storage of test data
- evaluation and analysis of test data

4.1.2.3 ASAM COMMON

ASAM COMMON standards are used in both areas, AE and CAT (Table 4.3).

Table 4.1 ASAM AE standards (see [1])

Standard	Title	Description
ASAM ATX	Automotive Test Exchange Format	Definition of an XML-based format for describing tests. Allows the reuse of test descriptions in different test automation systems. Covers activities such as test specification, test planning, test execution and test evaluation. Test specifications include test cases, sequences, steps, actions and corresponding meta-data. The format allows storing test values for inputs (stimuli values), outputs (expectation values) and execution conditions. Frequently used in conjunction with hardware-in-the-loop (HiL) test beds.
ASAM CC	ContainerCatalog	Definition of an XML-based format for describing development artifacts such as source code, compiled objects or documentation files, and associated meta-information about the objects such as creator, name or version. Primarily used for exchanging information between OEM and supplier.
ASAM CDF	Calibration Data Format	Definition of an XML-based format for storing calibration parameters and associated meta-data on acquisition and quality. Complements MCD-2 MC in that it provides the values of the calibration parameters described by MCD-2 MC.
ASAM FSX	Functional Specification Exchange Format	Definition of an XML-based format for describing the functional specification of software, primarily for the purpose of generating technical documentation. The format allows documentation using prose text in multiple languages, cross-referencing, linking and mentioning of requirements. Variations of the specification can be expressed. Complements MDX which contains the interface definitions of the functions.
ASAM HIL (new: ASAM xIL)	Hardware-in-the-Loop	Definition of an API (application programming interface) between test automation and HiL test bed components, particularly for ECU testing. Provides access to the simulation models, ECU-internal measurement and calibration data, diagnostics data and electrical error simulation. API is described as a technology-independent UML (Unified Modeling Language) model. Includes code examples for implementing the API in C# and Python.
ASAM ISSUE	ISSUE Exchange Format	Definition of an XML-based format for information exchange between different parties in a distributed development process. Covers change requests,

(continued)

Table 4.1 (continued)

Standard	Title	Description
		clarification requests, problem reports and further use-cases. Furthermore, definition of a workflow for handling the requests.
ASAM MBFS	Model Based Functions Specification	Definition of a block library for model-based design that contains typical functions needed in automotive control algorithm specifications. Includes the definition of multiple blocks with icons, pseudo code and test vectors. Also includes the description of a reference implementation in MATLAB/Simulink©.
ASAM MCD-1 CCP	CAN Calibration Protocol	Definition of a communication protocol between master and slave controllers on a CAN 2.0B network. Includes transfer of calibration data to slave controllers, continuous data acquisition from slave controllers and the execution of generic control functions. The standard is also known as "ASAP1."
ASAM MCD-1 XCP	The Universal Measurement and Calibration Protocol Family	Definition of a bus-independent communication protocol between master and slave controllers. Includes synchronous data acquisition and stimulation, read/write access on calibration data, memory page management, flash programming and further optional features. Transport layer specifications are defined for CAN, Ethernet (TCP/IP & UDP/IP), FlexRay, USB and SxI. Establishes an improved and generalized version of MCD-1 CCP.
ASAM MCD-2 D	Data Model Specification for ECU Diagnostics	Definition of an XML-based format for describing ECU diagnostic, programming and related vehicle interface data for data exchange between the ECU and external test equipment. If compliant to this standard, external testing equipment does not require special programming to process diagnostic data from an ECU. The standard is also known as "ODX."
ASAM MCD-2 MC	ECU Measurement and Calibration Data Exchange Format	Definition of calibration parameters (CHARACTERISTIC) and measurable variables (MEASUREMENT) via a non-XML format. Such data resides inside the memory of the ECU. The definition format makes the data accessible through application systems. Also contains a description of the ECU's hardware interface for device driver configuration of the application system. The standard is also known as "ASAP2."

(continued)

Table 4.1 (continued)

Standard	Title	Description
ASAM MCD-2 NET	Data Model for ECU Network Systems	Definition of an XML-based format for describing messages and their timing for automotive communication buses. It is the preferred description format for FlexRay and MOST, but also supports CAN, TTCAN, LIN and Ethernet. Is used for design, configuration, data logging and simulation of communication on the bus. The standard is also known as "FIBEX" (Field Bus Exchange Format). Content is harmonized with the AUTOSAR System Template.
ASAM MCD-3 ASAP3	Automation/ Optimization and ECU Calibration System Interface	Definition of an RS232 protocol between the test automation system software and a measurement and calibration system connected to an ECU. The standard describes an obsolete technology. MCD-3 MC should be used instead.
ASAM MCD-3 D	Application Programming Interface for MVCI Diagnostic Server	Definition of an object-oriented API for an ECU diagnostics server, which provides bus- and protocol-independent services for communication between client applications and ECUs. Includes a mapping of an object-oriented API to C++, Java and COM-IDL with example code. Is complemented by MCD-2 D which contains the actual diagnostic configuration for ECUs and vehicle networks.
ASAM MCD-3 MC	Application Programming Interface for Measurement and Calibration Server	Definition of an object-oriented API for a measurement and ECU application server, which provides bus- and protocol-independent services for communication between client applications and ECUs. Includes a mapping of the object-oriented API to COM-IDL with example code. Is complemented by MCD-2 MC which contains the actual measurement and calibration data description for ECUs.
ASAM MDX	Model Data Exchange Format	Definition of an XML-based format for describing interfaces of functions, their data (variables and calibration parameters) and scheduling in the ECU software. This allows the integration of such functions as object code into the overall ECU software without having access to the source code. The standard is complementary to FSX, which contains the functional description of functions. MDX is the predecessor of the AUTOSAR Software Component Template.

Table 4.2 ASAM CAT standards (see [1])

Standard	Title	Description
ASAM ACI	Automatic Calibration Interface	Definition of a client–server, object-oriented API for an automatic calibration system (client) for remote-controlling a test bed automation system (server). The standard includes: –Player service: controlling of test bed actuators for setpoint adjustment –Recorder service: recording of measured values (averaged or instantaneous) from the test bed –Watcher service: monitoring of out-of-bounds measurement values –Device service: further services such as ECU-specific and test bed-specific operations Client and server may reside on different host systems, communicate via TCP/IP and are suitable for static test execution. Includes a CORBA based middleware implementation guide, an interface definition file and a description for interface certification tests.
ASAM CEA	Components for Evaluation and Analysis	Definition of functional components for the evaluation and analysis of test and measurement data. Most commonly used for the development of applications for processing and the visualization of test data. Defined components are: Input, Worker, Output, Consumer, Producer and Viewer. Services are available for creating graphics, logging & tracing, unit conversion, undo & redo, inter-component communication and help. Furthermore, the standard defines a file format for CEA component description. Includes reference code for Java and . NET.
ASAM GDI	Generic Device Interface	Standard defines the connection of measurement and control devices of a test bed with a test bed automation system via a 4-layer architecture. It includes the specification of: –Layer 4: coordinator services for translating measurement and control operation requests from the test bed automation system to device driver commands –Layer 3: device drivers for uniform access to devices of different types –Layer 2: platform adapters for standardized interfaces to specific devices and operating system services –Layer 1: transport layers and communication types for communicating with devices via TCP/IP, USB, SoftSync or RS232 The standard specifies APIs for all layers and provides description formats for device functions and data connections. Includes companion standards for communication with chassis

(continued)

Table 4.2 (continued)

Standard	Title	Description
		dyno test beds, crash test devices, multi-channel DAQ systems and mapping of MCD-3 to GDI. Includes files for schema definition and C-headers as well as description format and implementation examples. Also available via ISO 20242 but without the automotive-specific companion standards and code.
ASAM ODS	Open Data Services	Standard focuses on the persistent storage and retrieval of data independent of IT architecture. Is primarily used in conjunction with test automation systems. –It includes the specification of: –a base data model –derived application models –a model for relational databases for the physical storage of data –one API to access stored data in the databases –one API to access meta-information about the application model implemented in the database –two file formats (one in XML) for file-based data exchange –There are application models for vehicle geometry, NVH testing, test bed calibration data, bus data and testing workflows. Includes files for schema definition, interface definitions and description format examples. –In 2015, ASAM started a project to create a new version of the standard.

Table 4.3 ASAM COMMON standards (see [1])

Standard	Title	Description
ASAM LXF	Layout Exchange Format	Definition of an XML-based format for describing layouts for graphical content used by data post-processing applications and automated document generators. Includes definitions for master layouts, canvases and graphic objects such as images, charts, shapes and curves. Can contain embedded formulas resolved during runtime. Used particularly in conjunction with CEA.
ASAM MDF	Measurement Data Format	Definition of a block-structured and channel-oriented binary format for storing measurement data including descriptive meta-information. Allows for synchronization of data by time, angle, distance and index. Allows efficient storage of measurement data in real-time. MDF files can be referenced by an ODS database.

4.2 Measurement Data Acquisition

Physical values from units under test (such as torque, speed, pressure, temperature, gear shift, etc.) generally pass through several processing steps on their way from the sensor to their evaluation, as shown in Fig. 4.3.

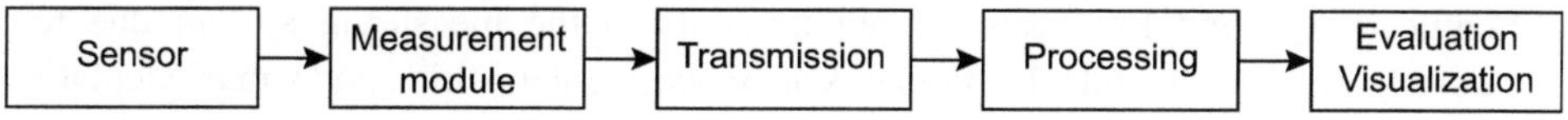

Fig. 4.3 Data flow in signal acquisition

Typically, the values are discretized as close to the measuring point as possible. With many solutions available on the market this already happens in the measuring module itself. Value discretization also entails time discretization. Depending on the analog-to-digital converter's measuring approach, this may be done either by sampling and holding the instantaneous value (sample-and-hold method) or by performing an integrating measurement by averaging the value over a sampling period. Value acquisition takes place either under conditions of chronological synchronism, angular synchronism or event-based.

To ensure correct representation and further processing, not only the digitized value but also its acquisition time t_s is of crucial importance. That is the reason why it is necessary to acquire the value along with its timestamp (time stamped data). Optionally, further properties can be included via modal criteria such as validity or availability. This information allows you to synchronize measured values from sources with differing sampling rates and values acquired on an event-driven basis.

4.2.1 Types of Data Acquisition

4.2.1.1 Time-Synchronized Acquisition

Critical to digital measuring systems is the ability to control the exact time the values are sampled. In time-synchronous acquisition, the time behavior of a defined periodical process (clock master) dictates the sampling rates. Periodic sources with maximum quality in terms of frequency stability and low jitter and with high reliability may act as clock master, for example a measuring module or the divided-down processor clock rate of the system PC.

A clock master has global validity across the entire measuring system. This is because the temporal consistency of data acquisition is essential to test bed systems, in particular, in order to represent causal relationships accurately. If the measurement has adequate time resolution (in other words, if the cyclic acquisition frequency is sufficiently high), it should equally be possible to represent events separated by time in the real world in the correct sequence and separated by time within the measuring system.

Depending on the measuring system's technical implementation and its analog-to-digital converters (ADCs), the system grabs a "snapshot" of a measurement quantity at a certain point in time (sample & hold) or measures the signal over a short measurement period and averages it (integrating ADC systems). Time synchronous acquisition, as applied by simple sensors on the test bed, for example for speed, torque, pressure or temperatures, has several advantages.

- Simple design and low resource requirements in the measuring system due to a carefully defined schedule that defines sensor access, acquisition and value calculation. These processes can be optimized when the system is designed.
- Allows deterministic procedure and therefore hard real-time.
- Simple detection of a failure in sensors, measuring modules, network nodes or computing steps as the time at which a new value occurs is known beforehand, making it easy to detect its absence.

Following the Nyquist-Shannon sampling theorem, a signal must be periodically sampled at twice the maximum signal frequency of the signals being sampled or higher. The maximum signal frequency is limited by implicit filtering (for example in the sensor) or explicit filtering (upstream of the ADC) (see also Sect. 4.3.3).

4.2.1.2 Angle-Synchronized Acquisition

In phase-driven processes, for example in-cylinder combustion in internal combustion engines, it is sometimes more effective to introduce phase-synchronous sampling and processing instead of measurement and calculation processes at equidistant time intervals. With angle-synchronized acquisition, for example, the crank shaft's top dead center triggers the start of an acquisition, which—unlike time-synchronized acquisition—leads to periods of varying time duration, depending on the speed.

This is how the acquired measuring quantity, e.g. the in-cylinder pressure, is acquired and represented (independent of speed) as a function of the crank angle, allowing it to be linked to the working cycle. The measuring system's acquisition and processing resources must be adequately designed to fit an acquisition rate that can handle the required maximum speed. For further details on ignition timing and indicating measurement, please refer to Sects. 3.3.11 and 3.3.16.

4.2.1.3 Event-Driven Acquisition

For some signal shapes, the time-synchronous acquisition method might be less suitable. Among these are events (aperiodic step functions) that do not occur periodically, such as gear shifting or a status message in the event of an error. Although in time-synchronous acquisition such events require high sampling rates (in accordance with the theorem) and consequently computing time and resources, there is no gain in information, which makes them highly redundant from an economical point of view. In this case, we recommend event-driven acquisition.

In contrast to time- or angle-synchronized acquisition, event-driven acquisition is acyclic, i.e. not in a previously defined periodicity, but instead triggered by an event. If variables are acquired driven by events, it must be possible to record the value itself as well as the event's timestamp to be able to process the data later on accordingly.

The unpredictability of events is a particular challenge when dimensioning the measuring system. In spite of this, there must be sufficient measuring and computing resources to ensure timely acquisition and processing, providing for time consistency in the measured values. For the implementation of the measuring systems, this means providing for sufficient acquisition and processing capacity to guarantee data consistency, even if events occur with a defined maximum temporal density (bursts).

4.2.2 Acquisition Time

To ensure correct representation and further processing, not only the digitized value but also its acquisition time t_s is of crucial importance. This time reference is made in various ways:

- Absolute time (also referred to as wall-clock time)
- Relative time referring to a certain event
- Sequential number in an ordered series of measured values

Assigning the time at which the value occurred to the cycle number usually requires cyclic sampling, i.e. periodic sampling in constant time intervals. For example, if you want to record the exact time a certain value occurred, it is sufficient to add the product of the value's cycle number and its cycle time to the time at which the recording started.

If two or more cyclically measured values are linked e.g. for calculating the mechanical power as product of speed and torque, for example, it is essential to ensure the use of values from the same cycle in order to prevent phase errors. A key factor in real-time data acquisition systems is to ensure this relationship even if the acquisition chains have varying lengths, for example if the measured values enter the system via different bus systems, or if they undergo different kinds of processing stages such as filtering or scaling.

In general terms: If two cyclically acquired quantities X and Y are used to cyclically calculate Z in accordance with:

$$Z[n] = F(X[n_1], Y[n_2]) \text{ where } n, n_1, n_2 \ldots \text{ cycle number} \quad (4.1)$$

Equation 4.1 Cyclic acquisition of quantities

One must ensure that it is true for every value Z[n] that it was calculated from input values where $n_1 = n_2 = n$.

Where the acquisition is represented over time, bear in mind that the assignment to a cycle number typically takes place in the central acquisition system. For this purpose, automation systems often use the clock master tick (CMT), i.e. a globally available, strictly monotonically increasing number of system cycles of the highest supported frequency since system start. If the signal is not captured at the highest but at a derived frequency instead, it is at all times possible to perform a conversion based on the division ratio.

It follows that the time at which a measured value occurs in an automation system results from:

$$t = (\mathrm{CMT}-\mathrm{CMT0}) * f_{\mathrm{ACQ}}/f_{\mathrm{CM}} + t_{s,\,aqc} \tag{4.2}$$

CMT	Registered clock master tick of the current value
CMT0	Registered clock master tick at the first acquisition of the value, i.e. at $t_{s,aqc}$
$t_{s,aqc}$	Absolute starting time of the first acquisition of the value
f_{ACQ}	Cyclic sampling frequency of the analyzed value, e.g. 1 kHz
f_{CM}	Frequency of the system's clock master, e.g. 10 kHz

Equation 4.2 Time of occurrence

The age of a measured value ("data age," also referred to as latency period) is the time difference between the actual acquisition in the measuring module and its availability in the system.

Data age is the accumulation of processing time in the measuring module, transmission time in the digital network (for example over a fieldbus) and transmission time in the central system, in other words until the value is available for further calculations, for example by its name.

4.2.3 Synchronization

System-wide temporal consistency is easiest to achieve by a uniform acquisition methodology and use of a global system clock. However, if different acquisition triggers are active within the same system, for example the concurrent acquisition of angle- and time-synchronized data, and if such data have to be calculated together, special methods are required for sampling, filtering or interpolation over time.

For this reason, high-quality data acquisition systems provide the capability of synchronizing all of these steps, from the start of the analog-to-digital conversion in the measuring module to the bus clock on the transmission medium (e.g. on the IEEE-1394 bus) or even the clock master tick (CMT) in the real-time system. The source providing this clock signal is called the clock master. As mentioned previously, the clock master is used as a reference to synchronize all further processing steps, so it is critical to use a clock that provides maximum quality in terms of frequency stability and low jitter.

In complex test bed systems it is not uncommon to combine different acquisition systems by using different bus systems. To ensure minimum data age and correct calculation with respect to the recording time in the real-time system even in this particular combination, it is crucial to synchronize acquisition and data transport with the clock master. For example, the automation system can act as a master on PROFIBUS-DP networks to determine start and duration of the bus cycles, thereby synchronizing them with a clock master. CANopen is equally capable of synchronizing all connected nodes with this clock master by transmitting a special cyclic message (SYNC message). For further details on the mentioned bus systems, please refer to Sect. 3.5.

Where technically possible, the system synchronizes one clock master with another by continuously readjusting a cycle time—a process which effectively results in a common, global clock master. This kind of synchronization is achieved, for example, by using NTP (Network Time Protocol) via LAN.

Other methods require continuous readjustment of the cycle time of periodic processes, similar to the PLL (phase-locked loop) method, either at a frequency ratio of 1 to 1 or at a predefined division ratio. Real-time Ethernet systems use methods that are even more precise, such as IEEE1588, to synchronize their nodes to within 1 µs.

Systems synchronized this way also ensure data consistency, i.e. there is always exactly one value available per cycle. If this is not guaranteed and the measuring and processing cycles fail to run in sync, for example, cycles may occur in which two values are acquired, and others in which no values are acquired at all. In sequential recordings this inconsistency will lead to missing values or, misleadingly, to duplicates.

4.2.4 Modal Criteria

Aside from the measured value and its time reference (timestamp), there are further attributes that can be acquired and processed as well, for example information about availability or the measuring method. A measured value's availability or its state depends on factors/properties such as correct measurement range, calibration status, etc. Using additional attributes, the measured value's state can be transferred, processed and represented as channel status together with the channel value, and is also included in the decision on whether the measurement is valid.

For example, the result of a cyclically calculated formula for durations during which an input value is marked as unavailable may likewise be marked as undefined; the same applies to the data recording or further possible data usages. However, the time range in which a value is valid or of relevance may also depend on other information. For instance, it is possible for data recordings only to start if certain criteria (triggers) apply.

4.2.5 Data Preprocessing

Data processing that takes place very close to acquisition, and therefore usually still within the real-time system, is referred to as online data preprocessing. Unlike offline processing or post-processing, which is based on the stored result data and usually does not take place until later, online processing requires the observation of time-related criteria. For instance, in cyclic data acquisition the calculated value must be generated cyclically too, in other words, the calculation time must remain smaller than the cycle time. In the best of cases, the value is processed in the same cycle in which it was acquired.

4.2.5.1 Scaling and Linearization

Sensors or transducers are devices that convert a measurable physical quantity "X" (pressure, temperature, distance, force, etc.) into an electrical quantity "Y" (voltage, current, resistance, frequency, etc.). A sensor's transfer characteristic defines such representation of a physical quantity as an electrical quantity.

$$y = f_i(x) \tag{4.3}$$

Equation 4.3 Scaling

This characteristic curve can be determined either by a mathematical analysis or by experimental measurement. The former results in a mathematical function, whereas in an experimental measurement the measuring points are stored as a curve and the intermediate values are determined by employing suitable interpolation methods (piecewise linear approximation, polynomials or fractional rational functions of a certain order). Inverting this transfer function allows calculating the physical value present at the sensor from the electrical quantity measured in the measuring module (e.g. voltage in mV).

$$X = \widetilde{f}_i(y) \tag{4.4}$$

Equation 4.4 Calculation of a physical quantity

The calculation is often performed internally inside the individual measuring modules, i.e. the already preprocessed measured value "X" is transferred to an automation system. In less sophisticated systems, the automation system performs this conversion as the first step of data preprocessing.

4.2.5.2 Anti-Aliasing

The Nyquist-Shannon sampling theorem states that a continuous signal with a maximum frequency f_{max} must be sampled at a frequency that has at least twice the bandwidth of f_{max}. This is necessary to reconstruct the original signal from the obtained time-discrete signal without any loss of information. Before sampling, any obtained signal components with a frequency higher than half the sampling frequency must be filtered out of the signal with a low-pass filter, because otherwise there would be artefacts. The sampling theorem was formulated by Harry Nyquist in 1928 and proved by Claude E. Shannon in 1949. In accordance with this theorem, the quantization that takes place in a measuring module generally includes limiting the bandwidth of the measured value by implicit filtering (for example inside the sensor) or explicit filtering (upstream of an analog-to-digital converter).

4.2.5.3 Unit Conversion

A measured physical quantity is represented by its numerical value and its unit. ISO/IEC 80000–1 specifies the use of the metric units defined in the **International System of Units**

(SI System). These form a coherent system of units. To make them easier to read and handle, SI prefixes may be used (e.g. milli, micro, kilo, mega).

Many sensors and measuring modules already deliver their values in SI units. On the other hand, it is not uncommon for values and their units to be adapted to the measuring range (e.g. voltages in μV or mV). Pressure sensors frequently provide their measured values in bars. However, the SI unit for pressure is specified as pascal (Pa). A similar situation exists with temperatures, which (at least in German-speaking countries) are generally given in degrees Celsius (°C) and not in the SI unit kelvin (K). So to ensure consistent representation in the system, it may be necessary to provide cyclic unit conversion. The procedure can be automated if meta information such as units are available for the intended processing steps. For example, if the sensor description specifies that measured values are transmitted in mV, and if the assigned input in the automation system references a quantity with the unit V, the required conversion (multiplication by 0.001) can take place without user interaction.

It is also not uncommon for users to want their measured results displayed in country-specific units on the graphical interfaces of their automation system. In Germany, for example, users will want to have vehicle velocity shown in km/h while in the United Kingdom or the U.S. the preferred unit will be mph. In this case, neither of the above will use the SI unit m/s for velocity. Similarly, pressure is frequently displayed in mmHG or in bars instead of the SI unit pascal. These are further cases that require cyclic unit conversion.

4.3 Signal Processing

4.3.1 Signal Generators

Many tests require predefined signal curves. Automation systems therefore provide functions that meet this requirement:

- Generation of a signal curve by means of a mathematical function (e.g. sine function, ramp functions, step functions). The quantity's value undergoes a change over time based on this function. By adjusting the parameters, e.g. for amplitude and frequency, the given function may be adapted to the user's needs.
- Processing of values that are saved in a file or an ASAM-ODS measurement. These are often data from a real-world test drive that are to be transferred to the test bed.
- Processing of curves that are predefined using special automation system editors (see Sect. 4.6.1).

Most automation systems treat quantities that contain values from the signal generators as measuring quantities and are equally capable of reusing them in calculations or storing them.

4.3.2 Calculation

On the test bed, individually measured quantities are used to calculate dependent quantities during runtime. Accordingly, the product of angular velocity and torque is power. Such calculations require high precision and real-time processing. Automation systems provide tools for such purposes, which are capable of defining simple as well as complex calculations. These calculations are stored in individual formulas. The automation system then executes these formulas on a cyclic or event-driven basis. Most calculations are performed in real-time.

To define the formulas, script languages are generally available such as Visual Basic Scripting or Python. Predefined mathematical, logical or statistical functions facilitate handling (see Fig. 4.4). To ensure flexibility in formula management across the entire test facility, the different formulas are frequently managed in a library, where they can be assigned to testing tasks or test bed equipment in a structured fashion.

Apart from calculating measured quantities directly, there are also formulas for correcting and compensating signal delay times. This ensures that the quantities are provided on the same time base, regardless of the signal source. With steady-state measurements, formulas can also be used to pre- and post-process quantities. In response to this very common use case, many automation systems already offer corresponding functions.

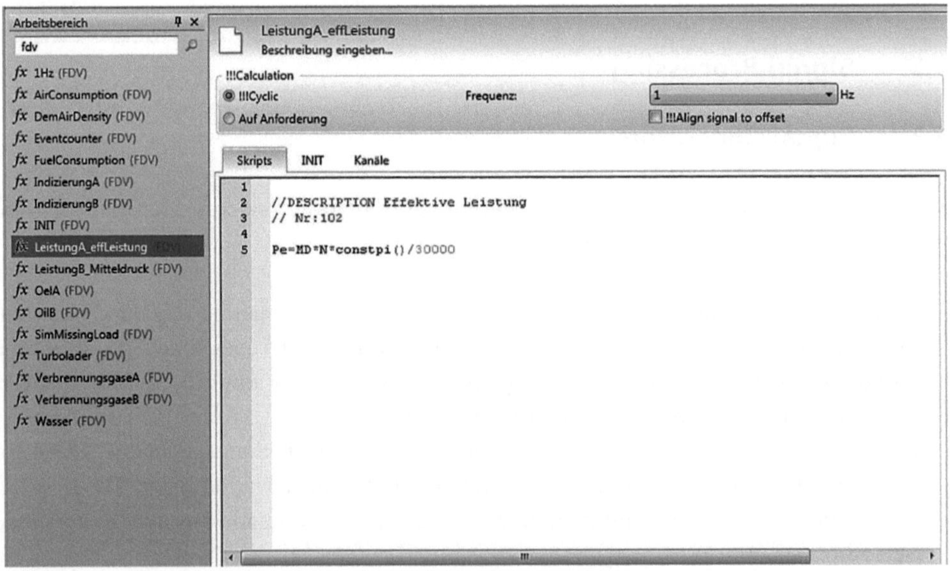

Fig. 4.4 Example of a formula creation editor in a test run library

4.3.3 Filtering

In accordance with the Nyquist-Shannon sampling theorem, the bandwidth must be restricted by applying adequate filters, even before quantization in the analog-to-digital converter. But even after digitization, it is occasionally necessary to apply a further filter, for example if you wish to visualize signals with high noise levels in order to evaluate the signal tendency, or when connecting two processing components with different sampling rates.

Digital filters are described by using what is referred to as the filter polynomial—the higher the order of the filter polynomial (also called filter order), the better the attenuation of signal components with frequencies in the stopband. On the other hand, higher-order filters are associated with longer calculation periods and a longer delay between source signal changes and target signal changes.

We distinguish between two main types of digital filters:

- IIR filters (infinite impulse response filters)
- FIR filters (finite impulse response filters)

Due to the anti-aliasing effect, an additional filter component is required for processing signals on different cycle frequencies. Figure 4.5 shows an example. Component A creates continuous values at a cycle frequency of 1 kHz; component B processes them, but at a cycle frequency of only 100 Hz. To avoid distortion caused by aliasing, a digital filter is inserted.

This can be achieved in an automated manner during the creation of the data acquisition chain by inserting a suitable filter component into the signal flow wherever the signal flow design leads to this kind of break in the frequency. Users who design explicit filters are additionally able to condition signals according to their specific requirements. But always bear in mind that filtering is only useful for continuous quantities. Discrete-value numbers, such as gear numbers in a transmission electronics system, must not be processed using filters.

4.3.4 Limit Monitoring

The high level of component integration makes it necessary for test beds to have a special monitoring system in place that ensures the safety of operating staff, unit under test and

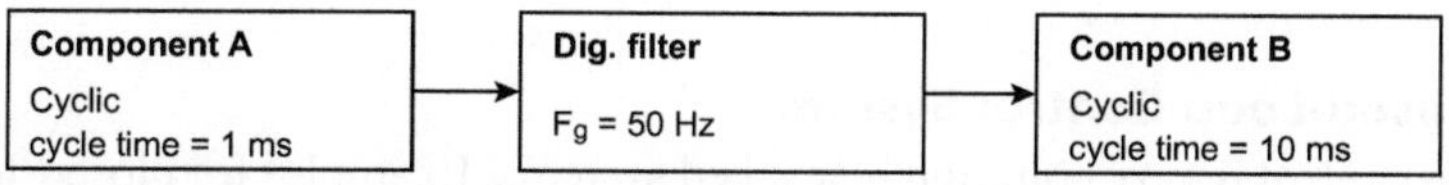

Fig. 4.5 Filter in the data acquisition chain

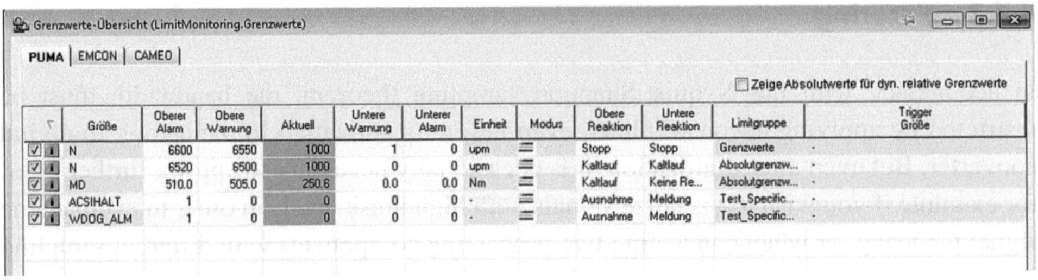

Grenzwerte-Übersicht (LimitMonitoring.Grenzwerte)

PUMA | EMCON | CAMEO

Zeige Absolutwerte für dyn. relative Grenzwerte

Größe	Oberer Alarm	Obere Warnung	Aktuell	Untere Warnung	Unterer Alarm	Einheit	Modus	Obere Reaktion	Untere Reaktion	Limitgruppe	Trigger Größe
N	6600	6550	1000	1	0	upm		Stopp	Stopp	Grenzwerte	-
N	6520	6500	1000	0	0	upm		Kaltlauf	Kaltlauf	Absolutgrenzw...	-
MD	510.0	505.0	250.6	0.0	0.0	Nm		Kaltlauf	Keine Re...	Absolutgrenzw...	-
ACSIHALT	1	0	0	0	0	-		Ausnahme	Meldung	Test_Specific...	-
WDOG_ALM	1	0	0	0	0	-		Ausnahme	Meldung	Test_Specific...	-

Fig. 4.6 Example of an online limit monitoring screen

test bed equipment. For this reason, a multi-level safety concept is a key feature in every automation system (see also Sect. 4.7).

Multi-level monitoring comprises three independent functions:

- Limit monitoring
- Test bed monitoring
- Post-mortem analysis based on a post-mortem recorder

We distinguish between limit values for the test bed, limit values for the unit under test and limit values for the test run. The specific “limit monitoring” function serves the safety of the unit under test and the test bed equipment. Measured variables (e.g. speed, oil temperature, etc.) are continuously compared with defined warning and alarm limits. Every time a limit value is violated, the system executes a defined event (e.g. the unit under test is stopped). Figure 4.6 shows the displayed limits for all monitored channels. Parameters such as the upper and lower warning and alarm limits, as well as the defined responses, are visualized. The user can modify these values online.

The Test Bed Monitoring function, as shown in Fig. 4.7, allows the monitoring of specific status bits on the test bed (e.g. door contact). As soon as an error is reported via a parameterized bit, predefined reactions can be initiated. Frequently this leads to the discontinuation of test bed operation for safety reasons.

A limit violation automatically activates the post-mortem recorder which saves the data of the previously selected measuring quantities for a predefined time, depending on the configuration. This means that the post-mortem recorder logs the limit value violation, helping to analyze the causes of test run failure.

4.3.5 General Controllers

4.3.5.1 Closed-Loop Control System

Closed-loop control is very common in test bed systems. Examples for closed-loop control are temperature control systems in fuel or coolant conditioning systems (see Sect. 3.2.4) or

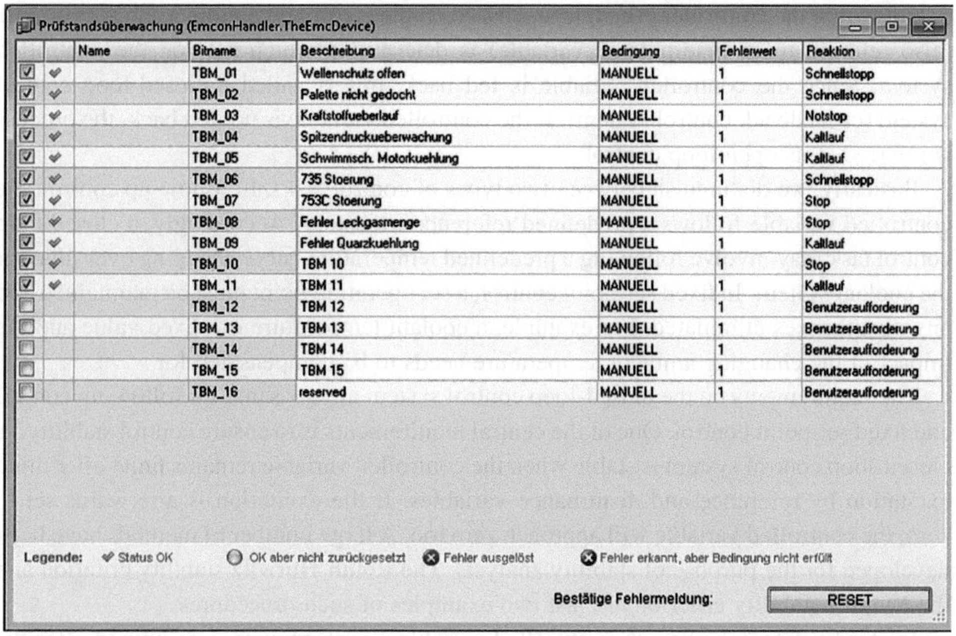

Name	Bitname	Beschreibung	Bedingung	Fehlerwert	Reaktion
	TBM_01	Wellenschutz offen	MANUELL	1	Schnellstopp
	TBM_02	Palette nicht gedocht	MANUELL	1	Schnellstopp
	TBM_03	Kraftstoffueberlauf	MANUELL	1	Notstop
	TBM_04	Spitzendruckueberwachung	MANUELL	1	Kaltlauf
	TBM_05	Schwimmsch. Motorkuehlung	MANUELL	1	Kaltlauf
	TBM_06	735 Stoerung	MANUELL	1	Schnellstopp
	TBM_07	753C Stoerung	MANUELL	1	Stop
	TBM_08	Fehler Leckgasmenge	MANUELL	1	Stop
	TBM_09	Fehler Quarzkuehlung	MANUELL	1	Kaltlauf
	TBM_10	TBM 10	MANUELL	1	Stop
	TBM_11	TBM 11	MANUELL	1	Kaltlauf
	TBM_12	TBM 12	MANUELL	1	Benutzeraufforderung
	TBM_13	TBM 13	MANUELL	1	Benutzeraufforderung
	TBM_14	TBM 14	MANUELL	1	Benutzeraufforderung
	TBM_15	TBM 15	MANUELL	1	Benutzeraufforderung
	TBM_16	reserved	MANUELL	1	Benutzeraufforderung

Fig. 4.7 Example of an online test bed monitoring screen

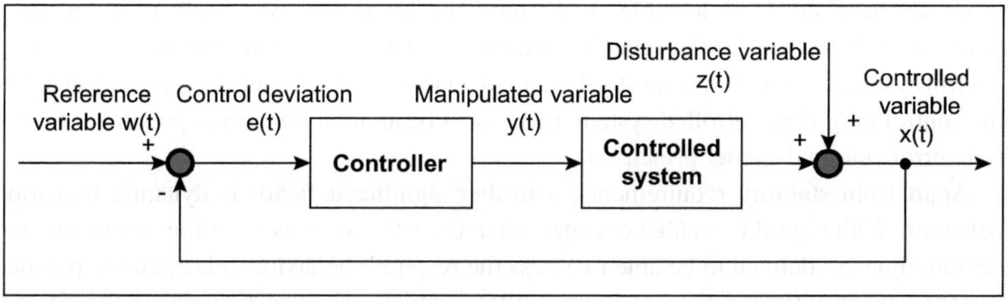

Fig. 4.8 Basic setup of a simplified closed-loop control system in accordance with DIN 19226

speed control systems in a test bed automation system. Figure 4.8 shows a schematic diagram of this type of closed-loop control system in accordance with DIN 19226.

A simplified closed-loop control configuration combines the actuator, the process to be controlled and the measuring element into the controlled system. The output variable of the controlled system may have a superimposed disturbance variable z(t), resulting in the controlled variable x(t). The controlled variable is fed back and compared with the reference variable w(t) by calculating the difference. The resulting deviation e(t) is the actual input variable for the controller.

Following the controller principle and based on the control deviation, a control variable (also referred to as manipulated variable) is determined which acts on the controlled system. Since the controlled variable is fed back, this is called a closed-loop control system (or feedback control system). If the controlled variable is not fed back, the process is referred to as open-loop control.

Basically, we distinguish between two types of control tasks. In follow-up control, the controlled variable follows a predefined reference trajectory. Accordingly, a closed loop control task may involve following a predefined temperature curve changing over time for the coolant system. In fixed set-point control, a set operating point must be maintained and any disturbances eliminated. For example, a coolant temperature is a fixed value and the impact of the changing ambient temperature needs to be compensated for.

The requirements on the closed-loop control system are the same for follow-up control and fixed set-point control. One of the central requirements is to ensure control stability. A closed-loop control system is stable when the controlled variable remains finite after finite excitation by reference and disturbance variables. If the excitation is afterwards set to zero, the controlled variable will approach zero too. A large number of methods have been developed for the purpose of stability analysis. The Routh-Hurwitz stability criterion and the Nyquist stability criterion are just two examples of such procedures.

In steady-state behavior, the controlled variable's performance is examined after the control processes have all subsided. For example, a sudden change in the reference variable initiates a control process for setting a new operating point. After the transient processes have subsided, a stable closed-loop control system will result in a stationary final value. If this final value matches the exact set reference variable, the closed-loop control system is said to have no steady-state deviation. For setpoint step changes in either the controller or the controlled system, there must be an integral term to prevent a steady-state error (internal-model principle).

Apart from stability requirements, a further significant factor is dynamic transition behavior. With regard to sudden changes, characteristics such as rise time, overshoot and settling time are defined to be able to assess the response behavior. It is common practice to resort to prototype curves of known filters, which are used to impress the desired behavior onto the closed-loop control system, based on the controller design. A controller's robustness is supposed to ensure that the closed-loop control system remains stable despite any changing conditions, for example due to aging or wear and tear. That is the reason why the design should always provide a safety margin for the controller. The keywords here are gain margin and phase margin.

Figure 4.8 shows the controlled-loop control system in the time domain. To analyze closed-loop control systems, the established practice is to transform the closed-loop control system to what we call the frequency domain using the Laplace transform. Systems described in the time domain by common differential equations are very easily expressed in the frequency domain using (linear) transfer functions. Many analysis and design methods are therefore defined in the frequency domain.

4.3.5.2 PID Controllers

PID controllers are very widely used in industrial control systems. As can be seen in Fig. 4.9, it consists of a proportional (P), an integral (I) and a derivative action (D) term. PID controllers are represented in two different ways, which can be converted into each other. If you have predefined controller modules in a test bed system, you always have to pay careful attention to the way the PID controller is implemented, so that no conversion errors are generated when transferring design parameters to the specific module of the automation system.

The top image results in the following equation in the time domain:

$$y(t) = K_P \cdot e(t) + K_I \int e(\tau)d\tau + K_D \cdot \frac{d}{dt} e(t) \qquad (4.5)$$

Equation 4.5 PID controller in the summation form

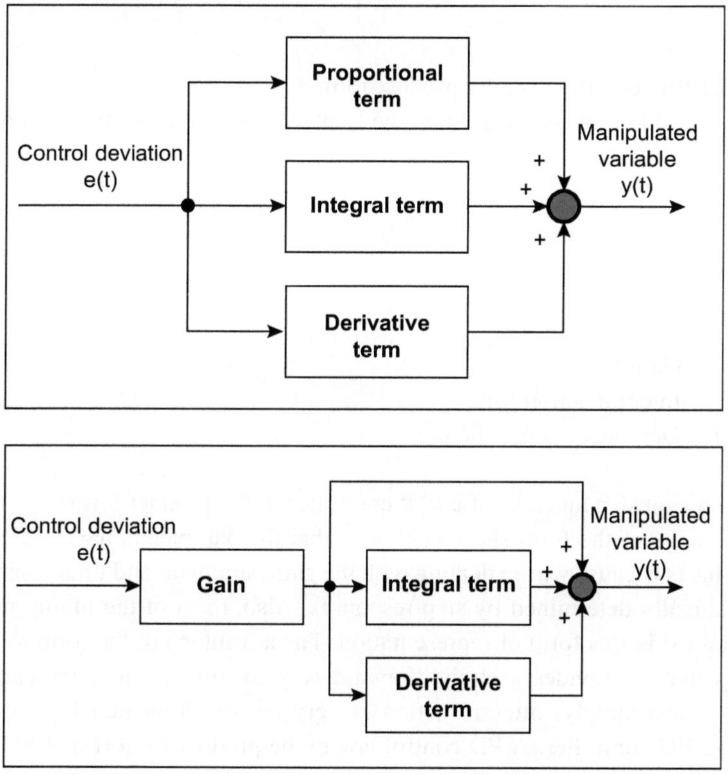

Fig. 4.9 PID controller shown in the two basic forms—summation form (*top*) and product form (*bottom*)

Equivalently, the transfer function in the Laplace domain of the PID controller is:

$$K(s) = \left(K_P + K_I\frac{1}{s} + K_D s\right) \tag{4.6}$$

where:

K_P Constant of proportional term
K_I Constant of integral term
K_P Constant of derivative term

Equation 4.6 Transfer function of a PID controller in the summation form

The bottom image results in the following equation in the time domain:

$$y(t) = K_R \cdot \left(e(t) + \frac{1}{T_I}\int e(\tau)d\tau + T_D \cdot \frac{d}{dt}e(t)\right) \tag{4.7}$$

Equation 4.7 PID controller in the product form

Equivalently, the transfer function in the Laplace domain of the PID controller is:

$$K(s) = K_R\left(1 + \frac{1}{T_I s} + T_D s\right) \tag{4.8}$$

where:

K_R Gain
T_I Integral action time
T_D(also T_V) Derivative action time

Equation 4.8 Transfer function of a PID controller in the product form

The advantage of the form shown below is that the parameters have a clear physical meaning. This is because we are dealing with the gain parameter and time constants which can be specifically determined by step responses. Also, most of the tuning rules lead to values contained in this form of representation. The advantage of the form shown above, however, is that it provides a straightforward way to implement modifications of the control law. Accordingly, integral action is very easily eliminated by setting $K_I = 0$, resulting in a PD controller. A PD control law in the product form (Eq. 4.8) can only be implemented approximately by adjusting the parameter T_I to a very large value.

Simply put, the different terms of a PID control law can be interpreted as follows: the proportional term takes into account the present, the integral term the past (based on the accumulation of the past) and the derivative term the prediction of the future (based on the rate of change).

For a PID controller there are a large number of empirical tuning rules to help you determine controller parameters. The most well-known tuning rules are named after their originators Ziegler and Nichols. If a system can be operated temporarily at the stability limit, the closed-loop control system is first closed using proportional-only control. The gain is steadily increased to the point at which, following a change in the set-point, the closed-loop control reaches its stability limit. The value of the gain parameter at the stability limit is the critical (or ultimate) gain K_R; the period of the sustained oscillation is the critical (or ultimate) period T_{crit}. The parameters for a PID controller can then be deduced from these two characteristic values.

$$\begin{aligned} K_R &= 0.6 \cdot K_{R,crit} \\ T_I &= 0.5 \cdot T_{crit} \\ T_D &= 0.12 \cdot T_{crit} \end{aligned} \tag{4.9}$$

Equation 4.9 Tuning rules for a PID controller using critical gain and critical period

Particularly if systems to be controlled are not oscillatory, it is sometimes possible to approximate the controlled system using a first-order system plus time delay, provided that the controlled system has a stable step response.

$$G(s) \approx \frac{K}{T_1 s + 1} \cdot e^{-T_t s} \tag{4.10}$$

Equation 4.10 Approximation of a controlled system for tuning a PID controller

These characteristic values can then also be used to determine the parameters of a PID controller.

$$\begin{aligned} K_R &= \frac{1.2}{K} \cdot \frac{T_1}{T_t} \\ T_I &= 2 \cdot T_t \\ T_D &= 0.5 \cdot T_t \end{aligned} \tag{4.11}$$

Equation 4.11 Tuning rules for a PID controller by approximation of a controlled system

Tuning rules are available for other characteristic step responses of a controlled system too and can be found in the relevant research literature (see [7]).

4.3.5.3 Real PID Controller

The PID controller according to Eq. 4.8 is, in terms of system analysis, a non-causal system because a pure derivative component cannot be implemented. That is the reason why a first-order low-pass filter is added to the derivative term.

$$K(s) = K_R\left(1 + \frac{1}{T_I s} + \frac{T_D}{Ts + 1}s\right) \tag{4.12}$$

where:

K_R	Gain
T_I	Integral action time
T_D(alsoT_V)	Derivative action time
T	Time constant of the low-pass filter

Equation 4.12 Transfer function of a real PID controller

The low-pass filter also ensures that any disturbance on the controlled variable is not amplified any further by the derivative term of the control law.

4.3.5.4 Discrete-Time PID Controller

If an automation system provides pre-manufactured controller modules, most of the required parameters are entered directly from a continuous design. Nevertheless, sometimes you will still have to implement the controller principles on a computer yourself. If you are provided with a controller from a continuous design, it is advisable to discretize it for a predefined sampling time T_S. This should generally be no problem if the sampling time is small enough compared with the closed-loop control system's critical frequency.

The easiest way to transform the continuous-time PID control law into a discrete one is by applying methods of numerical approximation. This involves approximating the continuous integral term by using the rectangle method (also called the midpoint or mid-ordinate rule) and the sum of sampled values and the derivative term over the difference quotient between two sampling values.

$$\begin{aligned} y(t) &= K_R \cdot \left(e(t) + \frac{1}{T_I}\int e(\tau)d\tau + T_D \cdot \frac{d}{dt}e(t)\right) \\ \frac{1}{T_I}\int e(\tau)d\tau &\approx \frac{T_S}{T_I}\sum\nolimits_{i=1}^{k} e(k) = y_I(k-1) + \frac{T_S}{T_I} \cdot e(k) \\ T_D\frac{d}{dt}e(t) &\approx T_D\frac{e(k) - e(k-1)}{T_S} \end{aligned} \tag{4.13}$$

Equation 4.13 Discrete-time approximation of integral and derivative action in a PID controller

This results in the discrete-time control law of an ideal PID controller.

$$
\begin{aligned}
y_I(k) &= y_I(k-1) + \frac{T_S}{T_I} \cdot e(k) \\
y(k) &= K_R \cdot \left(e(k) + y_I(k) + T_D \frac{e(k) - e(k-1)}{T_S} \right)
\end{aligned}
\tag{4.14}
$$

Equation 4.14 Discrete-time PID controller

Use of the slightly more precise trapezoidal rule to approximate the integral term results in the following approximation for the integral term:

$$
y_I(k) = y_I(k-1) + \frac{T_S}{2T_I} \cdot e(k) \tag{4.15}
$$

Equation 4.15 Approximation of the integral term using the trapezoidal rule

With this controller design, also referred to as a quasi-continuous design, there are further possibilities to derive a discrete-time control law from any given continuous-time transfer function. The most well-known transform is the bilinear or Tustin transform. For further details, please refer to the relevant literature (see [8]).

4.3.6 Evaluation

The productivity demands on modern test bed systems have risen sharply over recent years, due to the need for shorter development times at lower costs and raised quality requirements. The industry is responding to this trend by providing a high degree of automation, which in turn requires assurance of measuring data quality. In practice, however, it often happens in test facility operation that tests are faulty and have to be repeated due to the poor quality of the measured data. For example, it is not uncommon for a temperature sensor to fail or become disconnected. Another source of errors are drifting measuring signals. Given the growing number of required measuring points, it is becoming extremely difficult for a test bed operator to check and supervise the measuring equipment manually on a permanent basis.

4.3.6.1 Measured Data Plausibility Evaluation

As the early detection of useless or implausible measured data in manual and automated operation saves valuable test bed time, there is a need for continuous online data plausibility evaluation across all test bed measuring devices and subsystems during test bed operation—within reasonable time, with an acceptable amount of effort and without the need of having experts on site.

Some automation systems are extensible and allow the addition of special software components that are able to detect such erroneous measurements during test bed operation. Based on measuring signals they are capable of identifying states that lead to useless or implausible measured values. Such applications reflect expert knowledge and user experience.

The core of an integrated online diagnosis system is an integrated three-pillar model featuring the following groups of functions:

- System control
- Signal quality
- Signal plausibility

System control permits a function check of the connected measuring equipment. The second pillar, signal quality, deals with the signal curve and the statistical analysis of signal behavior (e.g. outliers, sudden discontinuities, noise, etc.). The third pillar examines the plausibility of the measured values. A plausibility check compares the measuring quantities with each other while checking the physical and user-defined rules (e.g. redundancies between identical or similar measuring quantities, temperature curves along an exhaust system, coherent mass balances, and many others). The plausibility pillar represents the system's ability to provide predefined plausibility methods based on expert knowledge and user experience gained in test bed operations over a long period of time (Fig. 4.10).

The three function groups cover all areas of test bed operation. Standstill diagnosis enables a quick check of individual channels even before the actual work has started. The results are presented in the form of a well-structured report. This kind of check is sometimes called "start-up check" or "good-morning check." The cyclic diagnosis function performs a continuous analysis on individual channels that are delivered by the automation system.

In case the system detects any contradiction between the observed measuring quantities, it immediately alerts the test bed operator, e.g. by a flashing warning triangle. More detailed information is provided by messages delivered in plain text. An integrated evaluation feature gives you a quick and easy way of visualizing the logged results in a clear and organized manner.

Some automation systems allow you to activate a measurement-synchronous mode while a steady-state step measurement is being performed at the same time. The purpose of this is to check the results of a steady-state step measurement for plausibility immediately after measurement completion.

4.4 Data Recording

An automation system acquires a wide variety of measuring quantities. These are used to calculate further quantities based on formulas, control algorithms, filters or simulation models. All of these quantities together contain the information that describes the way the unit under test performs along with the test conditions. For this reason, such values have to be saved and made available for post-processing.

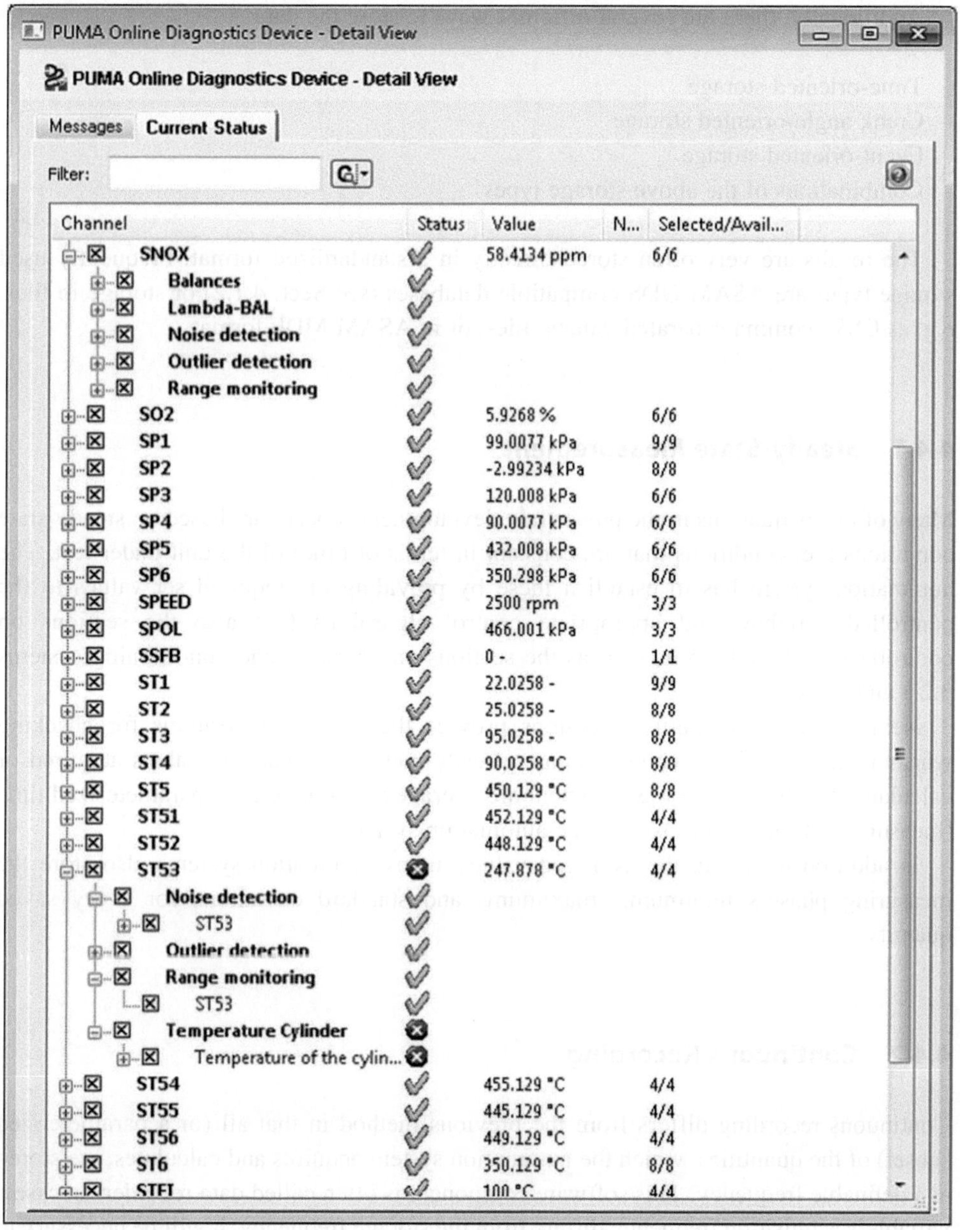

Fig. 4.10 Example of a detail view of the current diagnosis status

We distinguish between various different ways to record data:

- Steady-state measurement
- Continuous recording
- Post-mortem recording

Additionally, there are several different ways to store the data:

- Time-oriented storage
- Crank angle-oriented storage
- Event-oriented storage
- Combinations of the above storage types

The results are very often stored directly in a standardized format. Frequently used storage types are ASAM ODS-compatible databases (see Sect. 4.1.2) or storage in files, e.g. as CSV (comma-separated values) files, or in ASAM MDF format.

4.4.1 Steady-State Measurement

Many of the evaluations in the powertrain development process are based on steady-state conditions (i.e. conditions that are constant in terms of time) of the unit under test. The automation system has to establish these by providing the required set values to the controlled variables and appropriate control algorithms (see also the sections on controllers 4.3.5 and 4.5 as well as the sections on actuators and conditioning systems 3.2.3 and 3.2.4).

Next, the test automation functions provide the relevant algorithms for checking whether the steady-state conditions are already met. After that, the values acquired or calculated for this type of measurement are averaged over a specific, parameterized time (measurement time) and saved in the automation system.

In addition to saving the averaged values, many automation systems also store the measuring phase's minimums, maximums and standard deviations for every saved quantity.

4.4.2 Continuous Recording

Continuous recording differs from the previous method in that all (or a parameterized subset) of the quantities, which the automation system acquires and calculates, are stored at a definable frequency. This software component is often called data recorder. In case a variable's acquisition frequency differs from the storage frequency, it will be necessary to use the relevant anti-aliasing filters (see the section on anti-aliasing). Modern automation systems perform these anti-aliasing filter calculations automatically without the user having to parameterize them.

Many automation systems allow users to define when they want the recorder to start recording. Such systems distinguish between a start via a command in a test run, the occurrence of a certain parameterized event or immediate start at test start.

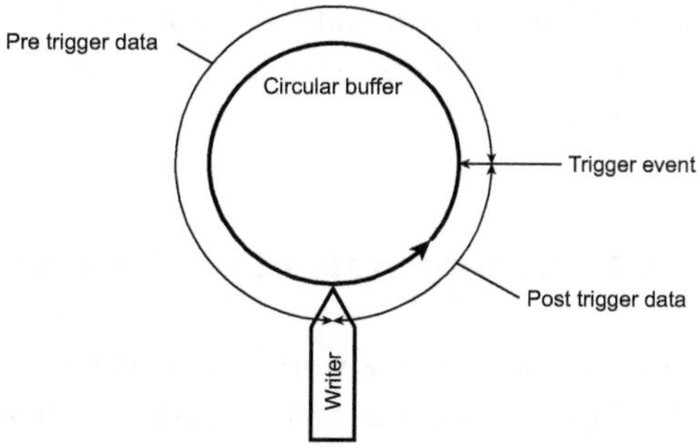

Fig. 4.11 Ring buffer recording of a post-mortem recorder

4.4.3 Post-mortem Recording

A special kind of relevance-oriented recording method is the post-mortem process as shown in Fig. 4.11. With this application, the decision to record and save cyclically acquired values before, during and after an incident is not triggered until a certain incident occurs, for example the unintended switch-off of a unit under test or the exceedance of a certain limit value.

Some automation systems enable post-mortem recordings by continuously recording selected signals in an infinite loop. Cyclically recorded measured values are copied to a ring buffer. Once the circular buffer's predefined length is full, the next, most recent value overwrites the oldest.

Upon the occurrence of the process defined as post-mortem trigger, the post-mortem recorder switches to the post-sampling phase. Unless the user has already stopped the process, the recorder continues to log data until the end of the post-sampling time. The contents of the ring buffer, consisting of pre- and post-trigger data, which indicate the preliminary events leading up to the incident and what happened immediately afterward, are stored as recorder results and made available for offline analysis.

4.5 Test Bed Control and Simulation

Control and simulation tasks have meanwhile become integral parts of a test bed automation system. To begin with, we will therefore address specific control concepts for different types of test bed configurations. Such concepts allow reproducing dynamic processes on a test bed that are (partially or fully) based on simulation. Some of the models used for such applications are quite simple. We will be dealing with some of these very straightforward vehicle models in detail. The reproduction of more complex driving maneuvers on the test bed demands a virtual test drive system for which detailed models of the driver, the vehicle and the environment are needed. For complete vehicle simulations, it is not uncommon to combine sub-models from a series of different domain-specific tools. We will therefore give a brief introduction to the concepts of integration platform and co-simulation. We will close our examination of this topic by discussing complex simulated communication structures based on residual bus simulation.

4.5.1 Control Systems on the Internal Combustion Engine Test Bed

Engineers all over the world regard the engine test bed as one of the standard tools in the development process of an internal combustion engine for performing development, testing and validation tasks. Work on an engine test bed aims at reproducing the load conditions for an internal combustion engine as realistically as possible. Vehicle testing involves the execution of frequently measured speed and torque profiles on the test bed. In such procedures, the values (speed, torque) measured during a road test drive are output to the test bed at sampling frequencies >100 Hz. Endurance strength testing based on this type of tests, in which speed and torque or speed and throttle position values, respectively, are applied to the unit under test, has been a well-established testing methodology for many years.

However, with units under test being smarter and more networked than ever before, this process is reaching its limits. For executing a gear shifting operation, for example, a transmission control unit uses information about driving conditions that it receives from other control units. The absence of a signal may lead to a change in the control unit behavior, which, in turn, will change the gear-shifting process and thus the test's boundary conditions. As a result, the required test run reproducibility is no longer ensured.

Distance-based testing with the aid of a virtual vehicle can help to solve this dilemma. For this purpose, a simulation model consisting of vehicle, environment and driver is used to calculate the internal combustion engine's load, which the torque setting device (dynamometer) then provides to the internal combustion engine (see Fig. 4.12). The aim of the simulation is to achieve as precise a reproduction of the load as possible, both in steady state and, most importantly, dynamic operation.

However, model-based testing on the test bed requires higher dynamics to handle the changes in the control variables. This can be achieved by combining a rigid shaft design between the unit under test and the dynamometer and a correspondingly adjusted control strategy.

4.5.1.1 Substitute Model

An engine test bed essentially consists of an internal combustion engine and an electric machine connected by an elastic shaft (see Fig. 4.13).

The internal combustion engine, the dynamometer's rotor and the elastic connection shaft constitute the mechanical components relevant to the control system. In simple terms, the system can be regarded as a dual-mass oscillator (see Fig. 4.14). With the exception of a dead time (or also called delay time), the electrical components, for example the dynamometer's electrical circuit, the converter and the torque and speed

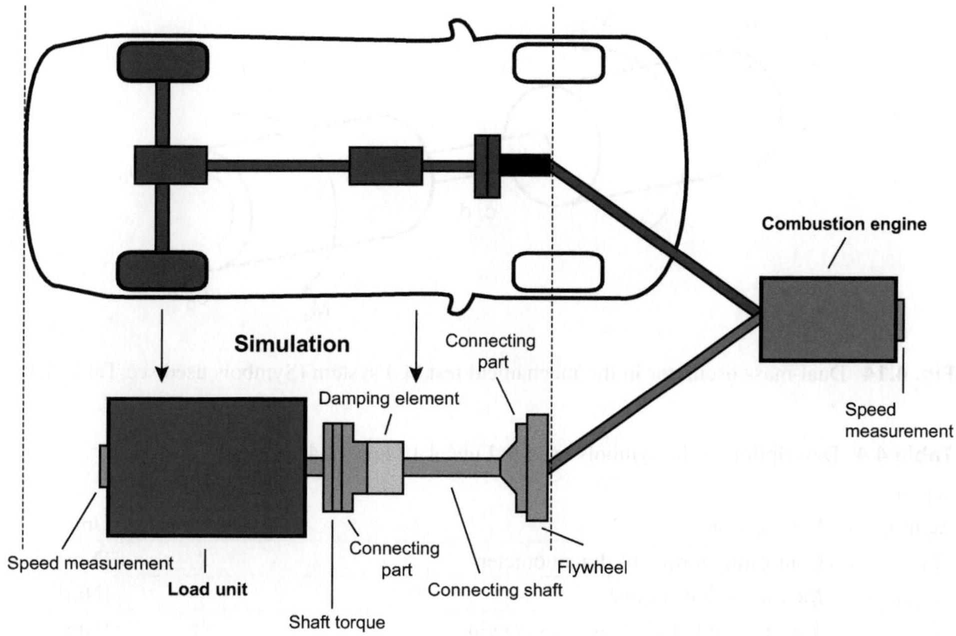

Fig. 4.12 Concept of an engine test bed

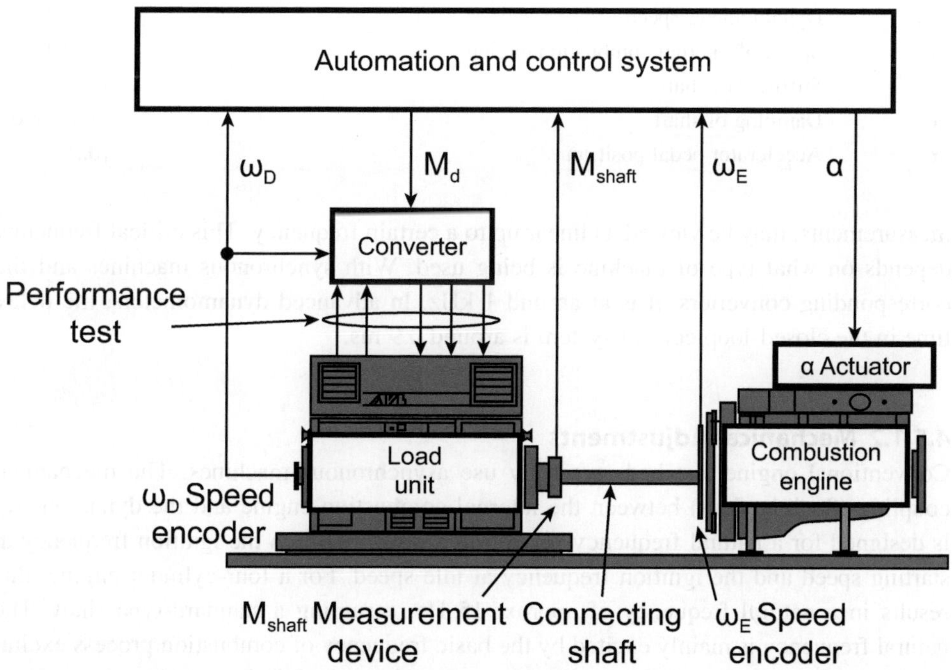

Fig. 4.13 Components of an engine test bed

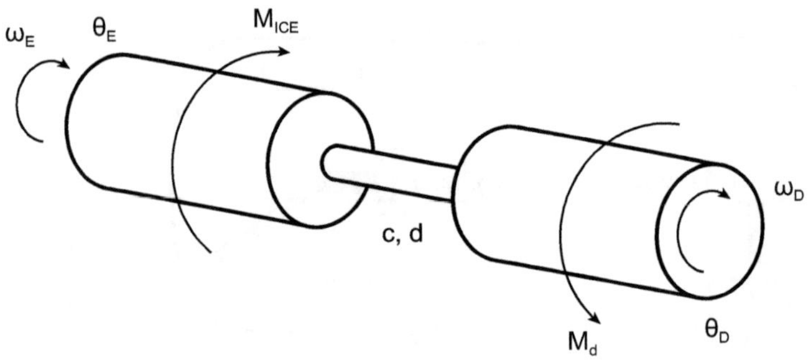

Fig. 4.14 Dual-mass oscillator in the mechanical test bed system (Symbols used see Table 4.4)

Table 4.4 Description of the symbols used in Figs. 4.13 and 4.14

Short term	Description	Unit
T_d	Controlling torque of dynamometer	[Nm]
T_{shaft}	Measured shaft torque	[Nm]
T_E	Torque of internal combustion engine	[Nm]
θ_D	Mass moment of inertia of the dynamometer rotor	[kg m^2]
θ_E	Mass moment of inertia of the rotating parts of the internal combustion engine	[kg m^2]
ω_D	Dynamometer speed	[rad/s]
ω	Speed of internal combustion engine	[rad/s]
c	Stiffness of shaft	[Nm/rad]
d	Damping of shaft	[Nm s/rad]
α	Accelerator pedal position	[0..1]

measurements, may be viewed as linear up to a certain frequency. This critical frequency depends on what type of machine is being used. With synchronous machines and the corresponding converters, it is at around 1 kHz. In advanced dynamometers, the delay time in the closed-loop control system is around 0.9 ms.

4.5.1.2 Mechanical Adjustments

Conventional engine test beds typically use asynchronous machines. The mechanical coupling (flexible shaft) between the internal combustion engine and the dynamometer is designed for a natural frequency settling in a range between the ignition frequency at starting speed and the ignition frequency at idle speed. For a four-cylinder engine, this results in a natural frequency of approx. 15 Hz, assuming a standard-type shaft. The natural frequency is mainly excited by the basic frequency of combustion process excitation. So with this design, the natural frequency is only excited during start/stop operations.

As a result, permanent (steady-state) operation is beyond the range of the overall system's natural frequency, thereby ensuring safe operation under normal operating conditions.

Given the very short period in which the system remains at natural frequency speed, and an additional damping element which is incorporated in the shaft setup, it is still possible to pass through the natural frequency speed during start/stop operations. However, being designed to allow a natural frequency of 15 Hz, the shaft does limit the bandwidth of the torque transferred to the internal combustion engine. To increase the bandwidth, it is necessary to raise the shaft rigidity ("stiff shaft") and thus the natural frequency of the test bed arrangement (see also Sect. 3.1.3).

Figure 4.15 shows the bode plots for the transfer functions from the dynamometer's controlling torque T_d to the shaft torque T_{shaft} for various shaft stiffnesses. Assuming a standard shaft, the bandwidth of a productively transferable torque ends below 10 Hz. In contrast, a stiff shaft allows transferring a torque dynamic up to 40 Hz. In this case, however, the test bed's natural frequency will shift into the internal combustion engine's operating range. This means that the internal combustion engine's ignition pulses will excite the test bed's natural frequency. Due to resonance effects, this can even lead to the destruction of the mechanical coupling system, particularly if the engine is operated in the natural frequency range at full load.

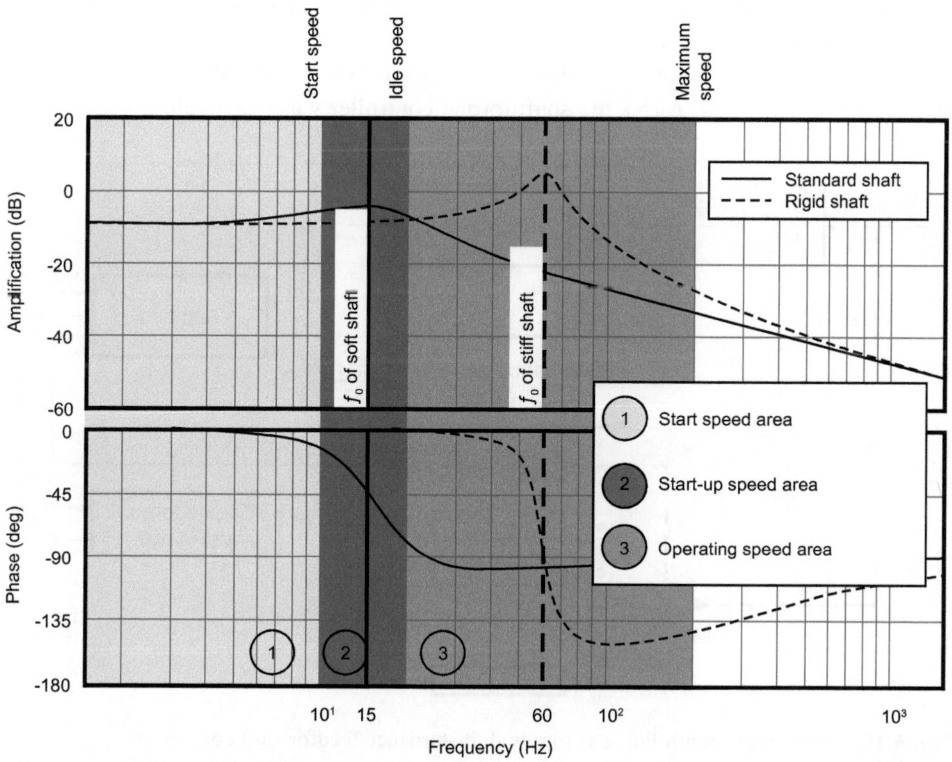

Fig. 4.15 Operating ranges of an engine test bed

An increase in the test bed's natural frequency places higher demands on the electric load unit's dynamic performance. The use of permanent magnet synchronous machines with a low moment of inertia allows accelerations up to 70,000 rpm/s, which are necessary to attenuate the resonance in the natural frequency range. It is additionally necessary to have a control strategy in place that is adapted to this type of operation.

4.5.1.3 Adapted Control Strategy

The operation of a test bed with a stiff shaft requires the use of a correspondingly adapted control strategy. As can be seen from Fig. 4.16, the control system essentially consists of a dynamometer speed controller, a combustion engine observer ($1/s \cdot \theta_{ICE}$), an engine torque estimator and a series of FIFO (first-in first-out) memories (see [9–12]). The goal is to apply the estimated and phase-corrected internal combustion engine torque to the shaft torque controller as a set-point. Disturbance feedforward control is based on a combustion engine torque $\widehat{T}_E$estimated by a Kalman filter.

By exploiting the periodical nature of the internal combustion engine's working cycle, it is possible to predict the disturbance variable in order to compensate for system delay periods. Since the system delay varies depending on the operating point, an estimation of the time necessary to calculate the disturbance variable is also required. This adaptation is implemented by using a series of FIFO memories with an adequate logic circuitry. The described method is protected by patents.

Figure 4.17 shows the result of shaft torque control on a real test bed at low internal combustion engine load. At 5 s the shaft torque controller was deactivated. The diagram

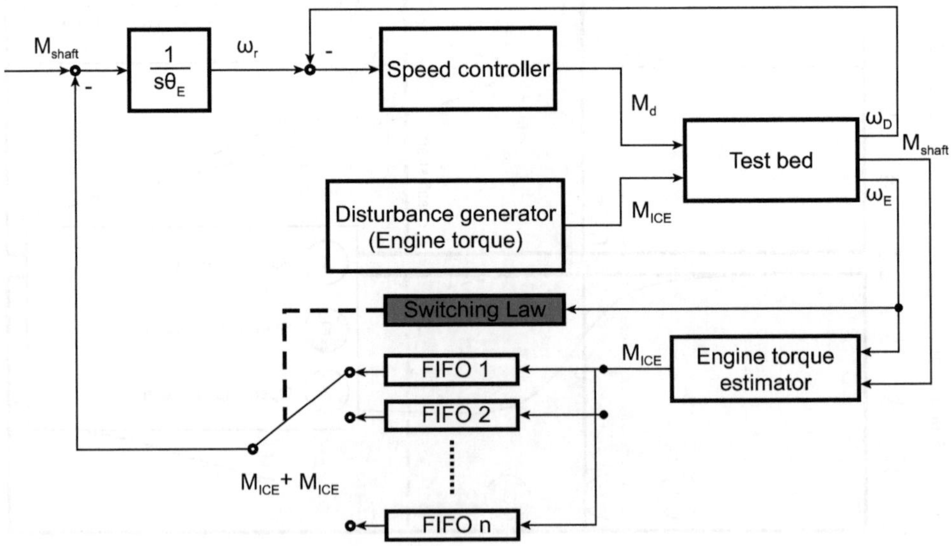

Fig. 4.16 Shaft torque controller structure with disturbance feedforward control

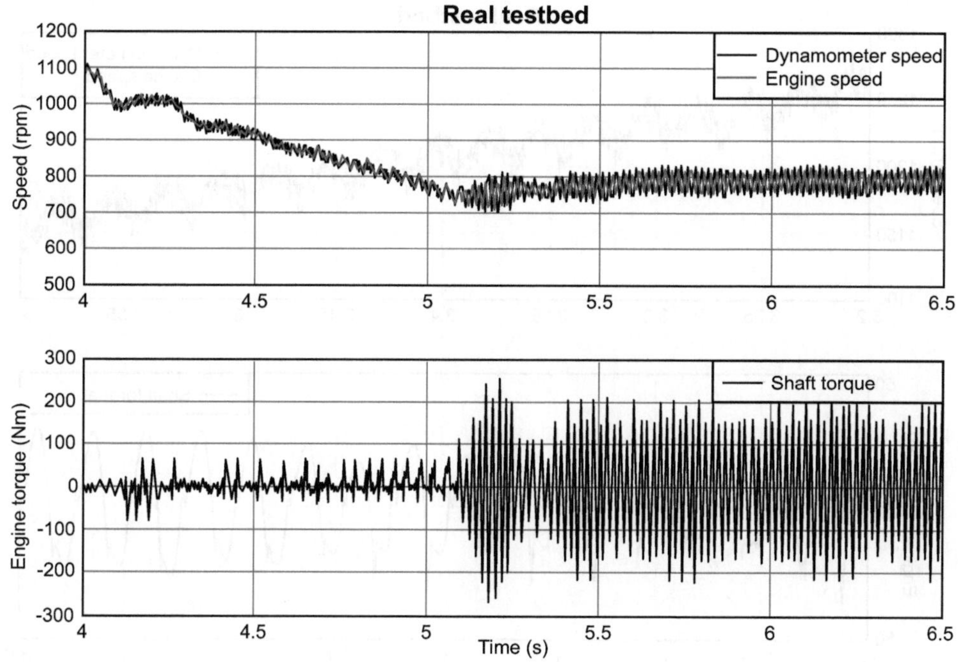

Fig. 4.17 Trend display for engine speed, dynamometer speed and shaft torque; at 5.1 s the shaft torque control is turned off

indicates a significant increase in the shaft torque because the speed of the dynamometer no longer follows the speed of the internal combustion engine. It is not possible to conduct this test under full load because resonance effects could lead to equipment destruction.

Figure 4.18 illustrates the shaft torque controller's transient response in close detail. The dynamometer's task is to follow the cyclic irregularities of the internal combustion engine up to a basic frequency of approx. 80 Hz (see the upper diagram in Fig. 4.18). This corresponds to keeping the angle of twist constant between the internal combustion engine and the dynamometer. A constant angular deflection means a constant shaft torque (see the lower diagram in Fig. 4.18). The shaft torque oscillates at a frequency of around ca. ±20 Nm. Assuming a shaft stiffness of approx. 11000 Nm/rad, this torque corresponds to an angle of twist of only ±0.1 deg. between the internal combustion engine and the dynamometer. For further details on control, implementation and application, please refer to [13–17].

4.5.2 Powertrain Test Bed Controllers

Typical test bed setups for powertrain testing can be seen in Figs. 2.50–2.53 (see Sect. 2.3). These test bed setups typically have a unit under test and up to four dynamometers

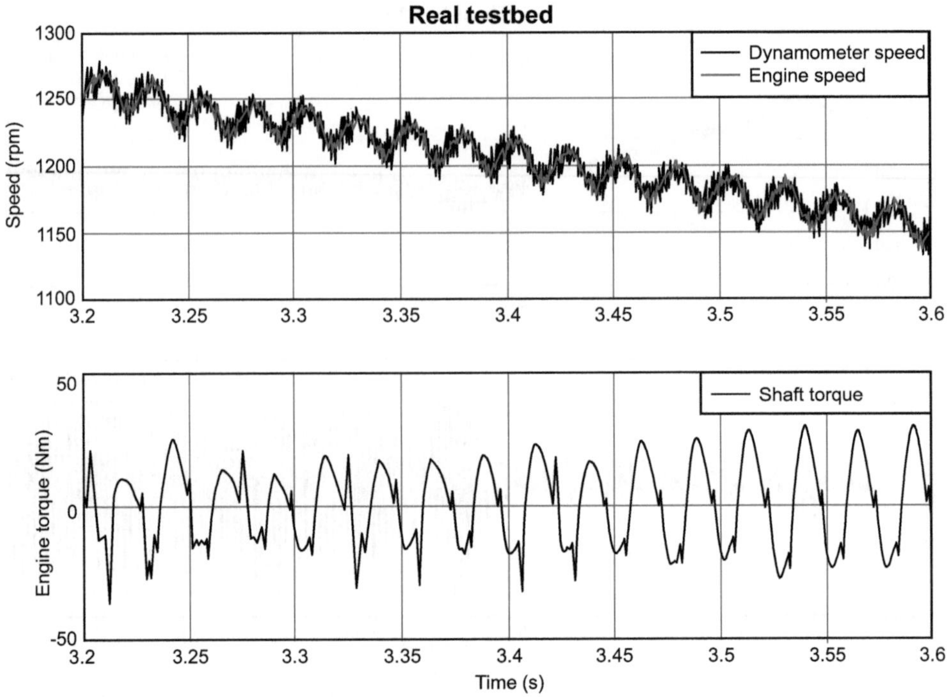

Fig. 4.18 Detail view of Fig. 4.17

Table 4.5 Control modes on the powertrain test bed

Load side	Drive side	Delta control
Speed	Pedal	Δn or ΔT
Torque	Pedal	Δn or ΔT
Speed	Torque	Δn or ΔT
Torque	Speed	Δn or ΔT
Speed	Speed	Δn or ΔT
Torque	Torque	Δn or ΔT

that apply loads to the axle shafts or the cardan shaft. To generate power, it is additionally possible to replace the original drive system (internal combustion engine or hybrid drive) by an electric drive unit (see Sect. 3.2.1). The major challenge associated with controlling these types of dynamometers is to achieve synchronicity in up to four individual machines to ensure that the powertrain under test is not subjected to unintended stress.

For this purpose, a powertrain test bed uses coupled control modes. The control mode determines which variables are controlled from the load side (dynamometers) and which from the drive side (internal combustion engine or electric drive unit). As can be seen in Table 2.1, a control mode consists of two sub control modes for load and drive side, as well as an optional delta control between the right and left wheel and/or between front and rear (Table 4.5).

4.5.2.1 Concept of Modal Control

Modal control is used particularly in multivariable control systems. This concept aims at transforming individual variables into independent variables by applying a suitable transformation. This coordinate transformation delivers decoupled control loops and thereby reduces the design of a multivariable control loop to that of several non-interacting individual control loops. On the powertrain test bed we call this a sum-delta-control structure.

The individual modes of modal control in the sum-delta-control structure are:

- Sum torque
- Center speed (average between front/rear and left/right side speeds)
- Delta torque
- Delta speed

When handling the difference values (delta values), the reference point must always be taken into account:

- Difference between left and right machine at the front axle
- Difference between left and right machine at the rear axle
- Difference between front and rear axle

Accordingly, a four-wheel drive unit under test with a fixed load distribution x between front and rear axle requires the following transformation rules:

$$\begin{aligned}
&\text{Sum torque}: T_{sum} = T_{f,l} + T_{f,r} + T_{r,l} + T_{r,r} \\
&\text{Delta torque front}: \Delta T_{front} = T_{f,l} - T_{f,r} \\
&\text{Delta torque rear}: \Delta T_{rear} = T_{r,l} - T_{r,r} \\
&\text{Delta torque front/rear}: \Delta T_{front/rear} = (1-x)(T_{f,l} + T_{f,r}) - x(T_{r,l} + T_{r,r}) \\
&\text{Center speed}: n_{center} = (n_{f,l} + n_{f,r} + n_{r,l} + n_{r,r})/4 \\
&\text{Delta speed front}: \Delta n_{front} = n_{f,l} - n_{f,r} \\
&\text{Delta speed rear}: \Delta n_{rear} = n_{r,l} - n_{r,r} \\
&\text{Delta speed front/rear}: \Delta n_{front/rear} = (n_{f,l} + n_{f,r}) - (n_{r,l} + n_{r,r})
\end{aligned} \tag{4.16}$$

Equation 4.16 Torque calculation

On the one hand, this transformation must be carried out for the modal controller's actual values in the sum delta structure. On the other hand, the set values of the modal controllers have to be transformed back to variables for the individual load units. Figure 4.19 shows the example of a basic control structure consisting of a main and delta controller on a four-wheel drive (4WD) test bed with two load units.

The advantage of modal control is that the couplings between the drive and load side are only reflected in the main control loop. That is the reason why the main controller's

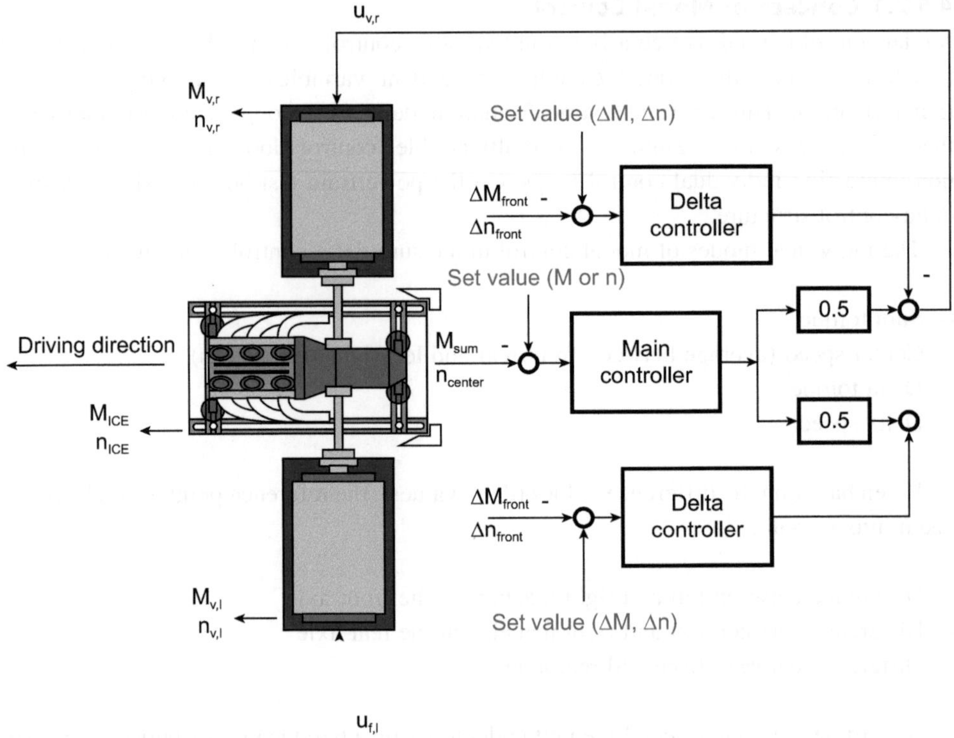

Fig. 4.19 Example of controller structure in a 4WD test bed

design is of particular significance. The delta controller has the task of controlling the desired load distribution ΔT or the desired delta speed Δn, for example when cornering.

4.5.2.2 Dynamic Follow-Up Control

The tasks in powertrain testing are seldom steady-state tests. The tests usually executed are predefined dynamic profiles for speed and torque. In such cases, the control structure has to be designed for dynamic follow-up control. It is common practice to use PI (D) controllers for this purpose (see Sect. 4.3.5). This control structure ensures that there is no steady state error during fixed set-point control. In tracking control, however, the lag error results in a dynamic deviation, which can be reduced or compensated for by using suitable additional control measures. Often, a control structure with main control and additional disturbance feedforward control is used.

The controller structure shown in Fig. 4.20 is able to improve the dynamic performance of the closed-loop control system without affecting stability or robustness. Disturbance feedforward control compensates for the impact of speed and torque from the drive side (internal combustion engine or electric drive unit) on the load side. The precontrol essentially serves to compensate for the lag error.

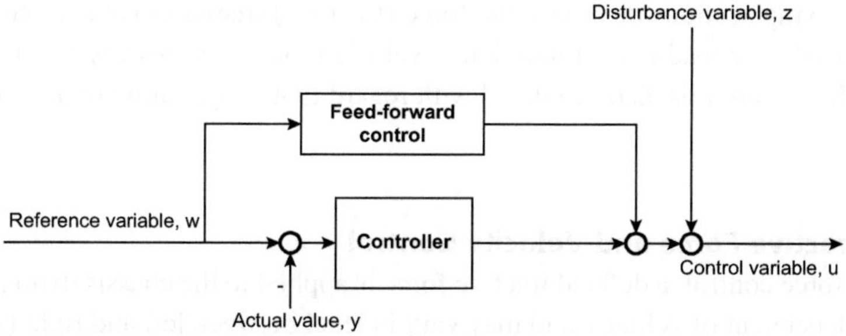

Fig. 4.20 Controller structure with disturbance feedforward control

Table 4.6 Control modes on chassis dyno test beds

Name of control mode	Test bed control	Unit-under-test control
RG/Alpha	Road gradient simulation	Alpha (accelerator pedal)
RG/n	Road gradient simulation	Engine speed
RG/v	Road gradient simulation	Roller velocity
RG/Driver	Road gradient simulation	Real driver
F/Alpha	Roller tractive force	Alpha (accelerator pedal)
F/n	Roller tractive force	Engine speed
F/v	Roller tractive force	Roller velocity
F/Driver	Roller tractive force	Real driver
v/Alpha	Roller velocity	Alpha (accelerator pedal)
v/Driver	Roller velocity	Real driver
n/Alpha	Engine speed	Alpha (accelerator pedal)

4.5.3 Control on the Chassis Dyno Test Bed

On a chassis dynamometer (chassis dyno test bed) the entire vehicle is physically present. However, as the vehicle on the chassis dyno test bed is not in motion, the simulation environment has to cover the effect of acceleration, uphill and downhill driving or even some of the environmental conditions, such as temperature, air pressure or humidity.

For operation with a real driver or a driving robot, chassis dyno test beds have three basic control modes:

- Force control
- Velocity control
- Road load simulation

For special applications, e.g. calibration tasks, further control modes may be available. Table 4.6 gives an overview of control modes available on chassis dyno test beds.

The following sections deal with the force (tractive force) and velocity control (basic control modes). For road load simulation, a vehicle model is necessary that is introduced in Sect. 4.5.4, where it is also discussed with regard to its application on the chassis dyno test bed.

4.5.3.1 Tractive Force and Velocity Control

In tractive force control, a defined tractive force is applied to the chassis dyno rollers. This force is independent of velocity and may vary in level between left and right (and/or front and rear) rollers if there are more than one dynamometers present. With velocity control, the opposite applies. In this case, the velocity is set regardless of the tractive force. If there are multiple dynamometers, it is additionally possible to adjust either a global center velocity or different velocities for each wheel of the vehicle. As these two control modes are used to set generic demand values and there is no vehicle or road load simulation used, these are independent of the unit under test in view of calibration and parameterization.

4.5.3.2 Tractive Force and Velocity Measurement

For all control modes the measurement of tractive force and velocity is an essential requirement. Velocity in most cases is measured using incremental or sine-cosine encoders on the engine shaft (see also Sect. 3.3.7). From the velocity it is then possible to derive the acceleration. In most configurations, tractive force is measured using cradle-mounted machines in conjunction with a load cell (see Fig. 4.21). The challenge here is to determine the tractive force at the contact between the tires and the surface of the chassis dyno rollers.

The force is measured directly using a load cell (F_{meas}). Based on the corresponding ratios of the levers and the radii it is possible to project this onto the force where the tires make contact with the rollers (F_{tract}). For static cases, this is sufficient. In transient, dynamic cases, however, further corrective forces need to be considered. On the chassis dyno test bed, for instance, there are velocity-related (in climate chambers also temperature-related) losses (F_{losses}), which are generally described in the form of a second-order polynomial. When the roller accelerates or decelerates, further inertia losses

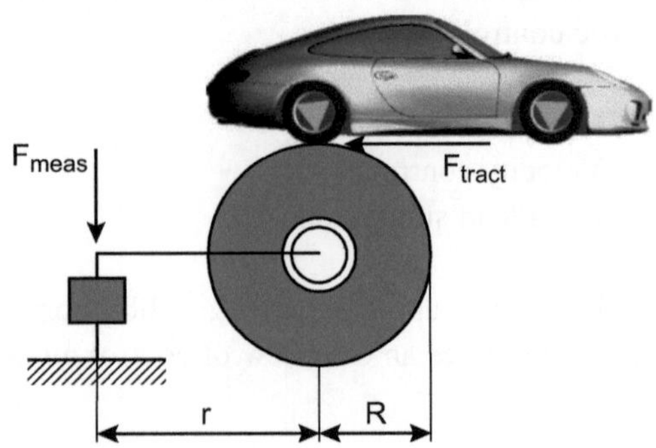

Fig. 4.21 Schematic diagram on tractive force acquisition

need to be taken into account ($F_{B,roller}$). For the calculation of tractive force in tire-roller contact, this results in the following equation:

$$F_{tract} = F_{meas} \cdot \frac{r}{R} + F_{losses} + F_{B,roller} \tag{4.17}$$

Equation 4.17 Tractive force

For further information, please refer to [18–20].

4.5.4 Simple Vehicle Model

Meaningful mathematical vehicle models allow simulations and parameter studies long before the first vehicle prototype has been engineered. Combining real components with simulation models ("hardware in the loop") allows the operation and testing of units under investigation at a very early development stage.

One of the first steps in modeling is the selection of a suitable substitute system. In the simplest of cases, substitute systems may consist of individual mass points, but they may also range from spatially extended rigid bodies or linked multi-body systems to finite-element systems.

By choosing the best suited substitute system for the problem at hand, engineers are able to break down complex structures and isolate central physical effects. The associated reduction of parameters and structures makes the system more manageable and often even accessible to analysis. These problem-specific models only open up very restricted or even local operating areas (e.g. straight-ahead driving, steady-state circular-course driving, etc.). A substitute system is well chosen if it exhibits a high effect in terms of a specific, chosen phenomenon.

A key attribute of simulation models interacting with real hardware components is their real-time capability. It must be ensured that the calculations are carried out in time, i.e. within a defined maximum response time, and also concurrently, i.e. quasi simultaneously using a real-time control system. Due to the limited computation resources on a test bed, the first models accepted by the community were very simple. For certain applications these simple models are still being successfully used today. Among these are the models "1D point mass model for longitudinal dynamics," "torsional vibration in the powertrain" and "slip simulation," which we will discuss in the following sections.

4.5.4.1 1D Point Mass Model for Longitudinal Dynamics

The model approach most widely used on powertrain and chassis dyno test beds is the so-called 1D point mass model for longitudinal dynamics. In this model, the vehicle mass is concentrated at the vehicle's center of gravity and the direction of movement is only described in one dimension (1D) for longitudinal dynamics. The equation for vehicle

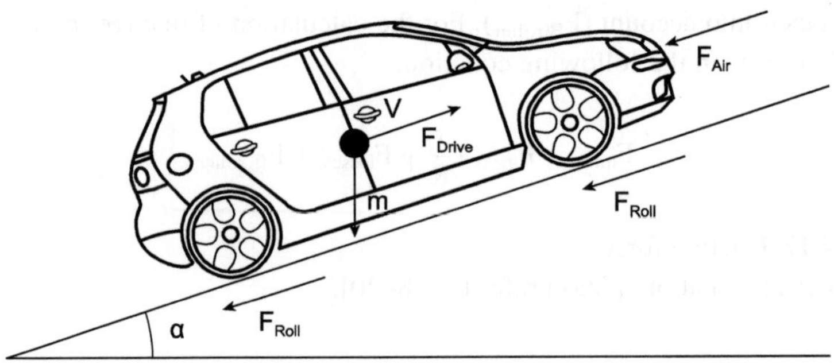

Fig. 4.22 Forces acting at the vehicle's center of gravity

simulation results from the conservation of momentum law, taking into account motive force, air resistance, rolling resistance and gradient resistance as shown in Fig. 4.22.

$$F_B = F_{Drive} - (F_{Air} + F_{Roll} + F_{Gradient}) \tag{4.18}$$

Equation 4.18 Equation of motion

The formulas for the aforementioned forces are listed below:

$$\text{Rolling resistance}\, F_{Roll} = f_r \cdot m \cdot g \tag{4.19}$$

where

f_r Rolling-resistance coefficient
m Vehicle mass
g Gravitational acceleration

Equation 4.19 Rolling resistance

$$\text{Air resistance}\ F_L = c_d \cdot A \cdot \frac{\rho_{air}}{2} \cdot v^2 \tag{4.20}$$

where

c_d Drag coefficient
A Projected frontal area of the vehicle
ρ_{air} Air density
v Relative velocity

Equation 4.20 Air resistance

$$\text{Acceleration resistance}\, F_B = m \cdot a + \Theta_{ges,R} \cdot \frac{\ddot{\varphi}_R}{r_{dyn}^2} \qquad (4.21)$$

where

m	Vehicle mass
a	Vehicle longitudinal acceleration
$\Theta_{ges,R}$	Inertia of all rotating vehicle parts (in relation to driving wheels)
$\ddot{\varphi}_R$	Angular acceleration of driving wheels
r_{dyn}	Dynamic tire radius

Equation 4.21 Acceleration resistance

$$\text{Gradient resistance}\ F_{gradient} = m \cdot g \cdot sin\,(\alpha_s) \qquad (4.22)$$

where

m	Vehicle mass
g	Gravitational acceleration
α	Gradient angle

Equation 4.22 Gradient resistance

4.5.4.2 Torsional Vibration in the Powertrain

Vehicle mass simulation has a fundamental impact on the results achievable on the test bed. In the past, flywheels on the powertrain or the chassis dyno test bed reproduced most of the missing vehicle mass. On the powertrain test bed, these flywheels were always coupled directly to the axle shafts of the vehicle under test. This ensured that the lowest dominant natural frequency (dual-mass oscillator engine-vehicle) on the test bed correlated with that of the vehicle.

However, where damping performance is concerned, the test bed differs from the vehicle because the components primarily responsible for damping—the tires—are missing. On the dynamic engine test bed or on a HiL test bed, however, there are multiple powertrain components that are not physically present. That is the reason why pure vehicle simulation is not sufficient. It is often necessary to reproduce the powertrain's vibration behavior, which is done by using a suitable dynamic model of the powertrain.

It is generally possible to represent the powertrain as a multi-mass oscillator. These masses may be, for example, internal combustion engine, clutch, transmission, cardan shaft, axle differential, axle shafts, wheels and vehicle. The masses are coupled to each other by linear and non-linear spring-damper elements (Fig. 4.23).

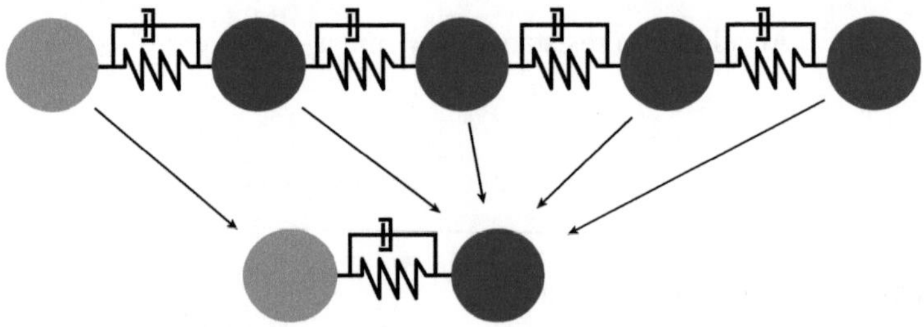

Fig. 4.23 Substitute model of the powertrain

Due to the model's real-time capability requirement, the multi-mass model is usually reduced to a dual-mass model with the dominant masses internal combustion engine and vehicle, as well as equivalent stiffness. This allows reproducing the powertrain's characteristic jerking oscillations (typically frequencies between 2 and 10 Hz) on a dynamic engine test bed. The equations of motion for the two masses are derived from the angular-momentum conservation law. The following equations are indicated for linear spring and damper principles:

$$\begin{aligned} \theta_1 \frac{d\omega_1(t)}{dt} &= T_{Input} + c(\phi_2 - \phi_1) + d(\omega_2 - \omega_1) \\ \theta_2 \frac{d\omega_2(t)}{dt} &= -T_{Output} - c(\phi_2 - \phi_1) - d(\omega_2 - \omega_1) \end{aligned} \tag{4.23}$$

where

Θ_1, Θ_2	Moments of inertia of the two flywheels
ω_1, ω_2	Speeds of the flywheels
c	Torsional stiffness
d	Damper constant
M_i	Drive and load torques acting from the outside
φ_1, φ_2	Angles of rotation

Equation 4.23 Equations of motion for a dual mass oscillator

4.5.4.3 Slip Simulation

Experience has shown that dynamic powertrain test beds with mechanical or electrical mass simulation and a simple model for longitudinal dynamics are not suitable for testing smart four-wheel drive systems, torque vectoring systems or hybrid drives using vehicle simulation. In these powertrain concepts, the combination of different kinds of control units for vehicle stabilization (ESP), transmission control (TCU) or brake pressure control (ABS) have a crucial influence on the speed and torque level for each wheel.

The application of delta speed or delta torque to the front or rear axle means that the test bed control system is acting against the unit-under-test control system. A driving dynamics model solves this problem. The multi-dimensional vehicle model with tire simulation calculates the tractive force for each of the wheels. The manner in which the tire makes contact with the road surface has a significant effect on each of the wheel's tractive force. This is described by the μ slip curve shown in Fig. 4.24.

The slip curve describes the transmittable tractive force in the wheel's longitudinal and lateral direction and contributes significantly toward driving stability, with longitudinal and lateral forces depending on each other. The resulting overall force is unable to exceed the maximum friction force. The individual load units apply the longitudinal forces to the powertrain while the lateral forces are only taken into account in the simulation model. For test drives on different kinds of road surfaces to be reproduced realistically on the powertrain test bed, the load units' polar moment of inertia must be comparable to that of the wheel. This ensures that the resonance behavior on the test bed is identical to that of the vehicle.

A model with six degrees of freedom in a vehicle-fixed coordinate system as shown in Fig. 4.25 is useful for vehicle simulation. The model describes the following vehicle movements:

- Longitudinal motion (x-direction)
- Transversal motion (y-direction)

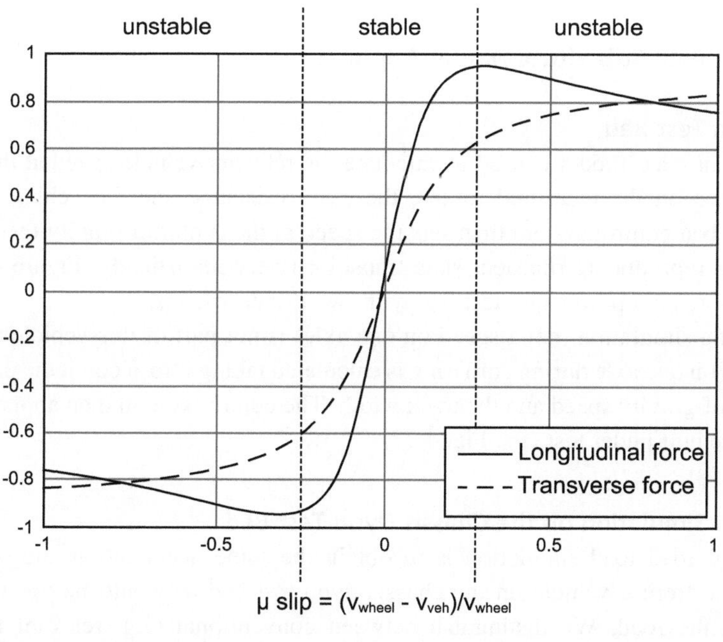

Fig. 4.24 Slip curve during vehicle acceleration

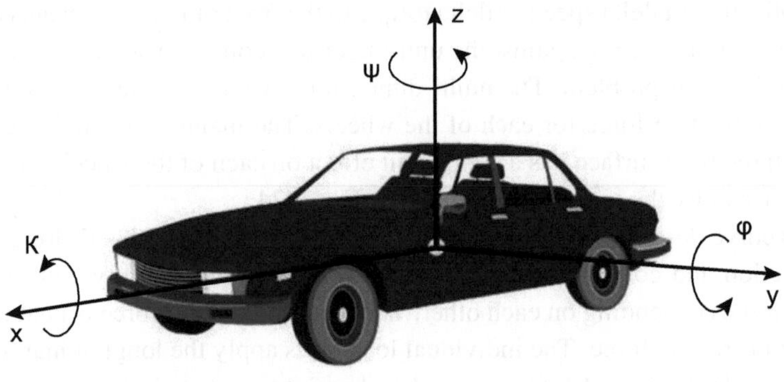

Fig. 4.25 Vehicle-fixed coordinate system

- Axial motion (z-direction)
- Roll (rotation around the x-axis)
- Pitch (rotation around the y-axis)
- Yaw (rotation around the z-axis)

A set of momentum balance equations for all six degrees of freedom with the forces and moments allows deriving a corresponding mathematical model [21]. A vehicle model that takes into account the tire-road contact is a prerequisite for maneuver-based testing on dynamic or high-dynamic test beds (see Sect. 4.5.5).

4.5.4.4 Vehicle Substitute Model Applications

Powertrain Test Bed

The 1D point mass model is used to calculate the relevant vehicle speed at the center of gravity based on the measured torque, the road resistance and the vehicle mass. The engine test bed control system then sets the speed at the center of gravity (center speed). This allows reproducing transient state transitions very realistically. Figure 4.26 shows this procedure on a powertrain test bed with two driven wheels.

Cornering simulation (delta speed on one axle) is not part of this vehicle model. The delta speed on one axle during cornering is calculated taking into account the curve radius, the center-of-gravity speed and the track width. The control system then applies this delta speed to the unit under test (see Fig. 4.27).

Road Load Simulation on the Chassis Dyno Test Bed

The goal of road load simulation is to obtain the same acceleration and deceleration performance from a vehicle on the chassis dyno test bed as would be obtained from a vehicle on the road. We distinguish between conventional (e.g. relevant for exhaust emission certification) and advanced road load simulation.

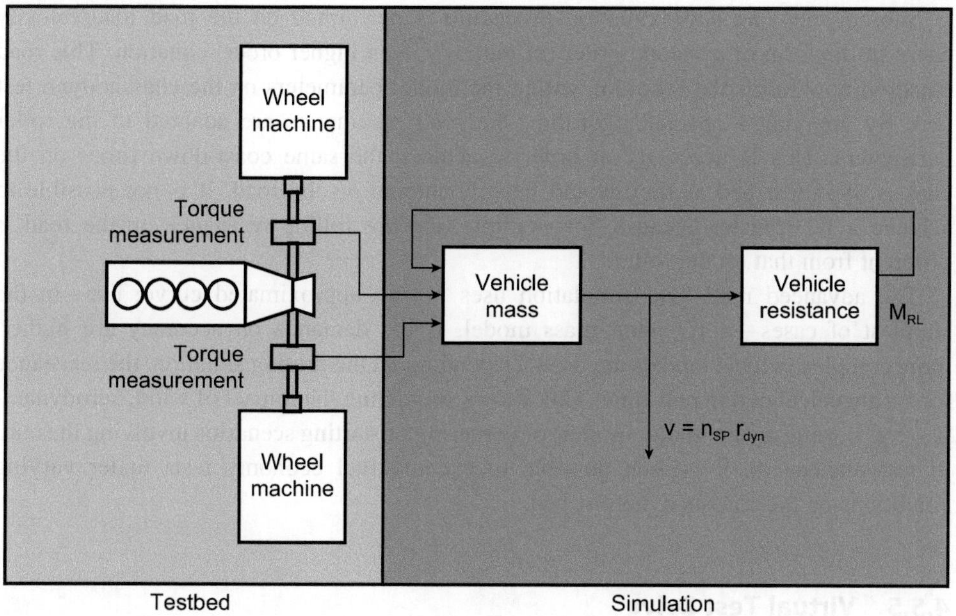

Fig. 4.26 Vehicle simulation on a two-wheel drive powertrain test bed

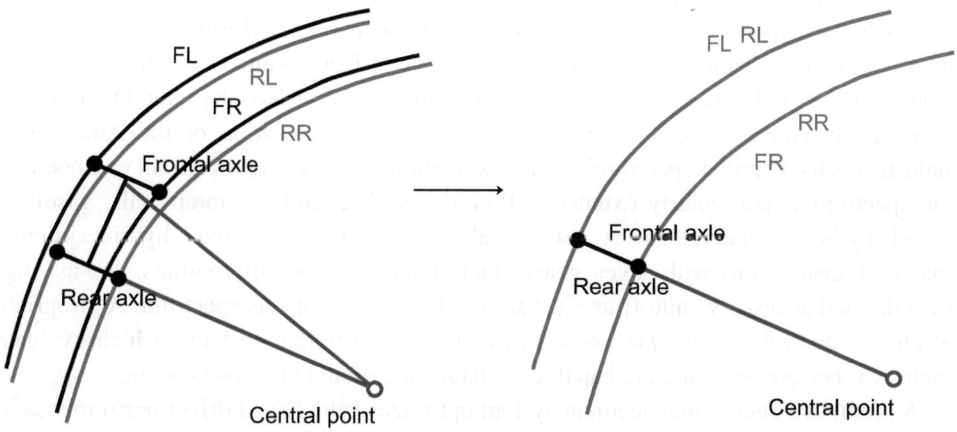

Fig. 4.27 Cornering in a one-wheel model

Conventional road load simulation requires fast parameterization of the simulation parameters. In other words, a fast and easy process to obtain the relevant simulation parameters is critical. For this purpose, a coast-down test is performed with the target vehicle on a real road under the defined conditions (level track, no wind impact, etc.). The test is recorded in a velocity-time diagram.

Subsequently, an approximation procedure is performed on the road load/velocity curve in the form of a second-order (or partially even higher order) equation. This road equation now forms the basis for setting the model parameters on the chassis dyno test bed. By applying a special algorithm, the road parameters are adapted to the roller parameters. This is necessary in order to achieve the same coast-down times on the chassis dyno test bed as they would be encountered on the road. It is not possible to achieve a 1:1 transfer because, for example, the tire rolling resistance on the road is different from that on the roller.

The advanced road load simulation uses not an approximated curve, but—in the simplest of cases—a 1D point mass model. If the demands on accuracy are higher, more complex vehicle models are used. Depending on the driving situation, the resistance forces are calculated in real-time. This allows simulating the impact of wind, aerodynamics (e.g. if using an automatic spoiler) or cornering or starting scenarios involving the curb in test maneuvers. It is then possible to execute fuel economy tests under varying conditions on the chassis dyno test bed.

4.5.5 Virtual Test Drive

Urban mobility is currently in the process of redefinition. Whereas previously people regarded their vehicle as a status symbol, they are now viewing it as a means of transport of which they expect environmentally and socially compatible performance. The question now is to identify which type of drive system is specifically suited for which boundary conditions. As nobody is able to predict this with any precision, the manufacturers are pursuing an approach which the VDA (the German Association of the Automotive Industry) calls a "broad-spectrum" or diversification strategy. In powertrain engineering, the spectrum is particularly extensive. Batteries with countless laptop cells, gasoline, diesel, hydrogen, natural gas, kerosene, salad oil or ethanol engines, hybrid systems, manual transmissions with seven gears, dual clutches, active differentials, downsizing, cylinder and accessory shutoff are just some of the facets of this spectrum. Appropriate electronic control systems are needed to control the different systems, which evaluate their own performance and are capable of diagnosis, should any errors occur.

A predictive energy management system optimizes vehicle and driver performance in the overall system of its environment: road and traffic. Even the Wankel-type engine is attempting a comeback as a quiet and light-weight drive system that extends the range in electric vehicles. The fact remains that nobody today knows which drive technologies will become established tomorrow. However, one thing is clear: the trend is going to produce a sharp rise in the diversity of vehicle variants (see also Sect. 1.1).

The vehicles of the future will be networked with each other and with the relevant infrastructure outside of the vehicle. While this makes the overall system safer, more agile, comfortable and efficient, the complexity and requirements with regard to costs, quality and development times will significantly increase. In conjunction with the diversity of vehicle variants, this will not only raise the development requirements, but, most importantly, the validation requirements in order to ensure safety, functionality and performance of the networked vehicle systems in the customers" real-world operating conditions. The interaction between the subsystems and the resulting effects on the performance of the overall vehicle is not always known, and the use of good engineering judgment, even by experienced experts, is no sufficient basis for SOP (start of production) release.

To protect themselves against legal disputes, manufacturers and suppliers have to adhere to the recognized and objective state of the art, which is not always required by the standards. Against this background, the traditional, component-based procedure is viewed critically, as it is impossible to verify functional safety at a vehicle system level merely by validating its components. The key focus is no longer on the unit under test which is merely one part of the overall vehicle system. This consideration calls for a shift toward consistent testing and validation processes in which virtual test drives play a decisive role, so that integration tests can be executed at an early stage—even if real vehicle components are missing.

4.5.5.1 Effects on Test-Bed Testing

What effects do the described boundary conditions have on modern powertrain development and testing tools? Let us recall the most generic definition of a test [22]: "*A test is an experiment to gain greater certainty as to whether a technical device or a process functions within the intended boundary conditions or whether or not certain properties are present.*"

If this were transferred to the task of "developing and testing the powertrain of the future," a modern development and testing environment for the "technical device" powertrain, for example, would allow us to gain greater certainty on the real fuel economy potential obtained by a predictive energy management function, considering the differing framework conditions of driving in cities or on mountain highways.

A further task arises from the question of how the traveling range of an electric vehicle is affected by different realistic use conditions. Likewise, evidence must be produced of whether the emission-reducing measures in real operating conditions are as effective as the law requires or as the end-consumer expects. The emissions from an internal combustion engine depend on a wide variety of parameters. In what kind of environment is the vehicle operated (city or country, winter or summer, rush-hour or clear roads, rain, snow, fog, etc.)? What type of driver is operating the vehicle (sporty, defensive, economical)? To what extent is the vehicle loaded? How was the engine and how was the gear shifting strategy calibrated?

The project manager of a new vehicle generation is interested in the cause-effect relationship between these influencing factors and the target function of a low-emission overall vehicle in order to determine the optimal settings of these factors. The real test drive is not suited for defining this cause-effect relationship systematically and with the necessary selectivity. One of the many reasons for this is, for example, the barely reproducible ambient conditions and traffic scenarios. Tests in a verifiable and controlled environment, however, are fundamental for obtaining the optimal settings.

These examples show very clearly that the individual components and subsystems are always also part of the overall system "vehicle." It is crucial to perform the assessment of vehicle characteristics in the complete vehicle during the function development phase at as early a stage as possible. How does my component (unit under test) affect the attributes/functions of the complete vehicle, and is it possible to achieve the goals and requirements of the complete vehicle by obtaining a certain component behavior?

This is the reason why these questions have to be answered at an early stage in realistic use scenarios close to the customer, with optimal reproducibility in what we call "closed-loop" operation in virtual test drives. Closed-loop operation means that the drivers are embedded in the "closed loop" of their environment and interact with it. Consequently, former methods, such as driving profiles that were recorded in the real world, are no longer effective. This is because the moment any adaptations are made to the functions/parameter settings in the unit under test, the feedback with the environment will necessarily lead to different variables compared with the test drive data previously recorded (Fig. 4.28).

Fig. 4.28 "Closed-loop" principle: Driver—Vehicle—Environment (road/traffic)

The only way to obtain a higher level of certainty about whether certain functions are working safely and just as intended is by testing very realistic driving situations and driving maneuvers. In addition, it must be verified whether or not the complete vehicle has the desired attributes. For this reason, vehicle manufacturers and suppliers are interested in methods, tools and processes that are sophisticated enough to tackle these complex requirements. Only then will testing become an element in the development process that adds true value. In this respect, the virtual complete vehicle has a key function. As a "virtual prototype" it remains present throughout all stages of the development process, permitting testing, securing and optimizing of the overall system. Costly mistakes and undesired developments due to a lack of greater vehicle-system knowledge are so avoided, or at least reduced to a minimum.

4.5.5.2 Virtual Complete Vehicle

"Driving tomorrow's vehicle today"—this is where the virtual test drive comes into play, reproducing a comprehensive and realistic test drive in a virtual world. In a virtual test drive, a virtual (test) driver drives a complete virtual vehicle in a virtual environment consisting of a 3D road network, road infrastructure (e.g. signs, traffic lights, guard railing, communication), traffic flow (e.g. passenger cars, trucks, motorcycles, pedestrians) and the required ambient conditions (e.g. temperature, air pressure, sunlight).

To provide the full range of feasible mechanical and electrical component interfaces needed for a broad application, it is advisable to use a 3D complete-vehicle model as a virtual vehicle. This technology comprises a totally non-linear 3D driving dynamics model complete with body, axles, suspension system, steering mechanism, engine bearing system, powertrain, hydraulic brakes, tires and aerodynamics. To build such a model, the complete vehicle system is split up into subsystems, each described separately, and subsequently put back together again in a modular way.

For mathematical description, the engineers therefore often use multibody systems (MBS). These consist of a finite number of rigid bodies that are connected to each other and to the environment by force elements (springs, dampers, contact forces between the wheel and the road surface) and joints. Such approaches are also particularly well suited for mapping frequently occurring nonlinearities (for example relating to the tires or the complex wheel control system), the interaction between individual components or the complex contact forces between the tires and the road surface. Mathematically speaking, the model takes the form of a connected differential-algebraic system that can be solved by applying the appropriate numerical methods.

Another aspect requiring consideration is that the complete vehicle system increasingly has to be described not only by mechanical components, but also by electronic control systems as well as electric or hydraulic actuators. This results in a multi-domain approach, in which simulation models from a variety of applications have to be integrated to a complete vehicle model. Section 4.5.6 looks at two concepts in further detail.

A further essential factor required to reproduce a real test drive in a virtual test drive as realistically as possible is the virtual driver, in other words the mathematical, numerical model of a real driver. Real drivers implement specific driving tasks in different ways and are distinguished by diverse driving strategies. These range from defensive and economical driving styles to sporty and even aggressive ones. Each of these driving strategies has an impact on certain vehicle functions and the system behavior. Ignoring this effect in the virtual driver will lead to problems when determining fuel economy and emission values or when testing the states of charge in batteries. A more thorough analysis of such driving strategies shows that drivers are distinguished by widely differing strategies involving the choice of driving line, keeping the vehicle on course, steering, braking and accelerating.

To make valid statements in the virtual terrain it is therefore vital for these differences to be modeled as realistically as possible in the virtual driver. For this reason, the producers of simulation software for virtual test drives have been focusing their development work on the realistic reconstruction of driver input for many years. Driver models are equipped with a whole range of driving styles and driving strategies which the user can select from a graphical user interface. Models for traffic-to-follow functions are also becoming increasingly significant in this respect. They allow the virtual driver to "see" and actively follow another vehicle while the virtual vehicle's speed is adapted to the vehicle traveling ahead, the driver's own requirements, the track, the vehicle's capabilities, the road signs (mandatory and prohibitory road signs) and the traffic lights along the route. Temporal and spatial distances and the required following strategy are defined as required.

For example, this permits a change in driving behavior to produce an altered accordion effect in the traffic flow, which consequently causes a change in fuel consumption levels. As a result, a virtual driver allows the performance of comprehensive systematic tests and realistic assessments of integrated functions in view of a variety of driver types. Such functions include, for example, a predictive energy management strategy or efficient battery management. This makes it possible to determine the attributes of the future vehicle very realistically and long before the first physical prototype is available.

4.5.5.3 Test Strategies in the Virtual Test Drive

The necessary tests are generally divided into three different test categories according to their test contents. Table 4.7 illustrates this by the example of longitudinal torque control.

The tests belonging to the first category concern safety and are also called fail safe tests. These are essentially set up and performed as defined in the ISO 26262 functional safety standard. These tests are carried out to verify whether or not the system is working reliably, errors are correctly identified, the system is brought to a safe state in the event of an error, the driver is appropriately warned and the diagnosis function is operating as intended. Typical kinds of errors include mechanical or electrical sensor or actuator errors, detection errors or communication errors.

The second category of tests concerns function. In a defined driving situation (a maneuver), tests are run to verify whether or not specific functions have been correctly implemented and are performing according to their specification. These frequently include a whole bundle of questions: Is the function activated at all in certain driving

Table 4.7 Test categories: safety, functional and performance tests illustrated by the example of longitudinal torque control tests

Software safety tests	Function tests	Performance tests
Fail safe tests	Driving maneuver catalog	
Overvoltage/Undervoltage	Slip control	Traction performance
Sensor signal interruption	Wheel spin reduction	Acceleration performance
Sensor error: offset, drift	Torque distribution/torque transfer	Hill-climbing capacity
Actuator defect	Acceleration	ISO lane change performance
Torque error	Traction control	Slalom performance
Pump performance error	Brake control	Lap time performance
Clutch calibration error	Active yaw rate control	Lateral acceleration performance
CAN error	Automatic torque reduction	Braking performance
Short circuit	Understeering reduction	Understeering performance
Loose contacts	Trailer function	Stability characteristics
Diagnostic messages	Clutch disengaging request by ESC	Steerability criteria
Interfaces and communication tests involving other ECUs	Ackermann steering control	Nürburgring racing track performance
	Automatic clutch calibration	Fuel economy performance

situations? Is the torque intervention achieved in the right direction, i.e. with the right (positive or negative) sign? Are there any differences in the units used for the signals? In connection with control unit functions, this can have disastrous consequences. A frequently quoted example is the loss of the Mars Climate Orbiter due to a simple unit error (imperial system versus SI system) in its navigation system [23].

In certain driving situations, it is even possible for functions that are actually desired to suddenly stop being useful and even become critical to safety. An example for this is the recuperation function of a hybrid vehicle with electric rear axle. When cornering on a road with a low friction coefficient, this can cause stability loss in the vehicle—with fatal consequences. This means that this particular driving situation requires that the recuperation and braking function is reliably deactivated. Such test cases have to be generated and documented in terms of systematic testing.

This still fails to answer the question of how efficient the functions are and what the driving or vehicle behavior is like. Answers to these questions are sought for in the third category, i.e. the performance tests. For example, these tests might deal with evaluating the achieved braking distances, the accelerating ability on ice, real-world fuel consumption on a predefined track or the range of an electric vehicle as a function of specific calibration parameters.

The examples make it evident that virtual test drives have to be incorporated into the test strategy. By doing so, the test bed operator of a component test bed is empowered to perform initial tests relating to a higher system layer, i.e. the complete-vehicle layer. The virtual test drive gives test bed operators a tool to verify functional safety (ISO 26262) and performance in a consistent manner and from the system point of view of a complete vehicle, allowing them to make an essential contribution toward making the development of future safety-related systems more efficient. In the future, this test methodology will even have effects on internal and external component suppliers beyond the confines of manufacturers and vehicle integrators.

There is much more to virtual test driving than just moving models. It means shifting the test driver's workplace from the road to the lab or office—along with test maneuvers and tracks, as well as diagnostic, measuring and application tools. This results in maneuver-based, event-based and track-based testing methods that can be combined according to requirements. As with a real-world test run, the (virtual) driver is given maneuver instructions that contain specific actions. These instructions might be vehicle control tasks, driver activities, system interventions and system operation, driving state monitoring, fault insertion or the operation of the measuring, calibration or diagnostics equipment.

State-of-the-art tools support the development of corresponding testing cases with the aid of interactive maneuver control, which transfers driving instructions to the virtual driver depending on events. The decisive factor here is that the maneuver has a generic parametric description, which can be automatically executed and used with any vehicle model across all configurations, largely unmodified and with different parameter sets. Open- and closed-loop mini maneuvers in longitudinal and lateral direction can be interchanged and combined as required.

Table 4.8 shows the assembly and disassembly of a virtual drive test based on the example of a braking attempt on a friction surface covered with micro gravel. The test is performed to assess a hybrid vehicle's attributes, including the integral function of its electric and hydraulic brakes, also known as torque blending, in view of the vehicle's stability, safety and performance.

In this maneuver- and event-based test, the vehicle is first accelerated to 100 km/h. This is done in closed-loop operation for longitudinal and transversal dynamics with the help of the virtual driver. "Marking reached" is an event that leads the driver to suddenly release the accelerator. In longitudinal dynamics, this controls the accelerator pedal in open-loop operation. In transversal dynamics, the driver continues to keep the vehicle on the desired course by making small steering movements. After that, by switching the driver to open-loop mode also in transversal dynamics, full braking also inhibits the controlled steering movement. As a measure for controllability, the virtual driver then once again performs the steering correction in closed-loop operation for the transversal dynamics. This is followed by another fault insertion before the vehicle is stopped.

Extending the maneuver- and event-based approach, track-based testing allows the analysis and evaluation of the complete vehicle's performance, as shown in Fig. 4.29. This includes, for example, fuel economy properties, emission performance or the range of electric vehicles in relation to different types of tracks, such as urban roads, rural roads,

Table 4.8 Maneuver- and event-based approach to a braking procedure on micro gravel to assess a hybrid vehicle

Mini-maneuver no.	Driver and action instructions	Longitudinal dynamics	Transverse dynamics
0	*Speed up to 100 km/h and maintain speed of 100 km/h* when approaching the micro gravel area.	Closed Loop VirtualDriver	Closed Loop VirtualDriver
1	*Marking:* Release accelerator quickly Coasting to assess stability achieved by drag torque control.	Open Loop	VirtualDriver
2	*Full braking:* Lock steering in position to determine the stability achieved by torque blending (yaw rate maximization in 1-s slot).	Open Loop	Open Loop
3	*Correct steering after 1 s:* Switch over to Closed Loop to determine the required steering angle as a controllability criterion.	Open Loop	VirtualDriver
4	*Fault insertion at 50 km/h:* Insert communication error between HCU (hybrid control unit) and BCU (battery control unit) to evaluate the error detection system and error effects, and test the diagnostics.	Open Loop	VirtualDriver
5	*Brake to standstill:* Read the fault memory to check whether the diagnostic entries are correct, delete the fault memory.	Open Loop	VirtualDriver

Fig. 4.29 Track-based approach with AVL InMotion powered by IPG/CarMaker

highways or up- and downhill drives, and when interacting with different driver types (sporty, standard, defensive or economical).

Using the track-based approach, the data of real 3D roads, for instance from a navigation system, map services or from one's own 3D track measurements, are imported into the simulation environment and given various road surface roughness profiles. The process results in customized home tracks or testing grounds as reconstructed reality, including any type of courses as part of the simulation (see Fig. 4.30).

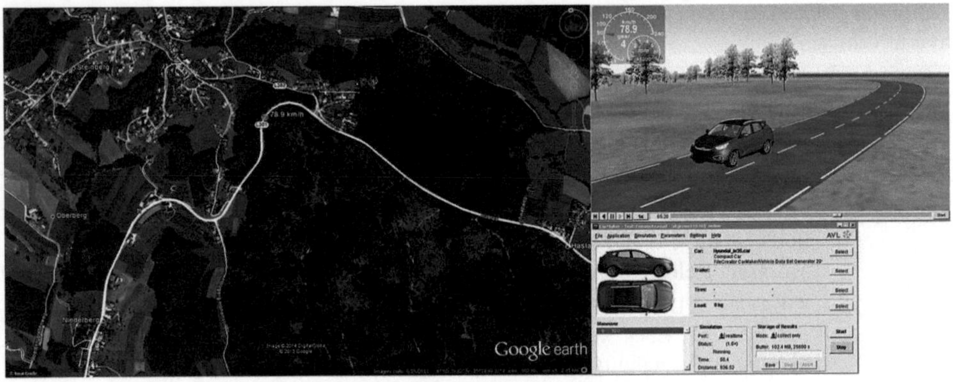

Fig. 4.30 Connection between the navigation system and the vehicle in the track-based approach

To allow for comparisons between the real test drive and the simulated results in various test environments, such as engine, powertrain or chassis dynamometer test beds, it is also possible to assign a track-based speed profile to a lead vehicle. The lead vehicle follows this speed profile with utmost precision in the simulation mode, the unit under test follows the lead vehicle on the simulated road. This allows results from different tests to be compared afterwards as time-based speed data from the test drive cannot be used, particularly not if the tracks are very curvy. One reason is that after a certain time period, time and distance in the simulation may drift apart compared with reality.

ADAS (advanced driver assistance systems) navigation data additionally contain information about the road infrastructure, such as e.g. gradients, road types, road signs, traffic lights, crossroads, number of lanes and lane widths. The quality in which such data are available is high enough to be used directly in the simulation environment: The virtual driver travels along the predefined route fully independently, according to a definable driving and course strategy, while always paying attention to the road signs.

Maneuver-, event- and track-based tests can be combined as needed in order to direct the vehicle into specified, reproducible driving situations, in which the relevant target function undergoes evaluation. It is additionally possible to investigate how this particular function affects the properties of the overall vehicle (see Fig. 4.31). For example, the specific introduction of traffic objects allows the virtual driver to follow the vehicle ahead or overtake it in order to assess state of charge (SOC) degradation or the effectiveness of the "sailing function."

The comparison below summarizes the different approaches to virtual test driving. In maneuver- and event-based testing, there is a fixed assignment between function, driving maneuver and the assessment criteria.

- The virtual driver receives instructions for a driving maneuver, defined actions and monitoring tasks which it then carries out.

Fig. 4.31 Combination of maneuver-/event-based and track-based approach

- The driver also receives driving recommendations from the human-machine interface, such as release of accelerator (sailing function), gear selection or brake pedal position, which it follows in different ways, depending on the set attributes.
- Further traffic objects, such as other traffic participants, are assigned driving maneuvers and monitoring instructions.

Using the track-based approach, i.e. when driving freely, integral assessments of a complete-vehicle attribute are made on long urban, rural or highway routes, depending on the different driver types and driving strategies. The examined attributes might commonly include emission levels, fuel consumption, proportion of all-electric driving or lap times.

- In such tests, the virtual driver makes the driving decisions freely according to the chosen driving style (sporty, standard, energy-saving, defensive, hectic, etc.). This influences the choice of course and speed, the use of the accelerator pedal and brake and the utilization of defined acceleration requirements.
- The virtual driver pays attention to mandatory and prohibitory road signs and obeys traffic lights.
- The virtual driver follows the vehicle traveling ahead or overtakes it according to its own decisions.
- The virtual driver also receives driving recommendations from the human-machine interface, such as release of accelerator (sailing function), gear selection or braking, which it follows in different ways, depending on the set attributes.

Based on the combination of free driving and maneuver- and event-based scenarios, situations and events are specifically inserted to be able to make an integral assessment on a complete vehicle attribute (e.g. emission levels, fuel consumption, proportion of electric driving):

- The virtual driver makes free driving decisions and pays attention to mandatory and prohibitory road signs and obeys traffic lights.
- Different kinds of traffic events are specifically initiated along the route, which are triggered by situations.
- The driver receives specific maneuver instructions at certain points along the track, or when defined traffic events occur (e.g. flooring the accelerator when overtaking).
- The driver receives driving recommendations according to the events, which the driver follows in various ways, depending on the set attributes.

4.5.6 Virtual Vehicle Integration

Over the past decades, a large number of specific simulation tools have become widely established in the automotive industry. For example, there are expert tools suited for signal-flow based description of control algorithms, multi-body simulation systems for describing powertrains and driving dynamics components, simulation tools for brake hydraulics or internal engine processes. These tools typically focus on very specific areas of expertise (also known as domains).

To develop state-of-the-art mechatronic systems, however, it requires a considerably broader approach. A holistic analysis must take into account the interdependences between subsystems from a broad spectrum of different domains. Bearing in mind the computing capacity available in real-time environments on test beds today, it is becoming increasingly important to integrate these usually very precise technical models into an overall concept.

Most of these efforts focus on two main approaches. A tool represents what is referred to as an integration platform and is capable of integrating and executing models from other tools via defined interfaces. Alternatively, a co-simulation environment is set up which supports heterogeneous simulation environments.

Both of these options have the further advantage of supporting the systematic distribution of different kinds of activities. Given the growing complexity in today's motor vehicle systems, this capability is becoming increasingly significant. For building models, domain-specific authoring tools are available, which can then be incorporated on the test bed while remaining seemingly invisible. Established user guidance systems and the human-machine interface are designed to reduce the complexity for the tester even further.

4.5.6.1 Integration Platform

Figure 4.32 shows the example of an integration platform.

All systems and components can be exchanged and extended as required, regardless of whether they are used as virtual model components, software, real control units or mechanical hardware. Pre-condition is the definition of standardized interfaces. Each

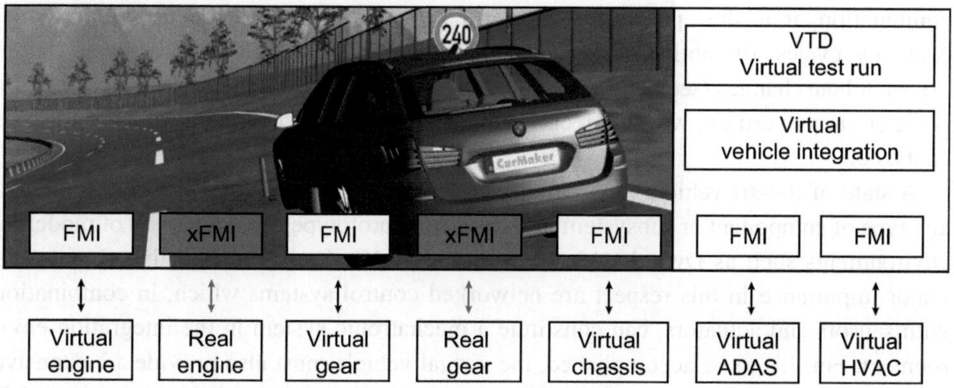

Fig. 4.32 Vehicle integration platform for virtual test driving

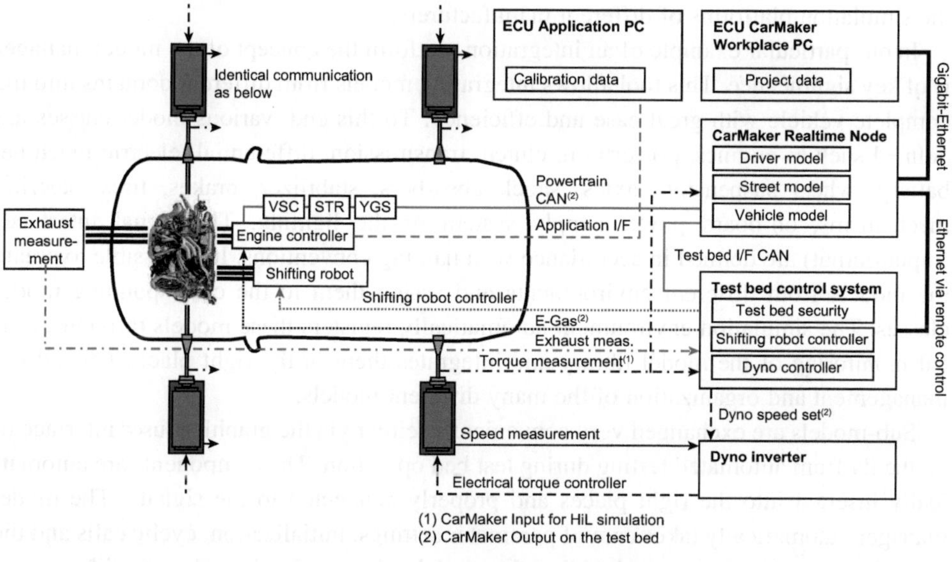

Fig. 4.33 Integration of a real powertrain in a complete vehicle by means of a powertrain test bed

component and subsystem can then be put in its predefined place and calculated according to a uniform and proven schedule.

The important thing is to ensure that different types of components—whether they are physical, virtual, mechanical, electrical, hydraulic or pneumatic—from different types of environments are capable of being integrated into the virtual vehicle. If they are, it is possible to build an environment for almost any test bed environments that allows performing a virtual test drive for the unit under test. Figure 4.33 shows the example of a real powertrain integrated into a virtual complete-vehicle environment. With this combination, it is thus possible to perform virtual test drives such as standard fuel economy cycles, drivability tests, acceleration or braking maneuvers or even slalom drives or lane changes (see Sect. 4.5.5). In this particular case, the unit under investigation is the entire powertrain, whose effects on the attributes of the complete vehicle can so be analyzed.

A state-of-the-art vehicle integration platform today is capable of integrating almost any type of component or subsystem into the virtual prototype, from a variety of modeling environments such as Dymola, Matlab/Simulink, AVL Cruise or as native C code. Of major importance in this respect are networked control systems which, in combination with sensors and actuators, can constitute a mechatronic system in the integration environment. For this to be accomplished, the virtual vehicle must also provide an extensive array of suitably performing sensors that can be freely positioned. Nowadays, models from different domains can also be integrated as functional mock-up units (FMU) based on the Functional Mock-up Interface (FMI) standard. This allows an exchange between the simulation platforms of different manufacturers.

In our particular example of an integration platform the concept of the model manager is of key significance. This tool allows integrating models from different domains into the complete vehicle with great ease and efficiency. To this end, various model classes are defined such as engines, powertrain, clutch, transmission, differential, electric machine, battery, wheel suspension, axles, shock absorbers, stabilizer, brakes, tires, steering mechanisms, on-board power supply system or the lighting. The signal interfaces (input/output) are defined in accordance with naming conventions. It is possible to create the models from different environments and assign them to the corresponding model classes. The simulation environment automatically registers these models through cyclical monitoring of the model library and integrates them in the right place. This allows management and organization of the many different models.

Sub-models are exchanged very conveniently, either via the graphical user interface or on the fly from automated testing during test bed operation. The components are automatically inserted into the right places and properly connected to the signals. The model manager automatically takes care of parameter settings, initialization, cyclic calls and the final clean-up. Apart from embedding the models, this mechanism also provides a very simple way of integrating hardware components and systems (for example, different types of control units, engines, the powertrain or the steering system) into the functional mock-up prototype, for example via a hardware-in-the-loop (HiL) environment. For the vehicle model itself, it is irrelevant whether a signal is exchanged cyclically with a model or whether the signals are transferred via the I/O interface of a control unit or an engine test bed [24].

The components, which may be virtual or real, can also be exchanged during simulation or, if required, operated in parallel. It is important to ensure that the interface is technically identical.

4.5.6.2 Co-simulation

The procedure described in the previous section often leads to simulation models being linked that were designed for different vehicle components and were created with different simulation tools. It is now possible to either translate such models into a single format for the integration platform or combine the relevant simulation tools during simulation by using a method referred to as co-simulation.

The introduction of co-simulation in the modern development process has made the task of complex mechatronic system development much more efficient. In the field of alternative drive systems, the focus is on integrating electrical and thermal vehicle components into existing (mechanical) drive concepts. Accordingly, the task of co-simulation is to take into account the complex interactions between the different simulation models in a suitable and, most importantly, correct manner (Fig. 4.34). The verified interaction between a large number of models (and thus also simulated components) enables a realistic virtual conception and validation of the overall system—consisting of vehicle, driver and environment (see also Sect. 4.5.5).

When coupling simulation tools (co-simulation) from different development areas, there are several important aspects to consider:

- The simulation models should be changed as little as possible, i.e. only input, output or control elements may be added for the co-simulation.
- The simulation models use their own specific numerical solution algorithms (integrators, solvers) and simulation step sizes.
- No changes are made to the solver settings, i.e. the settings defined by the model developer are also used in the co-simulation.
- The communication intervals between co-simulation platform and simulation tools are autonomous and adaptable.
- The co-simulation platform also enables remote control of the simulation tools.
- Coupling errors can be located in terms of time and topology.

For co-simulation, the (virtual) overall system is divided into component models. These can be individually decoupled with only the input and output signals (coupling variables) being connected to each other. The extrapolation of coupling signals at every calculation step, which is necessary for dividing an overall system into its components, leads to what we call a coupling error. To be able to ensure the accuracy of the

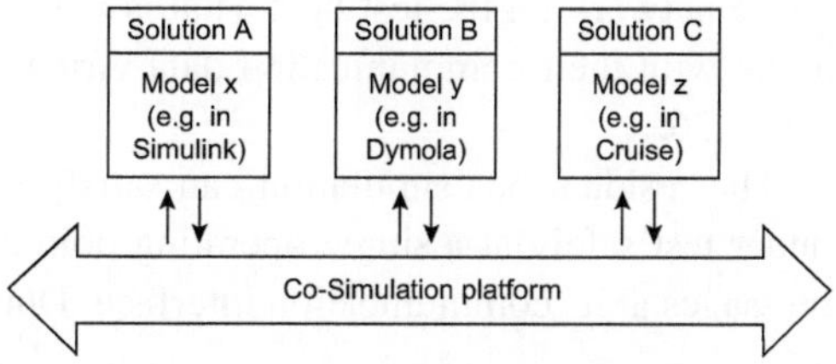

Fig. 4.34 Coupling of different simulation tools by means of a co-simulation platform

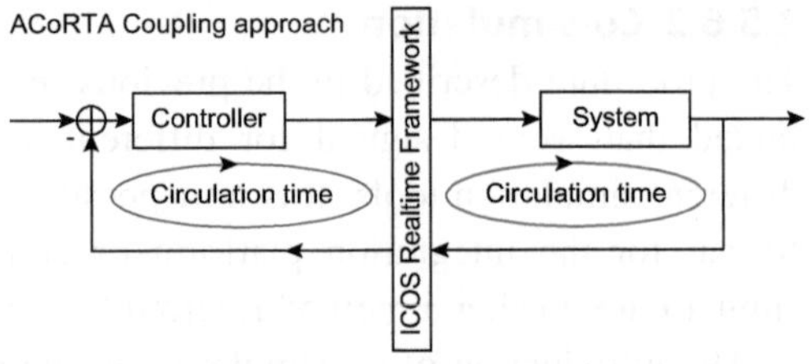

Fig. 4.35 Coupling of real-time systems or real-time with non-real-time systems

co-simulation, this coupling error has to be calculated and compensated for. To accomplish this, various methods can be used such as the energy preserving extrapolation method NEPCE (nearly energy preserving coupling element) developed by Virtual Vehicle (see [25]). In recent years, a standard called Functional Mock-up Interface (FMI) has become established for the integration of simulation models and tools (see also [6] and Sect. 4.1.2). FMI defines a uniform interface, which allows complex simulation models from different areas of physics to interact with each other. With a functional mock-up interface, both simulation models and simulation tools can be connected, making it far easier to operate and maintain co-simulation platforms.

A logical extension of the co-simulation approach is the integration of real-time systems into the system simulation. If one or more components are available as real hardware, these can be directly integrated into the existing system model. With the extension of the co-simulation into the real-time domain, additional challenges remain. Special coupling elements must ensure the time-correct coupling of the involved real-time and non-real-time systems. The round-trip times occurring across the entire system must be kept as small as possible in order to ensure the stability of existing control loops (see Fig. 4.35). In real-time systems noisy sensor signals can be a further problem. This topic is currently the focus of the research project ACoRTA (Advanced Co-Simulation Methods for Real-Time Applications) (see [26]).

4.5.7 Residual Bus Simulation

On the test bed, the unit under test is frequently linked to other smart systems via bus systems (see also Sect. 3.5). Unit-under-test operation is usually not possible unless communication with such nodes is fully ensured. However, due to unavailability, or for reasons relating to cost or complexity, it is often neither possible nor feasible to set up the entire network on the test bed. That is why it is possible to integrate the required nodes along with their communication data virtually in a residual bus simulation, as shown in Fig. 4.36.

The residual bus simulation can satisfy a variety of requirements. To operate a unit under test safely at a single operating point, it is often sufficient to supply it with static messages at its communication interface. During engine operation, for example, a control

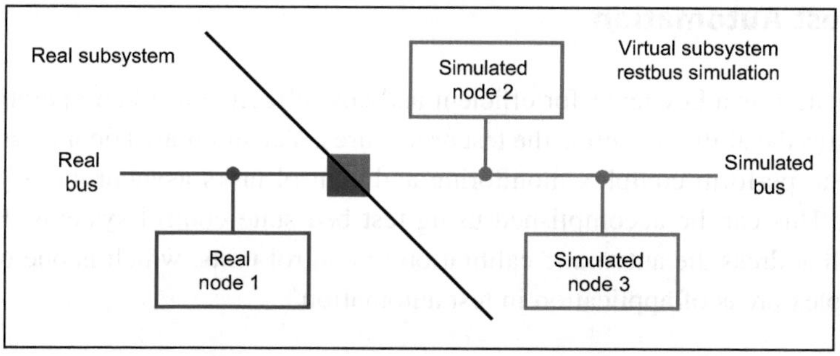

Fig. 4.36 Residual bus simulation with the aid of a virtual subsystem [2]

unit's bus enable signal may be required. This can be transferred by a communication interface without the need to integrate the actual physical control unit itself.

If a unit under test plausibilizes its own performance based on statements from other smart network nodes, a simple static value input will usually not be sufficient. In this case, a dynamic simulation of the network partner will be necessary to prevent the unit under test from encountering an error or, in a worst-case scenario, an emergency-stop situation.

Apart from ensuring the unit under test's operation, residual bus simulation can also be used to generate test cases for faulty communication within a network. In this use, the unit under test is deliberately brought into a fault situation by manipulated messages on its communication lines. This method is particularly suited to examine a tested unit's diagnostic and emergency stop functions. The technology is equally useful for reproducing errors in the bus physics, invalid messages or stress tests caused by bus overload situations.

For setting up a residual bus simulation, a variety of standards are available for storing the network's communication information. Accordingly, there are so-called DBC (Database CAN) descriptions for CAN; also A2L (ASAM-A2L) descriptions are available. With the FlexRay bus system, the so-called FIBEX (Fieldbus Exchange Format) description was introduced. By using this information, a number of tool producers also support users in setting up their residual bus simulation. Based on common programming and modeling languages, smart virtual systems can then apply important dynamic communication properties to the individual nodes. Standard tasks are usually already contained in the programming packages and are easy to activate by setting the parameters.

Critical for test bed operation is the seamless integration of the residual bus simulation, because normally no expert will be available on site. This is the reason why commercially available residual bus simulation tools are also operated autonomously based on configuration files. This permits an expert to design the required residual bus simulation in a tool of their own and provide it to the test bed operator as a "black box." The expert activity "residual bus simulation design" is so decoupled from the operation of the residual bus simulation. As a result, dedicated software interfaces to the automation system even allow remote control to a certain extent.

4.6 Test Automation

Test automation is a key lever for efficient and cost-effective test bed operation. First of all, it requires the ability to define the test procedures. Yet automated operation also means the ability to perform complex monitoring and control tasks asynchronously to the test procedure. This can be accomplished using test bed state control systems. Finally, this section will address the automatic calibration of control units, which is one of the rather more complex areas of application in test automation.

4.6.1 Test Procedure

Because of the very high level of complexity encountered in testing tasks, automated test procedures are a key factor in raising the efficiency on test beds. Among other things, this is down to the fact that productivity and cost considerations are not unimportant when judging the test strategy for a unit under test. In-vehicle tests are very expensive. That is the reason why auto manufacturers are interested in running road tests on chassis dynamometers and shifting chassis dynamometer tests to powertrain test beds or engine test beds. One result of this is that calibration tasks also require transient cycles that can be executed on high-dynamic test beds, for example. A certain number of physical tests are replaced by simulation or combined with real components, e.g. the simulation of a road profile and vehicle performance on a powertrain test bed.

While in the past the development focus was on the internal combustion engine, the attention today has shifted to exhaust gas after treatment, thermal management, the interaction between different control units (e.g. advanced driver assistant systems—ADAS) or the optimization of a vehicle's or component's energy management system. The control units with their differently optimized configurations and designs are a key factor in enabling reduced fuel consumption and even improved driving comfort and increased safety. For example, vehicles are equipped with a smart braking system to raise their level of safety.

To cope with this diversity of variants, modular and powerful tools are needed that are able to define and execute test procedures. These have to be capable of saving and managing large volumes of data that are generated by a variety of engine and transmission electronics and measuring devices. This necessitates real-time capability and smart synchronization mechanisms. In addition, consideration must also be given to the increased volume of data needed to design the test procedures, which requires efficient means for data plausibilization, data evaluation and data storage.

4.6.1.1 Basic Functions

Support of different types of testing jobs (e.g. definition of steady-state step sequences as shown in Fig. 4.37), execution of maps, definition of transient road profiles or maneuver-based testing are the core capabilities of every modern test run automation system.

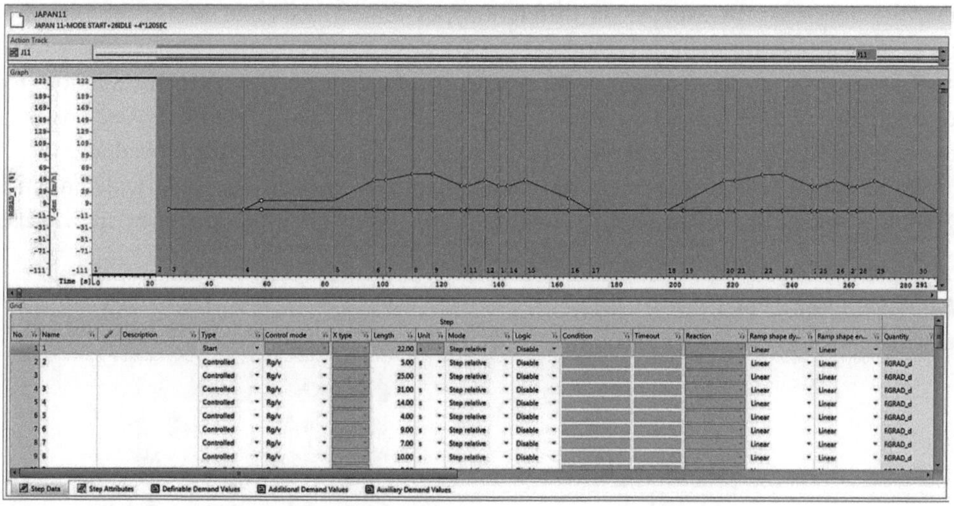

Fig. 4.37 Example of a parameterized step sequence

Unmanned test bed operation has significant effects on productivity and costs. Online limit monitoring and visualization capabilities are therefore a fundamental requirement for an automated test procedure.

Apart from that, automatic exception handling in the event of errors is absolutely essential for every test bed. These automated mechanisms ensure that the unit under test is brought to a safe operating state, even from unintended operating states, or, in the case that test continuation is possible, that the automated test run can continue at defined recovery points. Automation also requires advanced user interfaces with integrated debuggers that allow easy detection of errors in the test run design during test run development and facilitate the troubleshooting process.

The automatic test procedure must also provide an option to control subordinate systems, such as facility management, process control systems and other systems of relevance within the test facility context. It must be possible to respond to changes in the ambient conditions. Consequently, a system used to automatically execute test runs requires clearly defined interfaces to those subsystems. Freely definable state machines can help in this context to cover the diverse spectrum of application cases. A preferred means, in this respect, is a scripting interface, both in the test procedure and in the state machine.

To be able to separate frequently recurring testing tasks, one possible solution is to encapsulate function blocks in libraries. A mechanism like that allows you to reproduce tasks (such as warming up the engine, setting certain operating points or recording a full load curve) in library blocks which can be re-used in all test runs.

4.6.1.2 Test Creation

The simplicity and user-friendliness of test designs are critical to their acceptance at different user levels. Application engineers want to execute already existing step profiles that are very easy to execute on different types of test beds and re-use reproducible measuring results for their development work.

This calls for a tool that allows them to integrate all of the relevant information, for example demand values for the accelerator pedal position, road gradient, gear position, environmental conditions or elevation profiles, into the automated test procedure.

From predefined libraries, test run authors can now model their own test runs in a graphical operating environment (see Fig. 4.38). Being able to do things like re-use

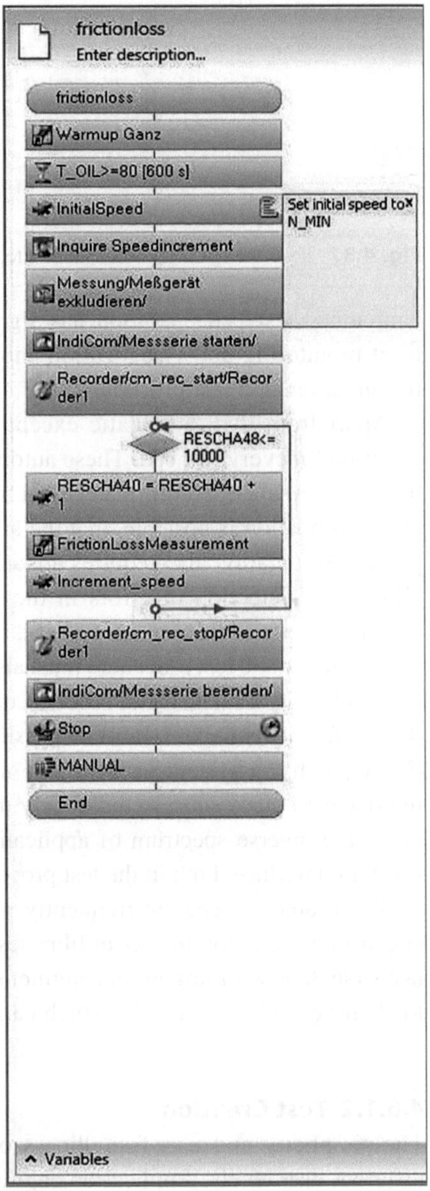

Fig. 4.38 Example of a graphical test run

predefined library blocks, define different types of measured value acquisition (steady-state or continuous measurement) or define storage data speeds up test procedure generation and makes it far more efficient.

4.6.1.3 Test Management

Versioning generated test procedures makes it easy to restore previous test designs. The central system that manages these test procedures is located on a host computer protected by various user rights. It enhances efficiency by providing support in maintenance jobs and ensures that only authorized users are able to modify or create a test design.

4.6.2 Test Bed State Control

When running automated test runs, a series of unchanging functions have to be executed that take place asynchronously to the test run. Examples are:

- Complex monitoring and control tasks to ensure the safety of units under test or unit-under-test components
- Control tasks to set defined ambient conditions or ones that are required by the test run or a simulation program by means of actuators or stimuli
- Complex responses to limit violations, for which defined sequences have to be met in order to bring units under test back to safe states. One example for this would be the shutdown of multi-dyno test beds in the event of a failure in a speed sensor.

Such asynchronous procedures can be programmed either in an external PLC (programmable logic controller) or in special test bed control function modules belonging to the automation system itself.

In case the PLC is external, the variables needed for the control tasks have to be exchanged between the automation system and the PLC on a continuous basis. The most common communication protocol for this purpose is the PROFIBUS protocol (see Sect. 3.5.3).

Where function modules in automation systems are concerned, state/event-based programming models have proved valuable. Such models allow the creation of any number of virtual devices within a test bed system, for which the required states can be defined. The state transitions are assigned to certain events, for example if a value for a test bed quantity is exceeded, or if a digital bit changes. Every state change triggers the execution of a short automation sequence, which the user defines in the automation system in a particular programming language.

4.6.3 Automatic Control Unit Calibration

4.6.3.1 Motivation

Another test bed automation task, which has gained significance in recent years, is the automatic calibration of control units. Novel approaches in control unit calibration are inevitable. This is evident not only in light of the growing complexity of units under test introduced by the increased number of variation parameters in the many different control units (see also Sect. 2.3.6) for the latest powertrains and the exponential growth in the number of control unit labels to be calibrated (characteristic maps, curves, values). But new approaches are also needed to meet the higher expectations regarding shorter development times, cost reduction and quality.

Control unit calibration is accomplished in four steps:

- The unit under test, of which the control unit requiring calibration is in charge, operates in various different operating points. In doing so, measured values are recorded and stored.
- After that, the recorded measuring values serve as a basis to calculate mathematical models that deliver an approximate description of the unit under test's performance.
- In a next step, these models are used to find optimal parameters in the control unit.
- The last step involves verifying the optimization results. For this purpose, the parameters determined in the previous step are loaded into the control unit and examined in reference tests.

The main objectives of using new automation methods and optimization tools are:

- Reduced calibration and testing requirements through special methodology, e.g. DoE (Design of Experiments) or dynamic engine measurement
- A high degree of automation through use of smart automation strategies
- Enhanced quality of results through reproducible procedures and use of model-based methods

Different approaches are taken to achieve the above goals. Fundamental in the development of these approaches and methods is their applicability in the automotive industry environment, particularly in view of calibration- and test-specific processes, organizational structures at the OEM and the employees' professional skills. If tools and methods are to come into widespread use, they have to be based on the above points and be understood and accepted by a broad base of users.

Below we have provided a brief overview of the different approaches, and an evaluation based on their advantages and disadvantages. The main focus is on approaches that are already in widespread use among automotive manufacturers and have proved their value in day-to-day operation. For a more detailed analysis, please refer to the relevant publications on this topic.

4.6.3.2 Different Approaches in Automatic Control Unit Calibration

Among the most commonly used approaches we distinguish between:

- **Online** and offline optimization
- **Local** and global modeling
- **Steady-state** and dynamic measurement
- Steady-state and dynamic **modeling**
- **Operating point optimization and cycle optimization**

The procedures printed in bold type are currently the most widely used in real-world control unit calibration applications in the automotive industry. This is because, aside from their high level of sophistication, they also enjoy broad acceptance among users.

Below is a list of the basic components which all methods and tools have in common:

- Test design (Design of Experiments)
- Test execution with automation/data generation
- Data plausibilization
- Model building and plausibilization
- Optimization
- Verification
- Map calculation and data set generation for control units
- Analysis of optimization results based on reference tests

4.6.3.3 Differences Between Online and Offline Optimization

With regard to the procedure, we distinguish between online and offline optimization. Both approaches require appropriate methods and tools that support a model-based application approach.

In **online optimization**, a safe start point is defined in one operating point (speed/load), i.e. a combination of the different engine variation parameters at which the engine can be operated safely. Examples for gasoline engine variation parameters are ignition timing, injection quantity and injection timing as well as boost pressure. For a diesel engine, the method may involve injection quantity, injection timing, number of pilot and post injections as well as boost pressure. Apart from defining a target quantity (e.g. minimum fuel consumption or maximum torque), limits are defined to restrict the ranges, e.g. for the exhaust gas temperature (as a measure to protect components) or in view of undesired emission values (for example black smoke).

Finally, a parameter variation is executed in the immediate vicinity of the start point. After calculating the average value of the first three variations, an online calculation automatically delivers the mathematical models that satisfy the objectives and constraints. In a large number of cases, the mathematical models are generated using polynomials.

Based on the generated models, an **online optimizer** is used to adjust the variation strategy in the direction of the nearest optimum. Such optimizers are commonly based on a gradient method. Depending on whether a minimum (e.g. minimum fuel consumption) or a maximum (e.g. maximum torque) is being sought, the optimizer moves the variation parameters toward the next minimum (steepest descent) or maximum (steepest ascent). The major strength of such a method is that the result (in the form of the optimum) is available straight after the test run. However, the benefits are offset by some significant disadvantages, e.g. the optimizer only finds the nearest local optimum in relation to the start point. In combination with methods capable of selecting multiple start points, this disadvantage could still be bypassed, yet the resulting overall runtimes are often no longer acceptable.

In **offline optimization**, the data generation step takes place on the test bed, decoupled from model generation. Figure 4.39 shows the individual stages of offline optimization.

The first step is **test design**. This is where the technique of **Statistical Design of Experiments (DoE)** comes into use. Drawing on the user's prior knowledge of the targets' dependencies on the unit under test's variation parameters, a test design is generated in order to vary the engine's control variables within an operating point (speed/load). Typical types of test design procedures are specified in Table 4.9, along with their advantages and disadvantages. For more detailed information on each of the test design types, please refer to the relevant publications on this topic [27].

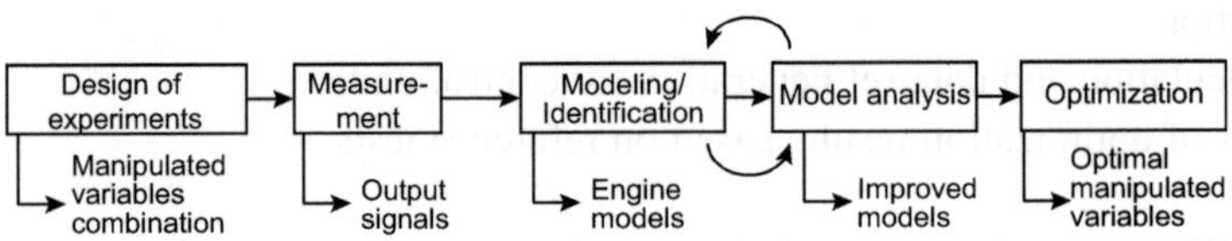

Fig. 4.39 Process stages of offline optimization [3]

Table 4.9 Examples for types of test design [3]

Test design	Model order/Advantages	Disadvantages
Box-Behnken	Second-order polynomials	Interaction not fully investigated
Central Composite	Second-order polynomials Fully identifiable	Symmetric design space needed
D-Optimal	Any order Direction-specific Any shape of design space Inclusions possible Any number of points	Computer support needed
Latin Hypercube	Fills the design space with points No prior knowledge necessary	No direction-specific differences possible Computer support needed
S-Optimal	Fills the design space evenly, even if the design space is extremely asymmetric	Computer support needed

The use of DoE aims at generating maximum information (in the model) with minimum measurement effort in the investigated variation range. In a second step, the **measurement** takes place on the engine test bed. Ideally, this step is fully automated in order to ensure maximum reproducibility.

Some of the particularly powerful control unit calibration tools allow the use of **adaptive DoE algorithms** as shown in Fig. 4.40. These make it possible, in a single automated test-bed session, to determine the drivable design space within predefined ranges for the operating points (**screening test**), calculate a corresponding DoE online and, in the same step, measure it (**adaptive DoE**). These sophisticated control unit calibration tools respond to possible limit violations by automatically adjusting the design space to the drivability limits. Test design types particularly suited for this purpose are D-optimal or space filling designs, as these support asymmetrically shaped design spaces.

Individual variation parameters can be extracted and automatically controlled to an optimal point for every operating state. To illustrate this procedure, we have included the

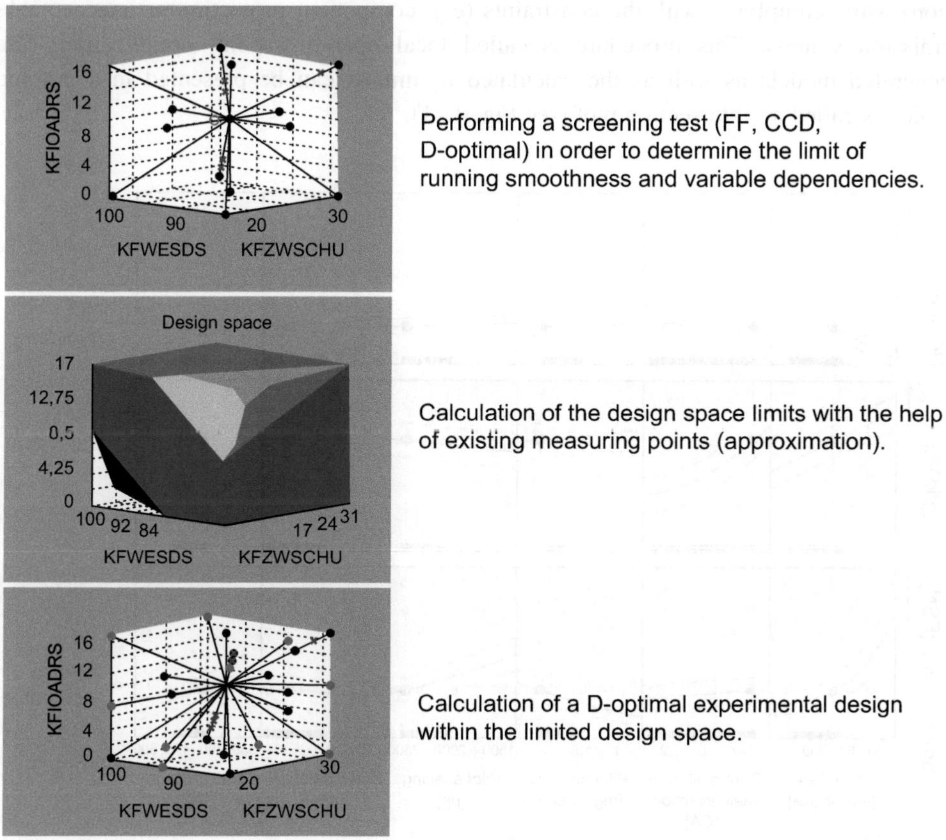

Fig. 4.40 Screening test with subsequent calculation of the envelope curve and D-optimal DoE measurement [28]

example of real-time ignition-timing control in a gasoline engine. In this example, ignition timing is controlled to the optimal center of gravity position, in other words to a defined safety margin in relation to the knocking limit. This reduces the number of the test design's variation parameters while, at the same time, adjusting the ignition time parameter to its optimal setting. Such control systems are applied in conjunction with secondary component-protection controllers [28].

The full measurement of the test design is followed by a statistical **raw data analysis and modeling process**. With the aid of statistical evaluations in the form of charts and characteristic values, the raw data undergo evaluation and modeling. Modeling takes place using a variety of approaches, with polynomial or artificial neural network models being most widespread. Here, too, please refer to the relevant literature [27].

After model generation and model verification using statistical methods, various methods are applied, e.g. gradient method or genetic algorithms, to optimize them. This is done by determining the best possible combination of parameters based on a target quantity (e.g. minimum fuel consumption) or target function (multi-objective optimization) while complying with the constraints (e.g. component protection or unacceptable emission values). This procedure is called local operating point optimization. The generated models as well as the calculated optimums can be presented in a graphic which is called an intersection plot (see Fig. 4.41).

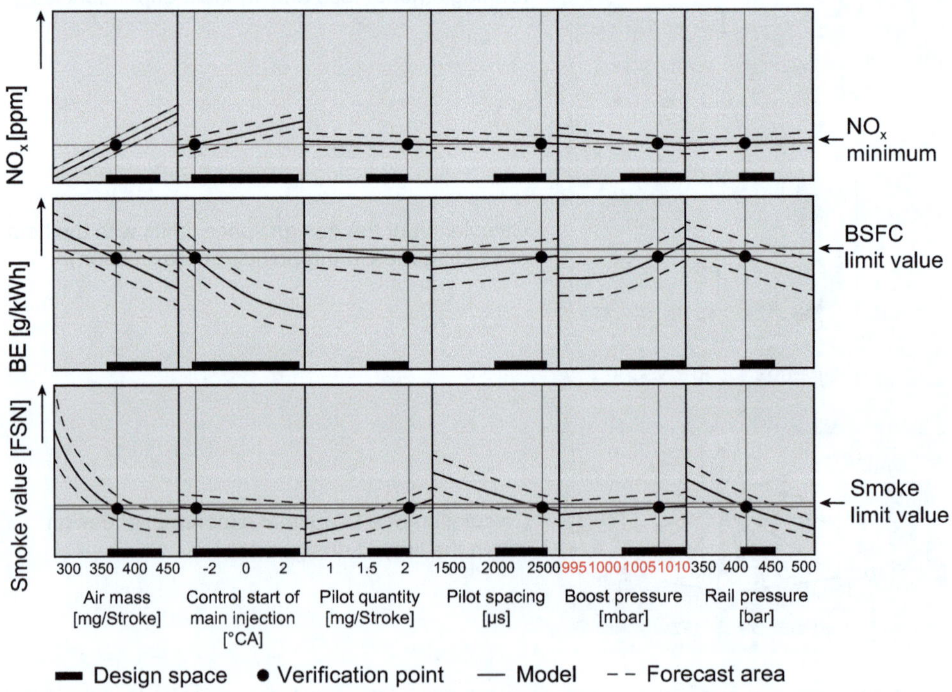

Fig. 4.41 Intersection plot—models for target and constraints across the engine's variation parameters [28]

After the optimization stage, the results undergo verification on the test bed. Based on the different optimums for each of the operating points, it is possible to use the related variation parameters to calculate the optimum settings for control unit maps using a map calculator, or even to derive an entire optimal data set for a control unit.

4.6.3.4 Differences Between Local and Global Models

Basically we distinguish between local (operating point-related) models and global models. Both approaches require different test design methods.

Local models always refer to operating points, that is, for example, a speed and a load. They are valid within the boundaries of this operating point. Nowadays, the information in the control units is usually stored in the form of maps with speed and load grid points. The grid points correspond to the operating points, making it easy to transfer the result to the control unit. To generate local models, the procedure is as described above. First, test designs are created and executed based on the operating points. Next, a plausibility check is run on the data, upon which the models are generated. The majority of models are based on the polynomial approach, as no irregularities are expected within the vicinity of the operating points. The following equation shows the example of a local second-order polynomial model with fuel consumption interdependencies:

$$\begin{aligned} \mathrm{BH} = {} & a_0 + a_1 \cdot S + a_2 \cdot S^2 + b_1 \cdot R + b_2 \cdot R^2 + c_1 \cdot A + c_2 \cdot A^2 \\ & + d_1 \cdot S \cdot R + d_2 \cdot R \cdot A + d_3 \cdot S \cdot A \end{aligned} \tag{4.24}$$

where:

BH	Fuel consumption [kg/h]
S	Injection start in [crank angle degrees before TDC]
R	Rail pressure in [bar]
A	Exhaust gas recirculation rate [%]
a_0 to d_3	Polynomial coefficients

Equation 4.24 Approximation of fuel consumption through a second-order polynomial including interdependencies [3]

Global models and the related test designs are not based on operating points. Instead, they contain speed and load, for example, as free parameters which, during the creation of the test design, are treated and varied like variation parameters. The results are models that describe the behavior in view of the target or constraint, going beyond speed and load. By fixing speed and load, global models can be used to determine local values for the operating points.

As the behavior of certain targets requiring to be modeled or of constraints may theoretically contain irregularities (e.g. sudden changes) that go beyond the operating point boundaries, and polynomial approaches are unable to represent such behaviors, automotive engineers generally use different model approaches. Among others, various types of neural networks come in useful that are capable of representing non-linear behavior in a suitable manner.

Whether a global model approach is to be implemented must already be taken into account in the test design stage. Prior knowledge, in particular, can be very useful, e.g. from pilot tests or relating to comparable predecessor engines. If there is no previous knowledge available, it may be helpful to conduct pilot tests. Such tests spread measuring points evenly throughout the investigated design space, e.g. using a Latin Hypercube test design (see Fig. 4.40). With the help of partitioning algorithms from artificial neural networks, the design space is divided into regions based on the acquired measuring data.

Regions with high nonlinearity are more finely partitioned by the algorithm than those with a linear curve. Regions in the map with high nonlinearity require more information, i.e. more measuring points, than areas with high linearity. This information is subsequently utilized to determine additional measuring points for regions with a high level of partitioning, i.e. regions with high nonlinearity, by using sub-test designs. In areas of high linearity there is usually no need for further information. In the end, the modeling algorithm is applied to smooth the sub-models and combine them into a global overall model.

4.6.3.5 Differences Between Steady-State and Dynamic Measurement and Modeling

Up until now, engines on test beds in basic applications were usually tuned in steady-state operation. But as actual driving conditions involve dynamic operation, the calibration engineer ideally optimizes the engine on a dynamic test bed and in the vehicle itself. The automated steady-state optimization approaches have already been discussed in detail.

Aside from these approaches there are methods to measure engines dynamically with subsequent dynamic modeling, which will be dealt with briefly below. For this purpose, special test designs are used to excite the engine dynamically. APRBS (amplitude-modulated pseudo random binary) signals are an example of such methods [29]. Determining the dynamic properties of an engine requires stochastic, dynamic excitation of all variation parameters across the entire operating range (Fig. 4.42).

A procedure like that results in data containing information about how the target values depend on the variation parameters in terms of time. In doing so, it must be ensured that any delay times of single measuring quantities, for example, that may be present in emission measurements due to the gas dynamic between exhaust gas sampling points and the measuring equipment are automatically aligned. These data can subsequently be used for dynamic modeling. For this purpose, artificial neural networks are used which, in addition to the already mentioned properties, are also capable of reproducing the system dynamics appropriately.

For the sake of completeness, we will also take a brief look at **transient measurement**. Transient measurement can be seen as an intermediate step between steady-state and dynamic measurement. In transient measurements, the variation parameters are changed

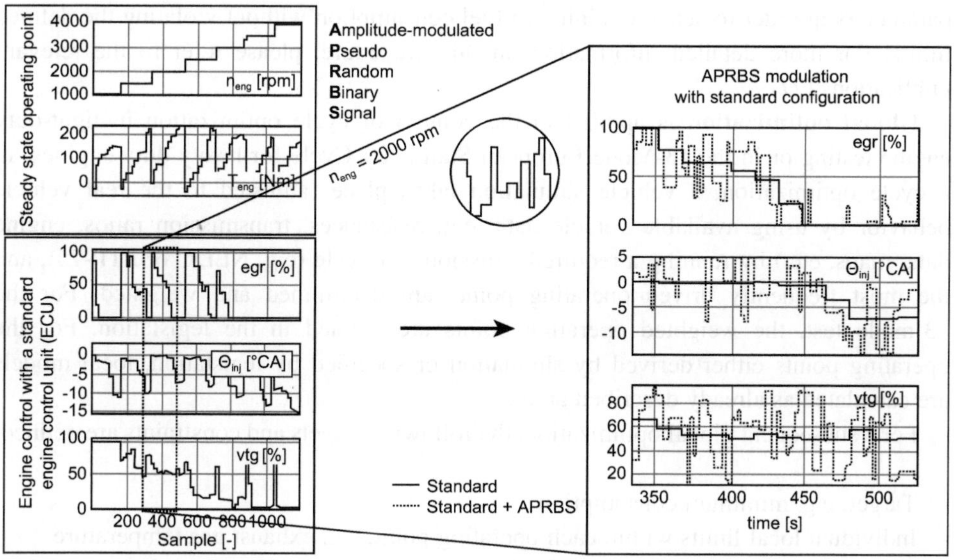

Fig. 4.42 Dynamic modeling procedure using APRBS signals [29]

on a continuous basis. But this must be done at a rate fast enough to keep testing times short and slow enough to prevent the system from being dynamically excited. During this process, the variation parameters are changed in quasi-stationary steps according to a ramp function so that the engine is practically operated in steady state. The procedure is also known as "slow dynamic slopes" [28]. The results are steady-state data. The method has the advantage of relatively short measurement times, however it is only suitable for specific variation parameters and calibration tasks.

4.6.3.6 Differences Between Operating Point Optimization and Cycle Optimization

With regard to the tuning procedure that follows model evaluation, we distinguish between two basic types of optimization:

– Local (operating point-related) optimization
– Cycle optimization (global optimization)

Local (operating point-related) optimization aims at calculating the best settings for the variation parameters within one operating point while complying with the local restrictions. The target may be minimum fuel consumption or maximum torque, for instance. With regard to the restrictions, we distinguish between legislative limits and engine-related limits. An optimizer, e.g. a gradient technique/method of the steepest ascent or an evolutionary method such as a genetic algorithm optimizes the variation parameters in order to achieve minimum fuel consumption without violating the defined limits. For more detailed information on this procedure, please refer to the relevant publications [27].

Global optimization is applied in the context of cycle optimization in light-duty engine testing or in the 13-Mode European Stationary Cycle for heavy duty engines. In a cycle optimization, a vehicle simulation takes place (adjusted to the real vehicle behavior by using available vehicle data, e.g. resistances, transmission ratios, engine data, maps, etc.) by running a required emission test cycle (e.g. NEDC or FTP-75), and the most frequently driven operating points are determined and weighted. For the 13-mode test, the weighted operating points are defined in the legislation. For the operating points either derived by simulation or specified by legislation, local models are calculated as already described above.

For a subsequent global optimization, the following targets and constraints are defined:

- Target, e.g. minimum consumption
- Individual local limits within each operating point, e.g. exhaust gas temperature
- Global limits, i.e. all operating points must satisfy these limits, e.g. emission variables such as nitrogen oxides, hydrocarbons, particles, etc.

Certain optimization algorithms, e.g. the Lagrangian method, enable determining the variation parameter settings that ensure minimum consumption while satisfying the local and global limits. It is critical that the local limits within an operating point are not violated. All global limits must be respected across all of the operating points, taking into account the operating points' specific weighting, however there may be violations within individual operating points. The results of a global optimization of this kind allow engineers at a very early stage of development, at a time where no vehicle is physically present, to draw conclusions on whether or not the vehicle will pass the cycle with the defined engine. Attention must be paid to the following boundary conditions:

- Many engine models are steady-state models and the driving cycle is dynamic.
- On the test bed, emissions are measured in mass per time; emissions in the cycle are defined in mass per kilometer.

The calibration engineer checks the plausibility of the generated models and the optimization results and verifies them on the engine test bed. Straight after their verification, the results, i.e. the optimal settings for the variation parameters within each of the operating points, serve as a basis for calculating new parameter maps for the engine control unit, which have to take into account transition behavior and the setting speed of the actuators. The parameter maps are calculated with the aid of a map calculator using specific calculation and smoothing algorithms. The parameter maps are then verified on the test bed and in the vehicle.

4.7 Measured Data Evaluation

The latest software tools permit conducting extensive assessments and analyses on the measured data in a task known as post-processing, i.e. a work step based on saved data, which is performed either alongside testing or afterwards. To conduct data evaluations, a number of options are available, ranging from simple data browsing to graphical interactive analysis or even complex applications with corresponding formula libraries.

These evaluations can be conducted on an interactive or automated basis. Adopting an interactive approach speeds up the generation of new evaluations significantly and makes the task more flexible. However if evaluations have to be repeated, the entire evaluation procedure has to be input all over again. In automated evaluations on the other hand, the initial effort is much higher, but the repetitive effort is a great deal lower.

4.7.1 Selection of Measurement Data

Every data evaluation starts with the selection of the measured data. Due to the continually improving methods of data acquisition, the amount of measured data available is continuously on the rise. This is why efficiency in measured-data selection is equally gaining in significance.

4.7.1.1 Interactive Selection

A high-performance data browser, like the one shown in Fig. 4.43, enables navigating the file structures and selecting the required data interactively, much like when using a file browser. Interactive selection is made possible not only by the filing structures themselves (folders) but also by the attributes of tests, test series or measurements.

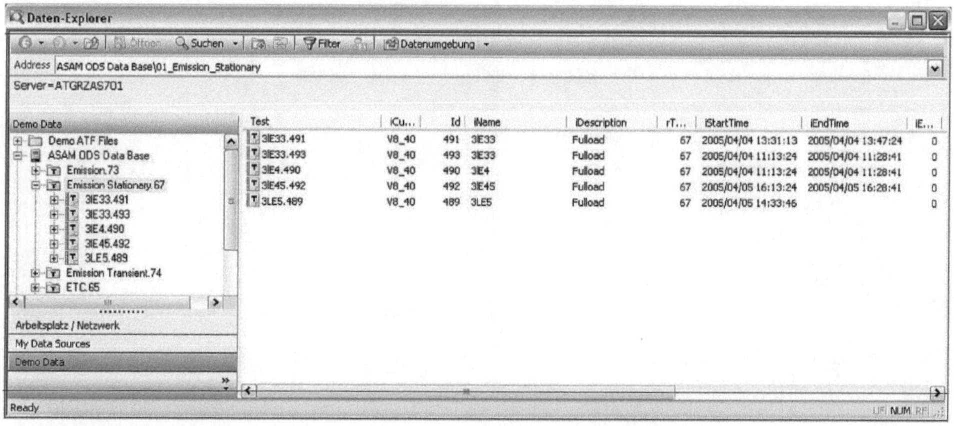

Fig. 4.43 Example of a data browser for displaying engine measurement data

The attributes are displayed in columns, allowing you to sort the test series, for example, according to the projects to which they are assigned. You can then select the located test series, which brings you to the next level. This is where you are once again able to sort the tests contained in the test series by certain attributes. In this way you can browse the data structures interactively.

Although both the ASAM ODS standard and the MDF format (see ASAM ODS in Table 4.2 and ASAM MDF in Table 4.3) have established themselves widely as data filing systems, the number of measurement data formats continues to increase steadily. Such measurement data formats have come to vary very widely in their structure. In one format, a file may correspond to a test, whereas, in another, a file may contain no more than one measurement channel. Other formats are based on databases, i.e. the structure is defined in the form of database tables and a data model.

The latest tools available for selecting measurement data display all of the measured data in one common logical structure, irrespective of their physical structure; the tool is preferably based on the logical structures of standardized filing systems such as ASAM ODS, for example. When dealing with test bed data, the test plays a central role. A test may contain several measurements, each of which contains the measurement channels belonging to each measurement. This logical representation, which is always the same, additionally enables an efficient comparison of measured data of different data formats and from different data sources (Fig. 4.44).

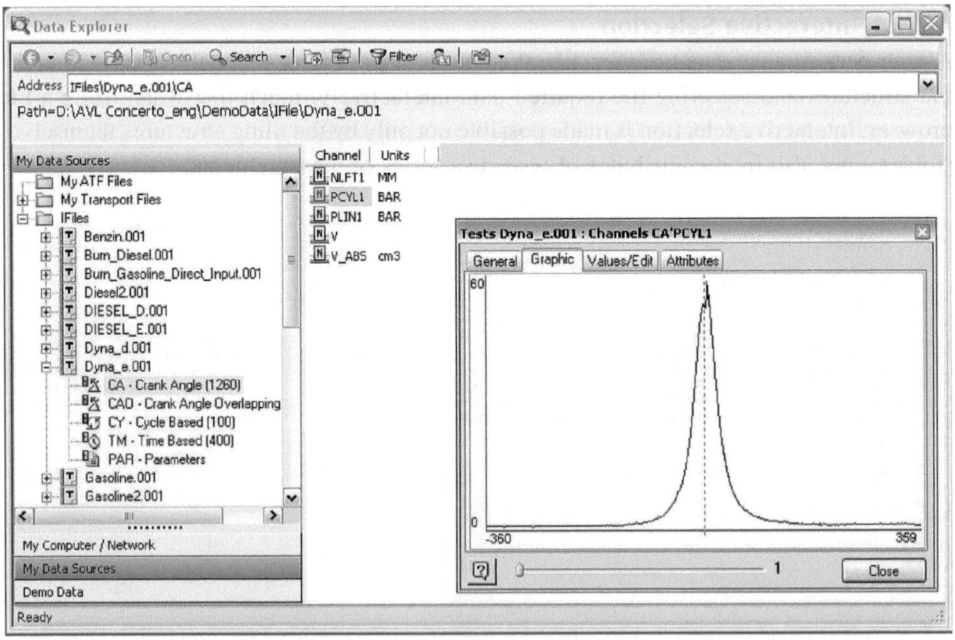

Fig. 4.44 Example of a measurement data browser for displaying indicating data

4.7.1.2 Data Mining

Beside interactive data selection, it is becoming more and more important to be able to employ powerful mechanisms for automatic data search. Ideally, attributes have already been defined in the filing system that can be used in the search process. This has the advantage that it speeds up the search, as the attributes can be read very quickly. The drawback, however, is that these attributes must have already been defined in the data model. Any addition of attributes requires a change in the data model. Accordingly, new attributes are only available in the more recent measured data (Fig. 4.45).

Another variation is searching by data contents. One of the benefits of this kind of search is that you can search for any data contents you like. As measured data are also vector data (unlike attributes), there are advanced search options, because you can search for values like maximums, minimums or means. The disadvantage with this type of search is that it takes longer because, firstly, the data contents have to be read and, secondly, there are vector data that have to be processed too. This kind of search demands much more of the evaluation software. Among other things, caching mechanisms play a key role.

In the age of index-based search engines, which we are familiar with from the Internet, this type of search is getting more and more attractive for measurement data. In this case, however, it is necessary to index the data first. Indexing is a process that must run automatically in the background, so that index data can be collected from newly measured data without any major delay. In contrast to Internet search engines that mainly index text contents, search applications for measurement data primarily have to index numerical data, while applications for vector data require additional statistical functions. The great strength of this search variant is the extremely high search speed (Fig. 4.46).

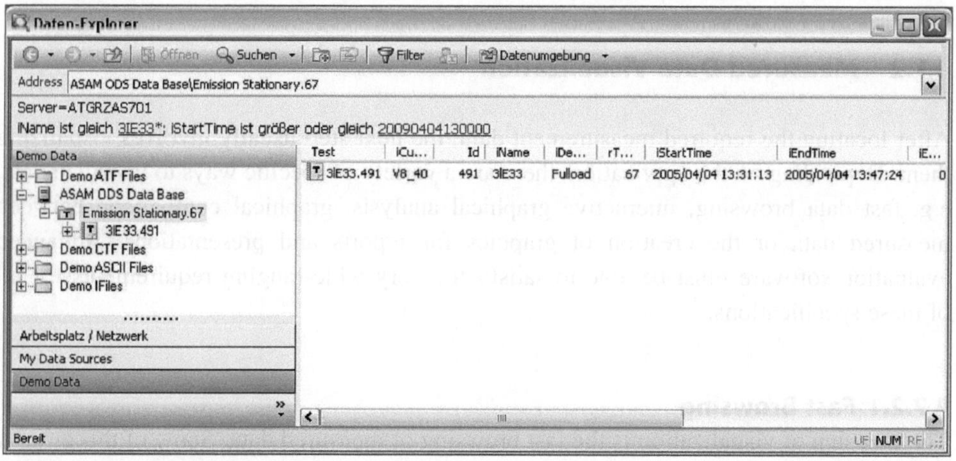

Fig. 4.45 Attribute-based data search

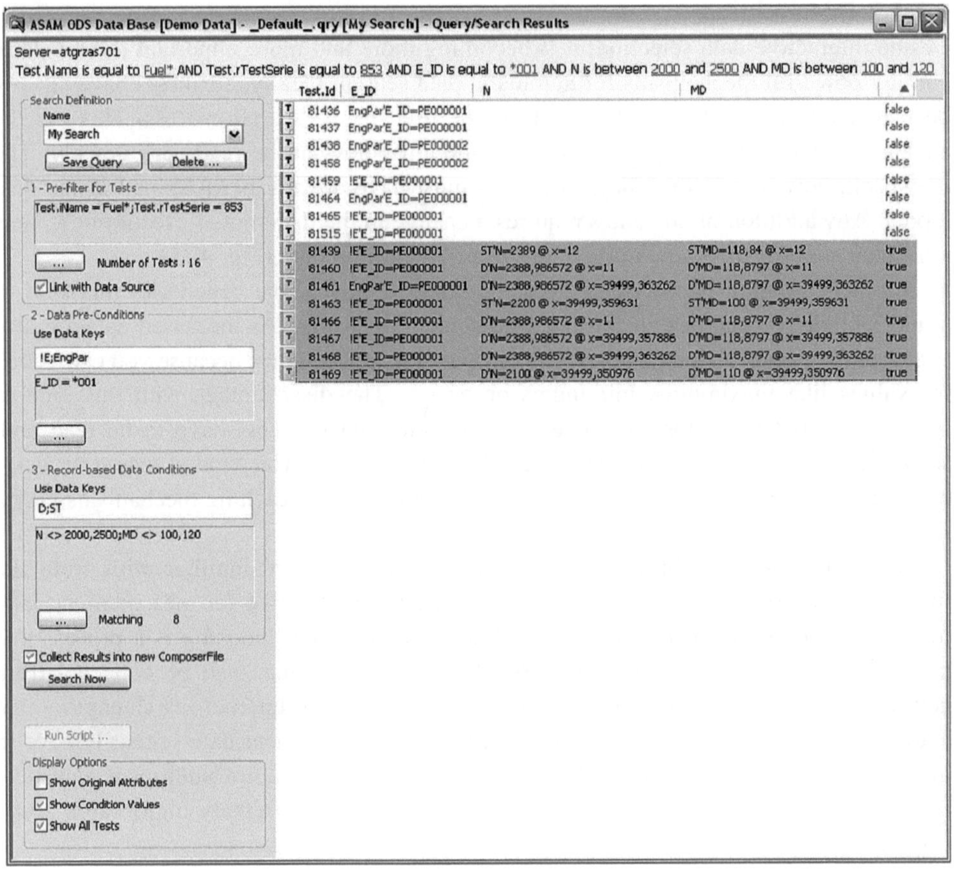

Fig. 4.46 Data search based on data contents

4.7.2 Measured Data Visualization

After locating the required measurement data, the next step usually involves visualizing them. Depending on the application, there are a variety of specific ways to visualize data, e.g. fast data browsing, interactive graphical analysis, graphical comparison between measured data or the creation of graphics for reports and presentations. Advanced evaluation software must be able to satisfy the very wide-ranging requirements for all of these specifications.

4.7.2.1 Fast Browsing

The first step of visualization is the fast browsing of measured data. Some of this already takes place when selecting the data if the data selection tool offers numerical or graphical preview features.

4.7.2.2 Interactive Analysis

There are various ways to visualize data for interactive graphical analysis. They might typically include charts with line graphics, bars or bands. These diagrams characteristically allow the free arrangement of almost any number of y-axes. It is even possible to display multiple data channels together on a single axis. A flexible legend is provided to identify the curves in the diagram. To measure the curves, these diagrams have one or multiple cursors that allow you to read individual values or differences between curves; they frequently also have extended statistical features. Further visualization options include numerical lists and tables (Fig. 4.47).

To create reports in presentation quality, it is necessary to be able to customize your diagrams according to your exact needs. This requires a comfortable graphical editor with an intuitive user interface. Each of the graphical objects must offer a sufficient amount of setting options to enable the desired customization. The use of standardized templates facilitates the creation of reports and ensures that they have the standardized appearance you require.

4.7.2.3 Map Visualization

Another requirement of the evaluation tool is that it must be able to represent characteristic maps. Maps are typically displayed as contour maps (Fig. 4.48) or 3D maps. In both cases it is necessary to use the more or less irregular measured data to create a regular grid

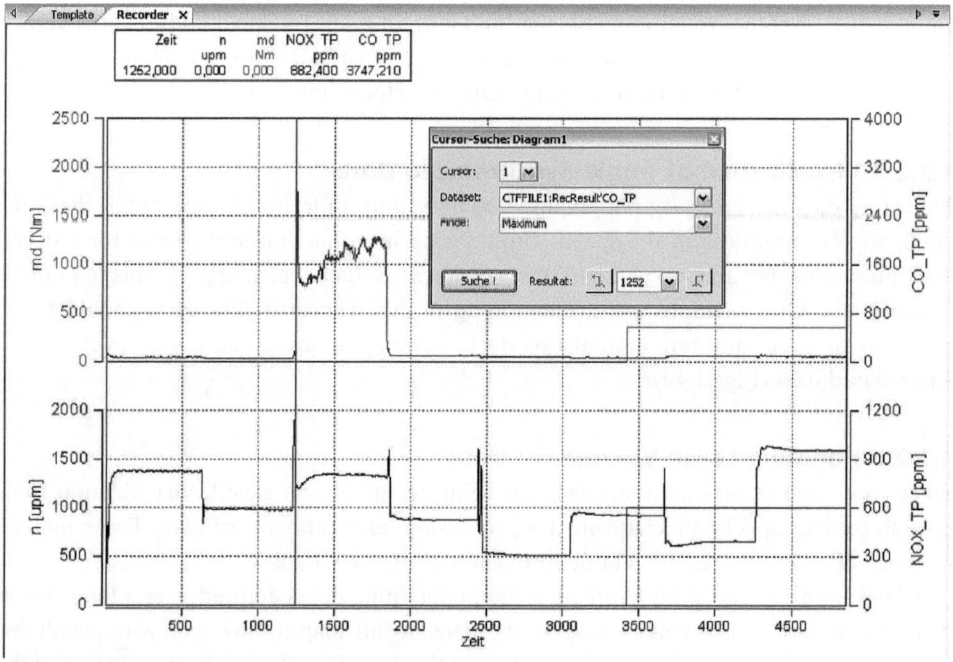

Fig. 4.47 Example of an interactive data analysis using a band cursor

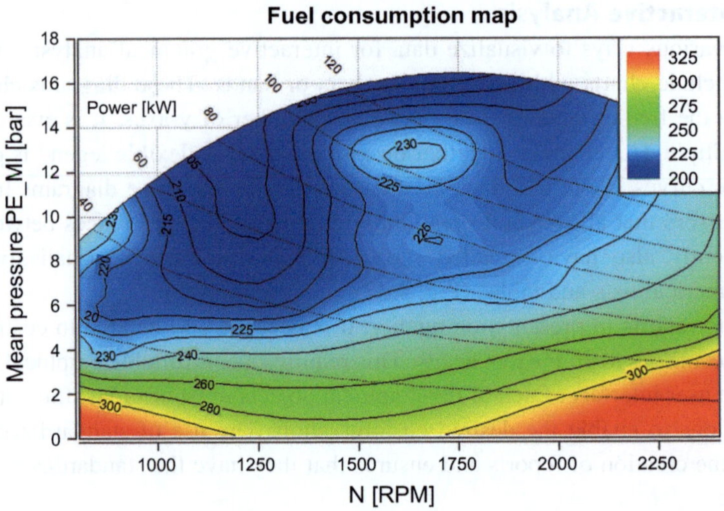

Fig. 4.48 Example for map visualization with contour lines for fuel consumption

as a basis for the visualization. This is done by the process of interpolation. Various algorithms are needed for the interpolation process in order to deliver useful results for differently organized data. For example, the algorithm for data that were captured on the test bed according to speed steps is not suited for creating maps for turbocharger compressors. Another requirement is the limitation of maps by upper and lower envelope curves. These envelopes are typically calculated automatically from the measured data's acquisition range. These require different kinds of algorithms too.

4.7.2.4 Visualization of Angle-Synchronized Data

Indicating data are typically not acquired equidistantly in terms of time but at the same crank angles regardless of the speed. Unlike generic evaluation tools, tools for test bed data must also be able to support the specifics of data occurring in this particular environment. One example worth mentioning in this respect is that there should be an easy way to select different combustion cycles when you are working with cyclic crank angle-based data (Fig. 4.49).

4.7.2.5 Handling Large Volumes of Data

Since more and more measured data are being recorded and saved, visualization tools need to provide special mechanisms for processing large amounts of data. These include methods of adaptive loading and the provision of preview data.

For example, if you want to create a chart containing data captured over a long period of time, the first step involves loading the data in full length, however with a reduced resolution. This gives users a good overview of the data. By effectively zooming in on the x-axis, it is now possible to enlarge a particular section. The software then automatically

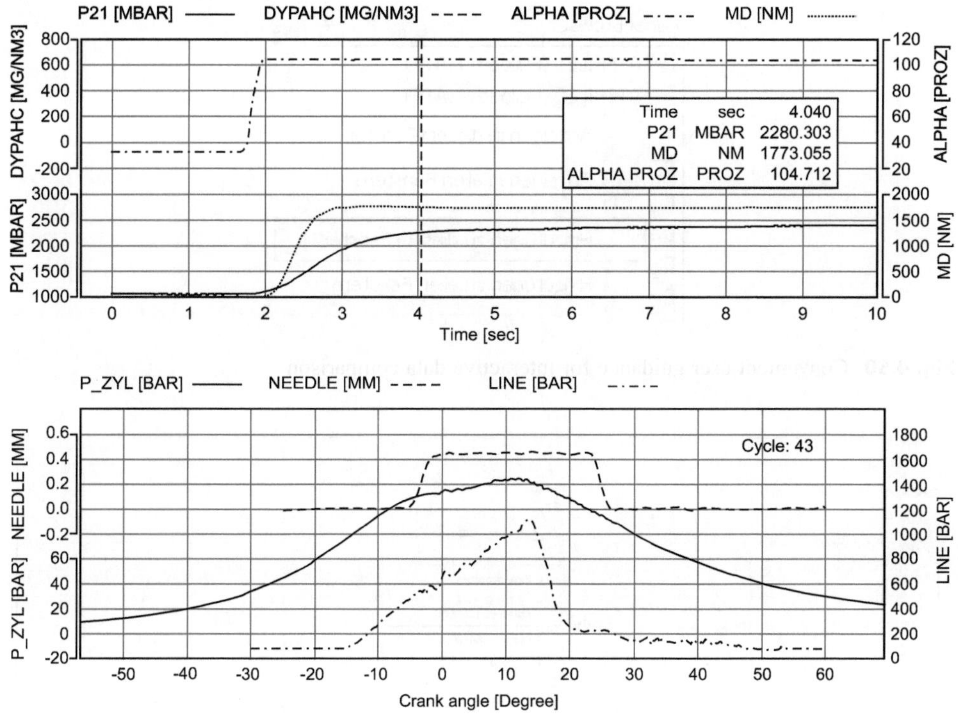

Fig. 4.49 Example of an interactive evaluation of indicating measurement data

switches to full resolution because now only a part of the data needs to be loaded. Here, too, caching mechanisms come into use, so that if the same section is reloaded, it is not necessary to load the data from a data server within the network all over again.

4.7.2.6 Data Comparison

A common application in the area of measured data visualization is the comparison of measured data. Typically, a diagram is initially created for measurement curves. Now if, for the purpose of comparison, you wish to superimpose the channels from another measurement on the channels in your diagram or even replace them completely, this should be conveniently possible with just a few simple interactions (Fig. 4.50).

This should also allow the automatic assignment of curve properties to the different measurements, e.g. line color, line thickness or line style.

4.7.3 Data Synchronization

Recordings on the test bed often involve multiple measuring devices. Many of these data are already synchronized via the automation system. But if measuring devices are involved that are not connected to the automation system and are only provided with a

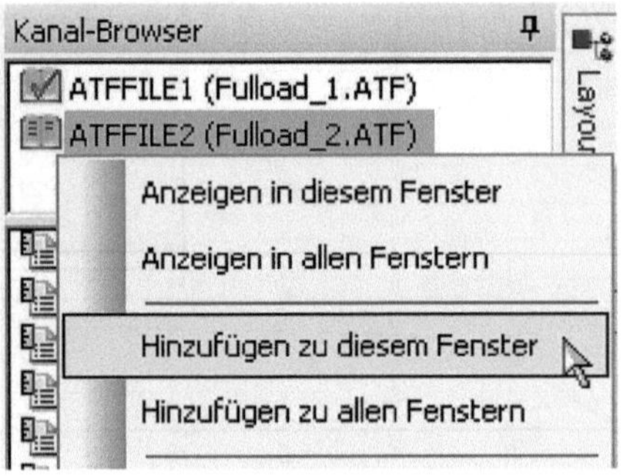

Fig. 4.50 Convenient user guidance for interactive data comparison

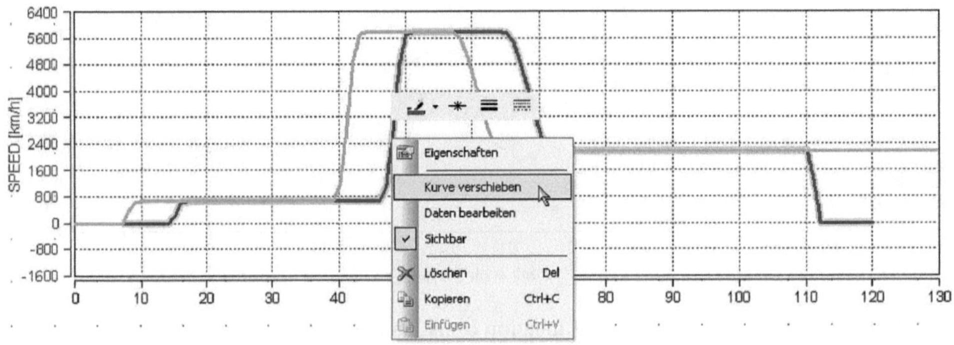

Fig. 4.51 Interactive shifting of measurement curves

recording of their own, the resulting measured data may not be synchronized in terms of their time base values. This means that there will be deviations with regard to resolution and starting times. For these data to be evaluated jointly, they first have to be synchronized to a common time base. Typically these devices will have also been used to measure shared signals or signals with similar curves (e.g. speed or velocity). Such signals can now be used for the synchronization. There are various ways to accomplish this task:

- Interactive time base shifting
- Automatic time base shifting

The synchronization system must meet further requirements if data recorded on a time-synchronized basis are to be represented together with data captured on an angle-synchronized basis, or if numerical data are to be combined with video recordings.

4.7.3.1 Interactive Time Base Shifting

The evaluation tool's graphical editor provides an option that allows you to display the curves to be synchronized. Then you can shift the required curves, either by dragging them with your mouse or by using specific dialogs (Fig. 4.51).

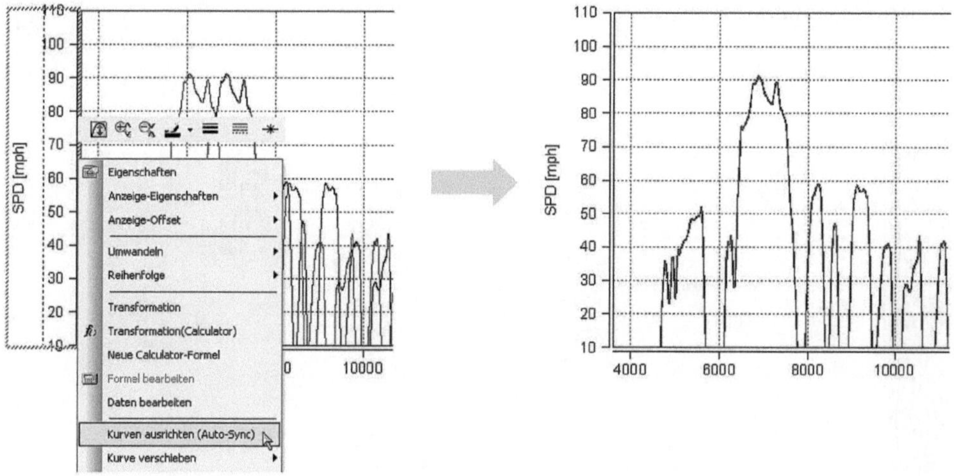

Fig. 4.52 Easy-to-use graphical user interface for automatic curve alignment

Once a distinctive curve has been shifted, the shifting process can be optionally applied to all channels of the measurement.

4.7.3.2 Automatic Time Base Shifting

In addition to manual shifting options, good post-processing tools should also offer the possibility to shift measured data automatically. Here, too, it is necessary to find curves among the measurements to be synchronized that are as similar as possible and have enough distinctive signal characteristics (Fig. 4.52).

The tool now employs special correlation algorithms which automatically calculate the shifting value, at which the two curves achieve maximum overlap. After the calculation, the result is immediately shown graphically, allowing its verification and manual fine-tuning, should this be necessary. With the shifting function being embedded in the graphical editor, it is also possible to select a specific range with the band cursor ahead of the automatic synchronization process. This makes the subsequent automatic synchronization process faster and more precise. Again it is possible to choose whether the shift is to be applied on the entire measurement.

4.7.3.3 Synchronizing Test Bed Data and Indicating Data

Apart from pure time-base synchronization, there are further applications for data synchronization. Imaginable are e.g. synchronizations between steady-state test bed data and data from indicating measurements running in parallel. The information needed for linking the data is saved together with the steady-state measurements already during the acquisition process. The evaluation software should be able to use this information to

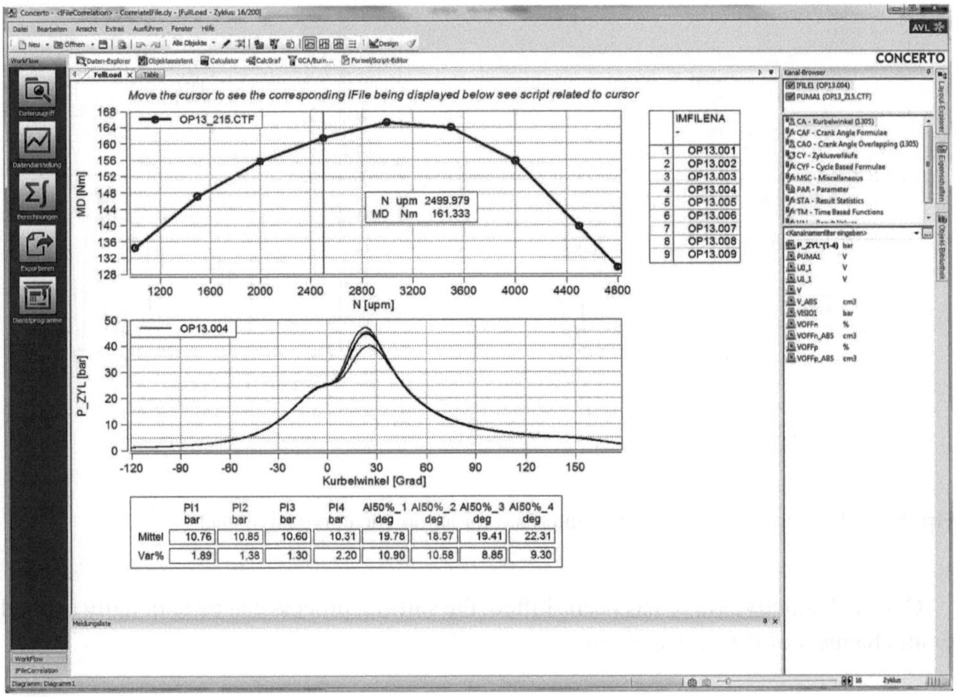

Fig. 4.53 Example of an interactive automatic synchronization of test bed measuring data and linked indicating measurement

automatically find the indicating measurement that matches a certain steady-state measuring point and display it (Fig. 4.53).

4.7.3.4 Synchronization Between Measured Data and Video Data

It is becoming an increasingly widespread practice to record image data in the form of videos alongside recorder measurements. Advanced evaluation tools should allow you to view and play measured data and video recordings synchronously. If you move the diagram cursor to a different section of the measured data, the video image should automatically adapt to the new position. If you go to a different position in the video, the diagram should adapt accordingly (Fig. 4.54).

4.7.4 Formulas and Calculations

In addition to visualizing the measured data themselves, it is also necessary to display the calculated quantities. Calculation typically takes place in a formula based calculation engine, which is capable of directly accessing the measured data as input quantity. The

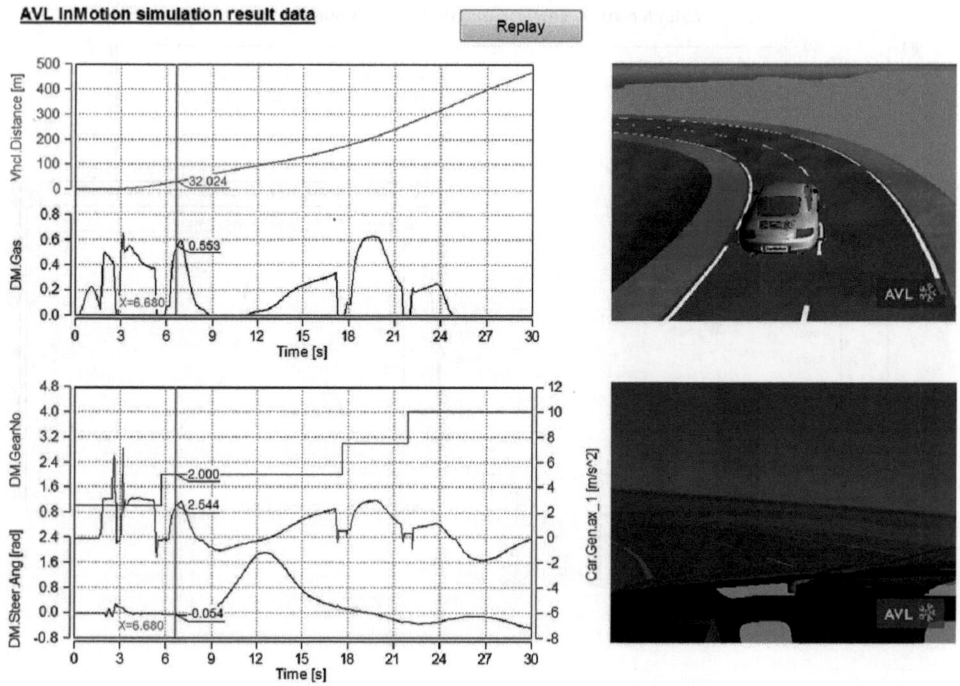

Fig. 4.54 Example of a synchronous interactive visualization of data curves and video contents

calculation results are ideally shown alongside the measured quantities, so that users can use the calculated quantities in exactly the same way as measured quantities. This formula engine is an integral part of the evaluation software. Calculation takes place automatically as soon as the calculated values are needed for visualization or further processing.

4.7.4.1 Tools

Depending on the application case, different tools are available that can be used to define the calculation. Simple calculations are typically carried out with a calculator tool; more complex calculations will require a simple programming language. Particularly convenient are graphical calculation editors where calculations can be defined by arranging and connecting function blocks. Data evaluation software is ideally provided with the full bandwidth of such calculation tools (Fig. 4.55).

4.7.4.2 Functions

Common Functions

Data analysis requires a large number of calculation functions. In addition to the basic arithmetic functions, all common functions, such as integration, differentiation, logarithmic calculus as well as trigonometric, statistical or logical functions, etc., are needed.

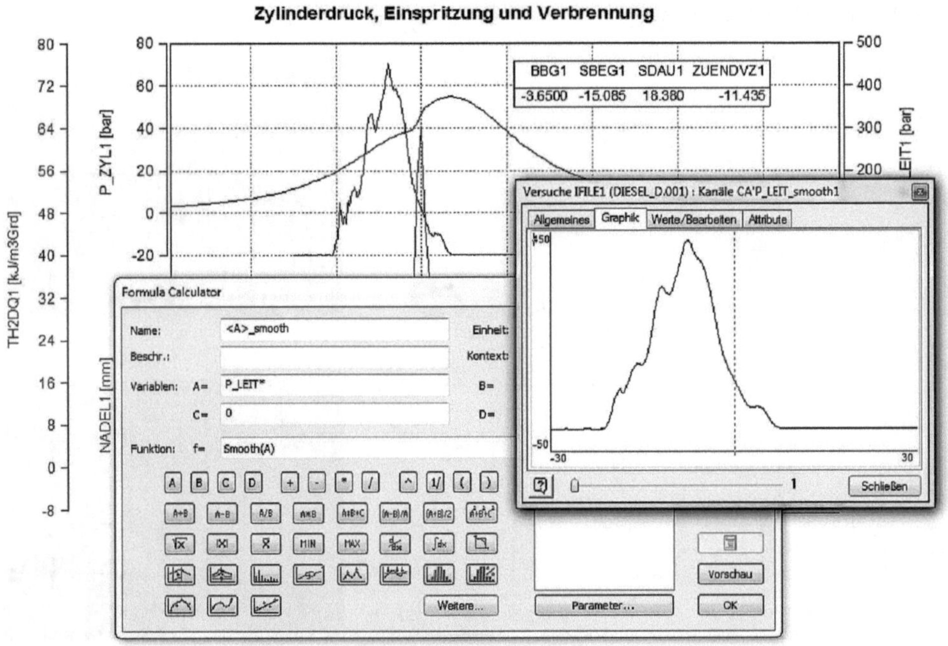

Fig. 4.55 Calculator—a simple tool for defining calculations

Furthermore, signal processing functions, such as smoothing functions, splines or approximation methods, are of essential significance. Every measurement data evaluation tool must also include a range of digital filters and frequency analysis techniques (Fig. 4.56).

Event Evaluation

Event-driven evaluation systems provide an efficient way to evaluate very large volumes of measured data which generally have been recorded over a very long measuring period. Typically, the portions of the measured data which are actually relevant for the evaluation only account for a fraction of the entire data. Very often these relevant portions can be identified by events, for example gear-shifting events. In such cases, only the data sections shortly before and after these events are of interest.

On the one hand, the tool should allow the fast identification of these events and, on the other, display the events on a time axis. In this respect, it is also very helpful if the events can be filtered, for example, if you want to limit your analysis to upshifting gear changing events. It should be additionally possible to use the identified events for navigation, allowing a convenient and fast further graphical examination of any interesting sections of data occurring before and after the event.

The upper diagram in Fig. 4.57, for example, shows all gear-shifting events. The diagram below only shows the selected gear-shifting events. If you move the cursor to a

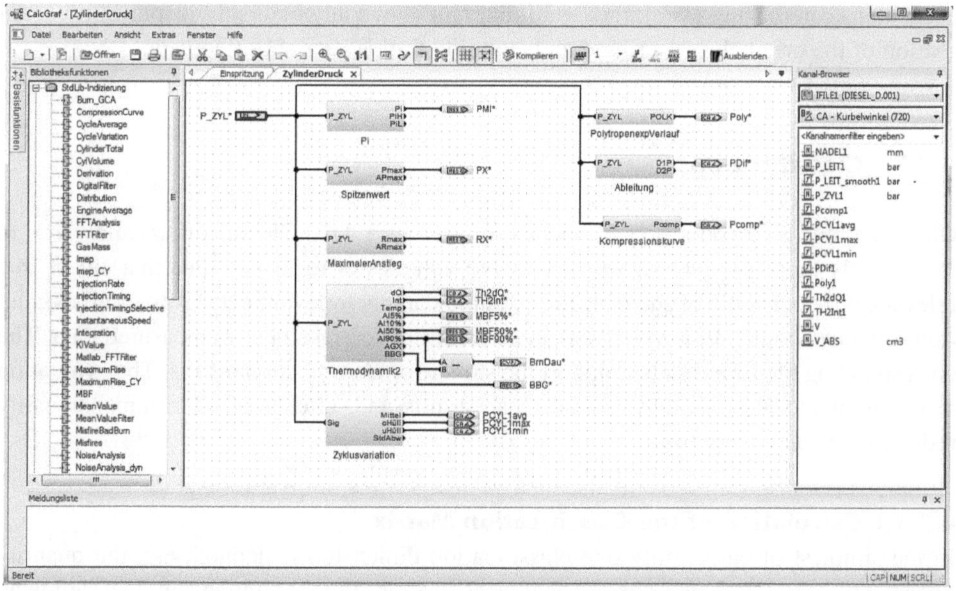

Fig. 4.56 Example of a graphical definition of calculations

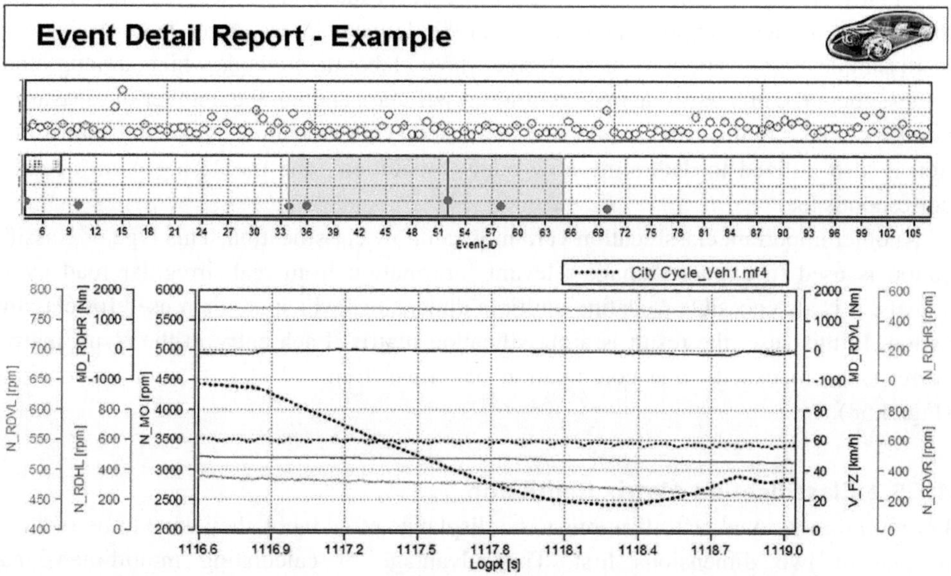

Event-ID	Begin [n]	End [n]	Eng_speed_after [Upm]	Eng_speed_before [Upm]	Gear_before [-]	Gear_after [-]	Shift_duration [s]	V_FZ [km/h]
52	335013	335747	2863.4	4442.5	1	2	2.4464	60.435

Fig. 4.57 Schematic diagram of the broad application area

particular gear-shifting event, the detail diagram above automatically displays the detail section of the original measuring data belonging to this gear-shifting event.

4.7.5 Classifications

Classifications give you a very efficient way to reduce data. Classifications allow you to extract only those portions of very extensive volumes of measured data that are of true relevance to your topic of interest, for example damage-relevant data. Advanced evaluation software should allow you to calculate classification results from measured data. The corresponding file formats also support the storage of classification results. The evaluation program should also be able to read these results and, after a possible dimensionality reduction, display them too.

4.7.5.1 Calculation of the Classification Matrix

In the simplest of cases, only one classification dimension is defined, e.g. the quantity Speed. This is called a one-dimensional classification. For this dimension, you define an area which you divide into classes, along with the width of each class. In the case of the simplest classification (referred to as sample classification), the next step is to determine how many measuring points are contained in each of the defined classes. The result is a distribution of measuring data across each of the defined classes. Aside from straightforward sample classification, there are further more elaborate methods which, among other things, count the number of actual revolutions per class and not the number of measuring points per class. In this case, in addition to the actual classification signal, speed information is also utilized to determine how many revolutions one measuring point actually corresponds to.

Another important classification variant is rainflow classification. This type of classification is used to extract damage-relevant information from real, irregular load cycle signals. It is also possible to define multiple dimensions which are then used for classification. In this case, the result is a classification matrix. Each entry in the result matrix corresponds to exactly one combination of values from all classification dimensions (Fig. 4.58).

4.7.5.2 Classification Matrix Reduction

For multidimensional classifications to be displayed effectively, they have to be reduced to one or two dimensions first. The advantage of calculating multidimensional classifications is that you only have to carry out the time-consuming classification calculation once. The task of reducing a multidimensional classification to a lower dimension accordingly requires no more than a fraction of the time and can now be done very quickly for any combinations of dimensions. This functionality should also be included in modern evaluation packages.

CO2persec_avg [g/s] EngineSpeed [1/min] Refresh

EngineTorque [Nm]	0	200	400	600	800	1000	1200	1400	1600	1800	2000	2200	2400	2600	2800	3000	3200	3400	3600
-20				0.0013	0.0080	0.1339	0.0623												
-10				0.1112	0.2593	0.3247													
0				0.2452	0.3215	0.4207	0.6051	0.9785											
10				0.3984	0.5699	0.5593	1.0862	1.0977											
20				0.5010	0.6563	0.7447		1.0254											
30				0.6137	0.7927	0.8928		1.3630											
40				0.6989	0.9049	1.0436	1.5732	1.6460	1.9415										
50				0.8389	1.0039	1.2268	1.7469	1.8153											
60				0.9556	1.1198	1.4012	1.9645	2.0366											
70					1.2350	1.5365	1.8019	2.1250											
80						1.6970	1.9735	2.3576	2.7839	3.0810									
90									3.0074	3.2817	3.6450	4.1344	4.6114	5.0169					
100													4.9774	5.2579	5.6339	6.2467	6.7635	7.2104	

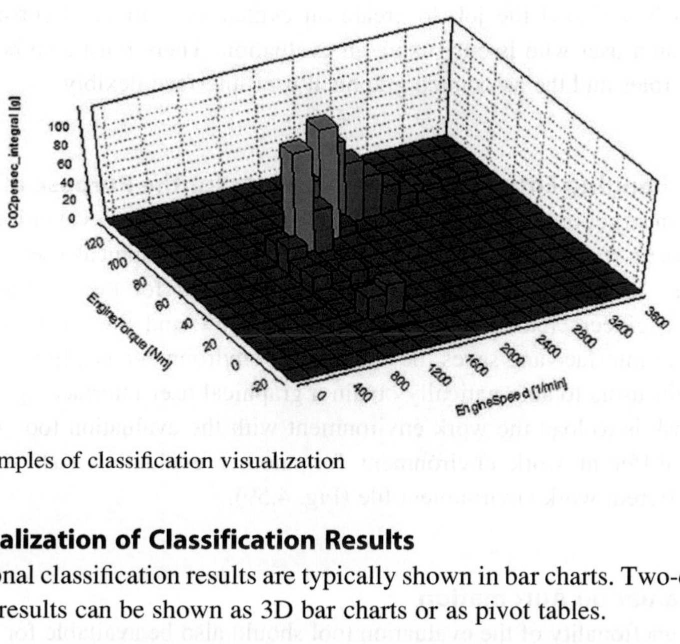

Fig. 4.58 Examples of classification visualization

4.7.5.3 Visualization of Classification Results

One-dimensional classification results are typically shown in bar charts. Two-dimensional classification results can be shown as 3D bar charts or as pivot tables.

4.7.6 Efficiency Enhancement in Data Evaluation

4.7.6.1 Layouts

Creating more complex views on measured data can take a great deal of effort. That is why it is essential that the common structure used for such views (the so-called layouts) can be easily saved and reloaded later. It should also be possible to re-use these layouts with any other measured data e.g. from other test runs as long as the names used for the measurement channels are identical.

4.7.6.2 Modeling Workflows

More complex evaluation tasks consist of several steps, which are typically processed according to a defined order. The evaluation tool should allow you to model specific workflows and customize the tool's graphical user interface to your workflow specifics. If necessary, any tool functions not needed should be able to be hidden, so as not to distract the user from the actual procedure.

4.7.6.3 Adjusting the User Interface to a Specific Evaluation Task

It should also be possible to customize the tool's graphical user interface to the user's role. A user who is assigned the job to create an evaluation will need considerably more functions than a user who is only using an evaluation. There must also be the ability to define these roles and the associated graphical user interface flexibly.

4.7.6.4 Exchanging Entire Work Environments for the Purpose of Evaluations

Particularly in larger organizations, there are a variety of complex evaluation tasks that are defined by their specific workflow and their customized graphical user interface. So it must be easy to exchange the entire work environment for the evaluation task. The application engineer creates all layouts, the workflow and the customizations to the graphical user interface and saves the entire work environment as a file. All that remains to be done for users to automatically obtain a graphical user interface optimized to their particular task is to load the work environment with the evaluation tool. Changing to a completely different work environment for another evaluation task simply requires loading a different work environment file (Fig. 4.59).

4.7.6.5 Evaluation Automation

The entire functionality of the evaluation tool should also be available for the automated evaluation of measured data. This means that all actions that are possible via the user interface should also be available programmatically in a script language. Macro recorders support the automation process by recording interactive steps and making them available as an executable script, for example. Such a recording can also serve as a basis for a complex automation routine.

This then makes it possible to execute evaluations fully automatically. For example, you can perform the same evaluations on all new tests, so that the same standardized reports are available for all tests afterwards. This means, the reports can be created either straight after saving the test or collectively for all tests in an overnight session.

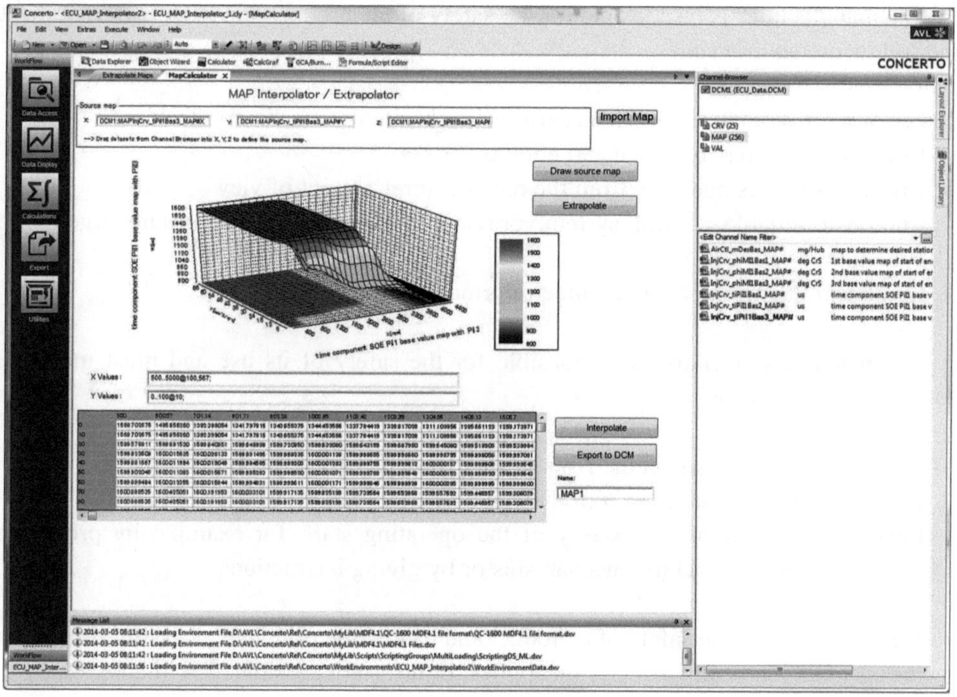

Fig. 4.59 Example of a work environment for map interpolation with customized workflow

4.8 Safety

4.8.1 Risk Analysis and Risk Assessment

As specified in the Machinery Directive 2006/42/EC, manufacturers of machinery are required to carry out a risk analysis in order to determine the risks associated with the machine. The EN ISO 12100:2010 standard "Safety of machinery—General principles for design—Risk assessment and risk reduction" specifies what is required for risk assessment. The standards EN ISO 13849-1 or IEC 62061 can be used to estimate the defined risks by safety category, performance level (PL) or safety integrity level (SIL). The standards specify design principles for the safety-related parts of control systems.

Risk assessment is based on the following criteria:

- Severity of the potential harm (serious/slight)
- Frequency or duration of exposure to the hazardous phenomenon (frequent/rare)
- Possibility of avoiding the hazardous phenomenon (virtually impossible/possible)

The manufacturer is responsible for the safety of the product and must therefore meet the following requirements:

- Perform a risk analysis on the hazards arising from the machinery itself
- Establish measures to reduce and eliminate risks
- Document the residual risk from the manufacturer's point of view
- Choose a suitable control system, provide safety systems and ensure ergonomic operation
- Document the machine and include information about residual risks

The machinery operator is responsible for the safety of its use and must meet the following requirements:

- Perform a risk analysis on the risks arising from using the machinery
- Document the residual risk from the operator's point of view
- Ensure safe operation and safety of the operating staff, for example by providing suitable barriers or locking mechanisms or by giving instructions

When it comes to the safety of a product and its use, the principle of risk minimization by means of an inherently safe design applies. If this is not possible, it is first necessary to implement technical protection measures. Providing appropriate user information is the last step in the event that design-related or technical protection measures are unable to be implemented.

4.8.2 Risk Analysis on Test Beds

Risk analyses have shown that the mechanical motions inside the test cell pose the greatest hazard to human life and health. Particularly hazardous are the rotating motions of the system (consisting of dynamometer, shaft and unit under test) on the engine or electric motor test bed and the rotating motions of powertrain components. High torques also present a great risk, particularly on multi-dyno test beds. Even at low speeds or at standstill (speed = 0), torsional stress in the powertrain may lead to material damage or breakage of axles or shafts. Broken rotating axles and shafts are dangerous even at low rotational speeds. The EN ISO 13849–1 standard pays particular attention to the risks described above. Risks arising from fire, contamination, transport by means of mechanical systems, and hazardous substances are dealt with separately. In test beds for electrical vehicle components, high direct-current voltages represent a new potential hazard. When dealing with fuel cell test beds, attention must be paid to escaping hydrogen gas (H_2).

The EN ISO 13849–1 "Safety of machinery—Safety-related parts of control systems—Part 1: General principles for design" describes the safety level of a safety function as a performance level (PL). Rotating parts on the test bed that can lead to serious injury or death therefore require the performance level "d." When using safety-related parts of control components, this performance level can be achieved by ensuring redundant structure, for example by a two-channel design. The systems have to be designed with a diagnostic coverage (DC) of ≥60 % and with high component quality (MTTFd >30 years). The rating PL "d" corresponds to the rating SIL 2 as defined in the IEC/EN 62061 standard, which describes the safety level of a safety-related function as safety integrity level (SIL). Figure 4.60 shows a risk graph in accordance with EN ISO 13849–1 to determine the performance level for rotating parts as they can be found, for example, in dynamometers, shafts and internal combustion engines or electric motors on the test bed.

The degree of diagnostic coverage DC is a measure for the effectiveness of the diagnosis and is determined based on the ratio of the failure rate of detected dangerous failures to the failure rate of total dangerous failures. The mean time to dangerous failure (MTTFd) can be used as a measure for component quality. An example of redundant architecture can be seen in Fig. 4.61.

The performance level has to be determined separately for every hazard (e.g. conditioning). After that, the safety chain is designed. The German Institute for Occupational Safety and Health of the German Social Accident Insurance developed a tool called SISTEMA that can serve to create computational proof. Safety circuits can be verified with the help of the Safety Calculator PAScal, for example, available from Pilz GmbH & Co. KG.

4.8.3 Safety-Relevant Systems

The term "safety-relevant systems" comprises all systems belonging to the test bed that present a danger to the life and health of people during operation and have to be safely

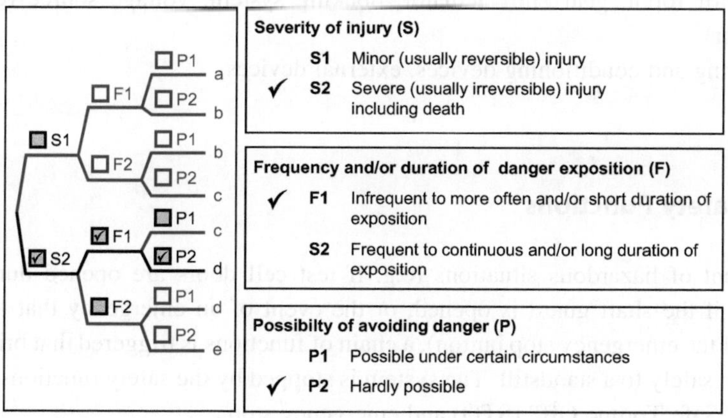

Fig. 4.60 Risk graph in accordance with EN ISO 13849–1 to determine the performance level for rotating parts

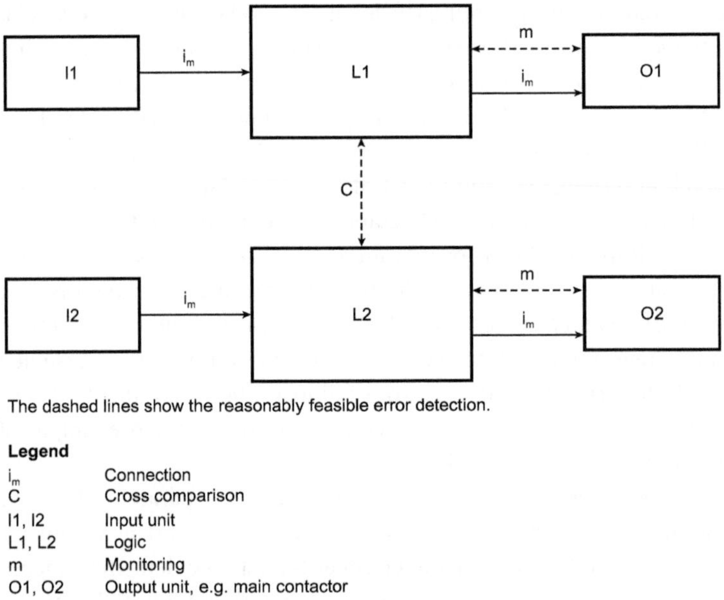

Fig. 4.61 Redundant architecture with cross comparison of two control units

brought to a standstill in dangerous situations (Safe Torque Off, emergency stop). The systems relevant to safety comprise:

- Dynamometer, electric drive unit
- Unit under test: internal combustion engine, electric motor, hybrid powertrain
- Shaft: connection between dynamometer and unit under test
- Source of electric voltage: battery, fuel cell, supercapacitor
- Gear shift robot, gear shift actuator, docking system, voltage source (e.g. battery emulator)
- Measuring and conditioning devices, external devices

4.8.4 Safety Functions

In the event of hazardous situations (e.g. if test cell doors are opened during testing operation, if the shaft guard is opened, or the event of an emergency that leads to the pressing of the emergency stop button), a chain of functions is triggered that brings the test bed system safely to a standstill. The system is stopped by the safety functions "Safe Stop 1" (SS1), "Safe Torque Off" (STO) and emergency stop.

In "Safe Stop 1" (SS1) the electrical power drive system or the external safety control system performs the following functions:

(a) Triggers and monitors deceleration of internal combustion engine or electric motor within pre-defined limits and triggers the STO function after the engine speed has reached standstill conditions, or
(b) Triggers deceleration of internal combustion engine or electric motor and, after an application-specific delay time, triggers the STO function.

This safety function corresponds to a controlled stop in accordance with IEC 60204–1 (Stop Category 1).

In the state "Safe Torque Off" (STO) the internal combustion engine or electric motor is no longer supplied with energy which could cause rotation (or motion in the case of a linear motor). The electrical power drive system delivers no energy to the internal combustion engine/electric motor that could generate torque (or force in the case of a linear motor). This safety function corresponds to an uncontrolled stop in accordance with IEC 60204–1 (Stop Category 0). This safety function is used where power removal is required to prevent an unexpected start-up or to limit the speed.

STOP, on the other hand, refers to a controlled stop where the energy supply to the machinery drive elements is maintained. This safety function corresponds to a controlled stop in accordance with IEC 60204–1, Stop Category 2.

The following requirements must be met on the test bed to enable compliance with the safety requirements:

- The test cell is spatially separated from the operating area. All potential danger sources are located inside the test cell: dynamometers, unit under test, shaft, gear shift robot, gear shift actuator, clutch actuator, docking system, battery simulator, measuring and conditioning units and external devices.
- Walls, floor, ceiling, doors and windows of the test cell are sufficiently solid, so that any possibly breaking parts ejected through the air—for example in the event of material failure—are unable to break through to the outside.
- All moving parts inside the test cell—most importantly those that present a danger to the life and health of persons (e.g. shaft, moving parts of the internal combustion engine/electric motor)—are covered for protection against contact (guard).
- All components needed for test operation comply with all of the applicable EC directives. The corresponding declarations of conformity and/or declarations of incorporation must be provided.

4.8.5 Safety Hardware

4.8.5.1 Test Beds for Internal Combustion Engines and Electric Motors

The figure below (Fig. 4.62) shows the basic setup of the safety hardware on an internal combustion engine test bed.

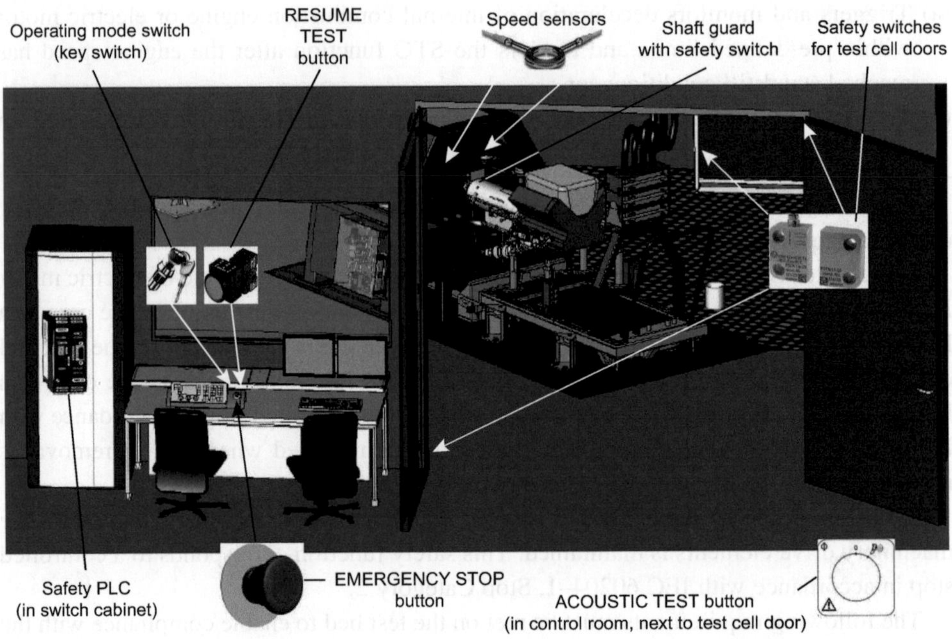

Fig. 4.62 Example of safety hardware for internal combustion engine or electric motor test beds

As the rotating parts on an internal combustion engine test bed are no different from those on an electric motor test bed, this arrangement can be used for either.

4.8.5.2 Multi-Dyno Test Beds

The safety hardware for multi-dyno test beds is somewhat more complex than for single-dyno testing facilities. It is shown in the following Fig. 4.63:

4.8.6 Setup of Safety Functions

The basic setup of the safety functions has the following characteristics:

- Where required, the entire hardware is based on a two-channel design.
- To achieve the safety level required by the standard, adequate hardware components are used.
- The safety switching device (safety PLC) has a response time that is fast enough to be able to activate a Safe Torque Off in the dynamometers (that ramp up fast) before any dangerous speeds have been reached. The abbreviation PLC stands for programmable logic controller.

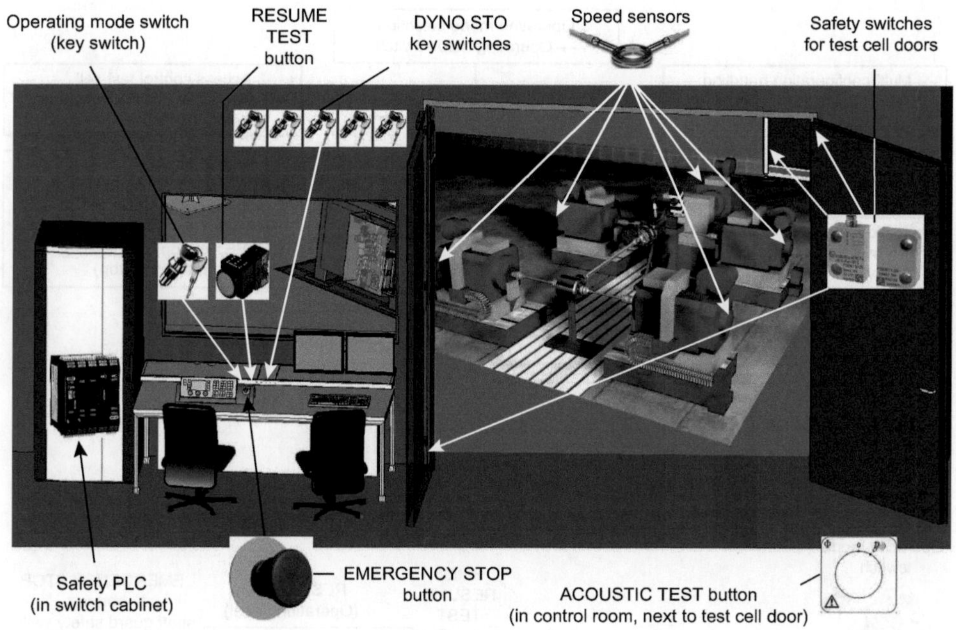

Fig. 4.63 Example of the safety hardware for powerpack or multi-dyno test beds

- The control units for hazardous motions (dynamometer, gear shift robot) have a two-channel design inside, too, in order to ensure the functions Safe Torque Off and emergency stop.
- The dynamometers must have a safety-oriented interface that enables activating Safe Torque Off and emergency stop.
- By means of functional signals, the respective operating modes and responses are also displayed and processed during automated operation. However, this functionality is not safety-oriented. Safety functions are performed by the safety hardware.

As shown in Fig. 4.64, the safety PLC monitors the following sensors and safety functions:

The safety functions are differently implemented, depending on the operating mode. They are described below.

4.8.6.1 Test Mode

Test mode comprises manual operation, automatic operation and remote operation. No person must be inside the test cell during normal test operation. The opening of test cell doors is monitored by door contacts to ensure safety. In this operating mode, test runs can be executed. Opening a door to the test cell triggers a safety-oriented switch-off of the system. After a controlled stop, the dynamometer is in the safe state "Safe Torque Off" (STO). If the door to the test cell is opened at standstill (e.g. end of test run), the

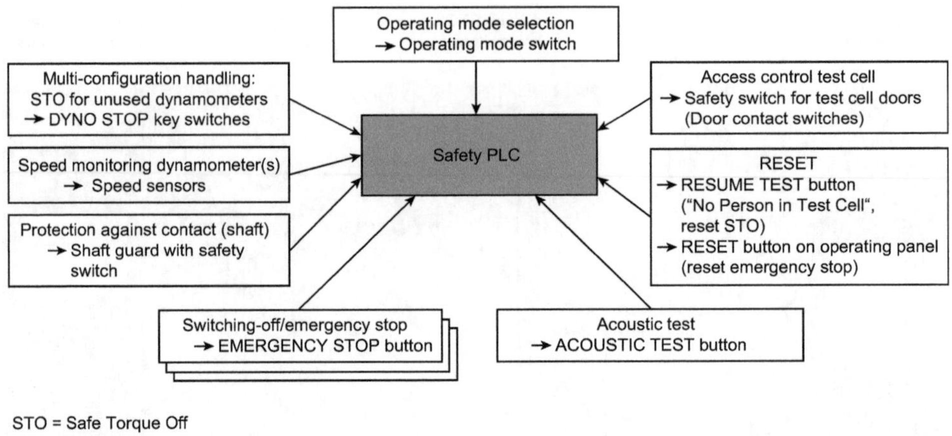

Fig. 4.64 Inputs (sensors and safety functions) for a safety PLC

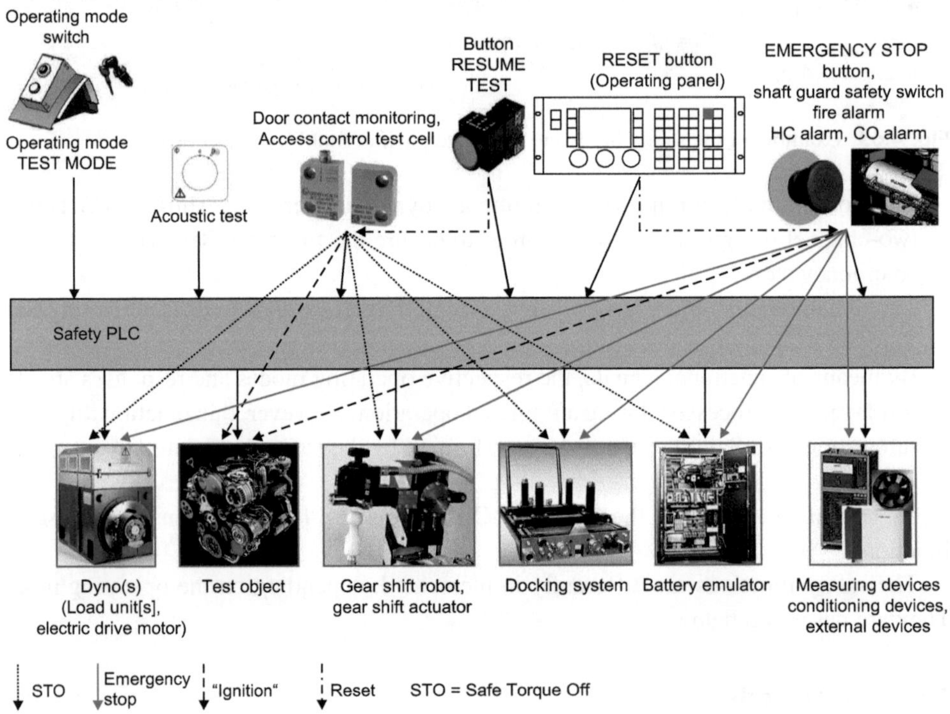

Fig. 4.65 Safe Torque Off and emergency stop in Test operating mode

dynamometer also changes over to the Safe Torque Off state. This satisfies the goal of protecting against unexpected start-up. A RESUME TEST button is also implemented in the hardware. Pressing this button enables resumption of the test run and confirms that no person is inside the test cell. Figure 4.65 shows how this could be designed.

4.8.6.2 Test Cell Service Mode

In TEST CELL SERVICE operating mode, all safety-related systems are switched permanently to the emergency stop state (see also Fig. 4.66). Dynamometers are switched off electromechanically. The unit under test's ignition system is switched off. In this operating mode, the test cell doors may be opened and remain open. People are allowed to be inside the test cell. Work on the mechanical system (unit under test, shaft, dynamometer, etc.) may be carried out. Before any jobs are performed on electrical systems, the main switch must be set to OFF.

4.8.6.3 Reduced Speed Mode

Internal Combustion Engine and Electric Motor Test Beds

In this operating mode, people may enter the test cell. The internal combustion engine or electric motor continues to run as long as the speed remains below the safe speed limit. When the speed limit is exceeded, the dynamometer immediately switches to the Safe Torque Off state (STO) and coasts down. At the same time, the unit under test is switched off. If the user triggers a stop and the system reaches standstill, the dynamometer also changes over to the Safe Torque Off state. This satisfies the goal of protecting against

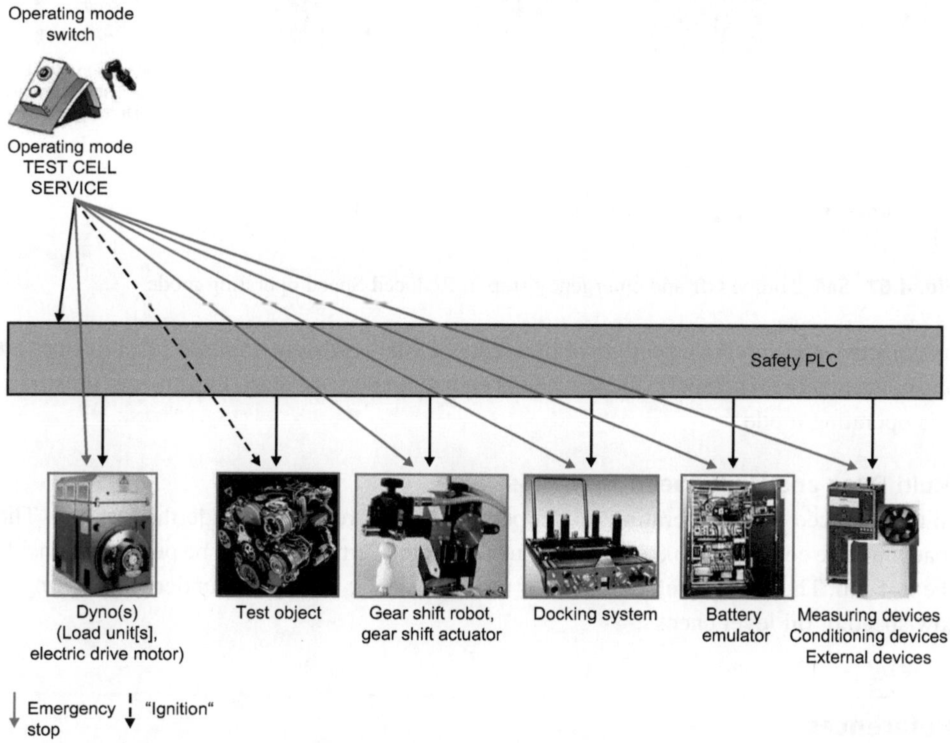

Fig. 4.66 Emergency stop in Test Cell Service operating mode

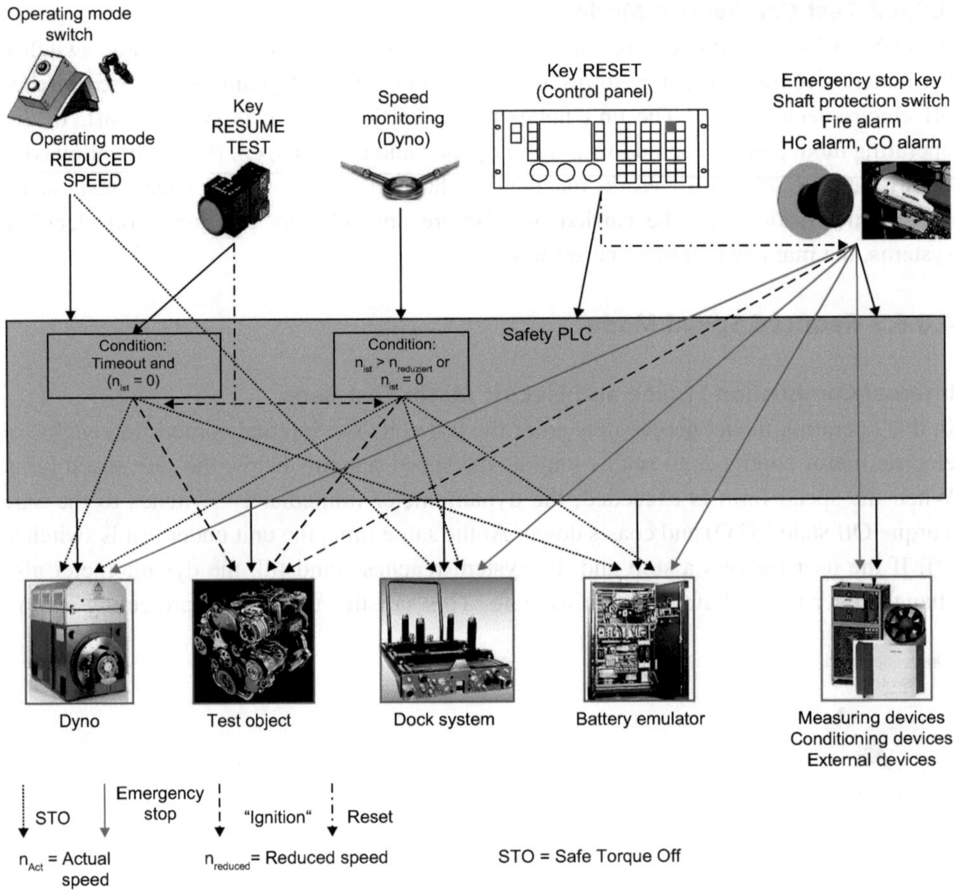

Fig. 4.67 Safe Torque Off and emergency stop in Reduced Speed operating mode

unexpected start-up. Push-buttons implemented in the hardware are used to enable restart and confirm that no person in the test cell is endangered by the start. Figure 4.67 illustrates this operating mode.

Multi-Dyno and High-Speed Test Beds

In the Reduced Speed operating mode, people are allowed to be inside the test cell. The load units are controlled to the Safe Torque Off state. Certain jobs can be performed inside the test cell. These may include visual or acoustic checks on the unit under test running at low speed or oil level checks.

References

1. ASAM e.V., ASAM Standards, 2013. [Online]. http://www.asam.net/nc/home/asam-standards.html. Accessed 28 Apr 2014

2. S. Albrecht, P. Decker, Schnelle Wege zur Restbussimulation. Automobil Elektronik **10**(3), 58–60 (2012)
3. G. Merker, C. Schwarz, R. Teichmann, Verbrennungsdiagnostik, in *Grundlagen Verbrennungsmotoren* (Vieweg + Teubner|Springer Fachmedien, Wiesbaden, 2012), pp. 309–378
4. OASIS, OASIS Advancing Open Standards for the Information Society—Home, 2014. [Online]. https://www.oasis-open.org/. Accessed 5 May 2014
5. OSLC, Open services for lifecycle collaboration—Home, 2014. [Online]. http://open-services.net/. Accessed 5 May 2014
6. Modelica Association, FMI—Home, 2014. [Online]. https://www.fmi-standard.org/. Accessed 5 May 2014
7. H. Lutz, W. Wendt, *Taschenbuch der Regelungstechnik*, 9th supplemented edn. (Harri Deutsch Verlag, Frankfurt am Main, 2012), p. 1471
8. J. Lunze, *Regelungstechnik 2—Mehrgrößensysteme, Digitale Regelung* (Springer Vieweg, Heidelberg, 2013). 7th revised edn
9. H. Kokal, Feed forward disturbance rejection by a multiple FIFO approach for varying system delay compensation, Dissertation, JKU Linz, JKU, 2013
10. E. Grünbacher, H. Kokal, M. Paulweber, L. del Re, M. Schmidt, Bandwidth extension of dynamical test benches by modified mechanical design under adaptive feedforward disturbance rejection, in *American Control Conference (ACC)*, Baltimore, 2010
11. P. Colanieri, H. Kokal, M. Paulweber, L. del Re, M. Schmidt, Feed forward disturbance rejection by a multiple FIFO approach for transient operation of an engine test bench, in *IEEE Multi-Conference on Systems and Control (MSC)*, Hyderabad, 2010
12. H. Kokal, J. Mayrhofer, L. del Re, M. Schmidt, Verfahren und Regelanordnung zur Regelung eine Regelstrecke mit sich wiederholendem Arbeitszyklus. EUR Patent PCT/EP2009/061009, 1 Sept 2008
13. M. Dank, D. Denger, S. Jones, H. Kokal, J. Mayrhofer, M. Schmidt, Prüfstandsanordnung. EUR Patent EP2264421 A2, 27 May 2010
14. D. Denger, A. Hirsch, G. Hochmann, J. Zehetner, The virtual vehicle at the engine test bed. MTZ **71**(9), 32–36 (2010)
15. D. Denger, P. Ebner, G. Hochmann, P. Hollerer, M. Schmidt, J. Zehetner, Effiziente Konzept Evaluierung und Fahrzeugvariantenkalibration am Motorenprüfstand, in *VDI Nachrichten*, Düsseldorf, 2009
16. M. Dank, H. Kokal, Model-based test of starter motors with a synchronous machine. ATZ **115** (7–8), 28–33 (2013)
17. D. Denger, S. Dragaschnig, G. Hochmann, H. Kokal, J. Mayrhofer, J. Schmidt, M. Schmidt, Prüfstandsanordnung. EUR Patent EP2264422 A2, 1 Jun 2010
18. F. Le Rhun, F. Pfister, C. Schyr, Effizientes Testen für mehr Energieeffizienz—Der Rollenprüfstand als mechatronische Entwicklungsplattform. ATZ **111**(11), 847–852 (2009)
19. A. Albers, M. Behrendt, T. Düser, R. Hettel, A. Schwarz, Extended flexible environment and vehicle simulation for an automated validation, in *Proceedings of the FISITA 2012 World Automotive Congress*, Beijing, 2012
20. C. Beidl, T. Düser, C. Gietzelt, T. Weber, Realer Kraftstoffverbrauch und manöverbasiertes Testen. ATZ **115**(4), 290–295 (2013)
21. D. Schramm, M. Hiller, R. Bardini, *Modellbildung und Simulation der Dynamik von Kraftfahrzeugen* (Springer Vieweg, Berlin, 2013)
22. Wikipedia, Test, 2014. [Online]. http://de.wikipedia.org/wiki/Test. Accessed 5 May 2014
23. Wikipedia, Mars climate orbiter, 2014. [Online]. http://de.wikipedia.org/wiki/Mars_Climate_Orbiter. Accessed 5 May 2014

24. F. Pfister, B. Schick, Die Zukunft hat einen Sensor—Location Awareness meets Powertrain Controls, in 4th *International Symposium on Development Methodology*, Wiesbaden, 2011
25. M. Benedikt, A. Hofer, D. Watzenig, J. Zehetner, A nearly energy-preserving coupling element for weak-coupled problems and co-simulations, in *V. International Conference on Computational Methods for Coupled Problems in Science and Engineering*, Ibiza, 2013
26. M. Benedikt, G. Stettinger, N. Thek, J. Zehetner, Extending co-simulation to the real-time domain. SAE Technical Paper 2013-01-0421, 2013, pp. 1–8
27. W. Kleppmann, *Taschenbuch Versuchsplanung—Produkte und Prozesse optimieren* (Carl Hanser Verlag, Munich, 2008)
28. M. Bollig, M. Büchel, F. Henzinger, R. Leithgöb, Methodik zur effizienten Bedatung modellbasierter ECU-Strukturen, in *1st International Symposium on Development Methodology*, Wiesbaden, 2005
29. R. Isermann, A. Schreiber, M. Weber, Modellgestützter Entwurf von Versuchsplänen zur dynamischen Vermessung von Verbrennungsmotoren, in *1st International Symposium on Development Methodology*, Wiesbaden, 2005

Software Perspective: The Test Facility 5

Following the detailed discussion of the three lower layers of the generic test bed architecture, as shown in Fig. 5.1 and as also mentioned in the introduction to the preceding chapters, this chapter deals with the layer of test facility-wide data processing. In this architectural layer, the main focus is on efficiency enhancement in large-scale development centers within the automotive industry. Accordingly, the goal in this layer is to analyze and optimize not only logistics processes, such as regular quality assurance management for measuring equipment, but also matters pertaining to test facility capacity management. In these processes, savings potential is achieved through economy-of-scale effects.

The layer additionally provides support to development teams all over the world by standardizing the ways they access data (measuring results or control unit parameters), which they require for the development process, making access fully independent of the actually used measuring and testing equipment. Consequently, this layer helps considerably to increase efficiency in the powertrain development process, meeting a basic requirement of all OEMs in the automotive industry.

5.1 Introduction to the Test Facility

5.1.1 Classification

A development project requires the execution of a wide range of tests with vehicles or vehicle components at a variety of different ambient conditions (partial load, full load, heat, rain, snow, etc.). The testing task therefore assumes a key position in the vehicle development process. The applied testing methods have a direct impact on the vehicles' quality in terms of energy consumption, noise behavior, safety or drivability criteria.

M. Paulweber, K. Lebert, *Powertrain Instrumentation and Test Systems*, Powertrain,
DOI 10.1007/978-3-319-32135-6_5

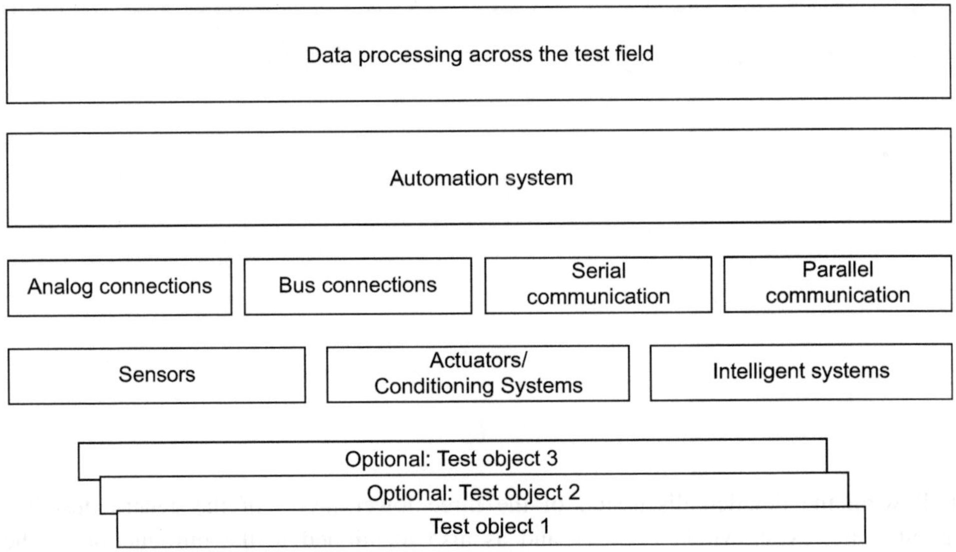

Fig. 5.1 Generic test facility architecture

Testing also greatly affects the market launch times of new vehicles, which automotive manufacturers regard as a major success factor.

State-of-the-art test facilities consist of the following fundamental components:

Test beds (hardware, automation software)
Measuring and conditioning systems
Data management solutions
Process and logistics systems

5.1.2 Challenges

Globalization has continued steadily over recent years, intensifying the competition among automakers even further. The increased sales possibilities outside of the domestic markets put additional pressure on car manufacturers. To remain competitive, they are forced to bring a greater variant diversity to the market in short development times. This has had a direct impact on the test facilities, which, likewise, have to be prepared to respond to the changing framework conditions with great speed and efficiency. Car manufacturers are therefore constantly on the lookout for new testing methods and flexibly extensible test systems. High cost effectiveness automatically led to globally utilized testing capacities and to test facilities being shifted to emerging economies. As a result of this trend, test facilities today are increasingly being used globally, requiring the necessary support through suitable solutions.

Competitiveness also calls for the development of novel testing methods, capable of raising the quality of test bed operation and enhancing its efficiency. To satisfy the complex testing requirements, widely differing technologies have been incorporated into powerful overall solutions. Sophisticated software solutions have led to a steady increase in the degree of automation in testing operation. Over the last years, OEMS have additionally been making an effort to support their test facility processes in the best manner possible through the use of dedicated software solutions (similar to SAP in business applications). One of the core issues of advanced test facilities is thus the capability to manage all test data with maximum efficiency.

The emphasis, in this respect, is not only on the management of (result) data but also on the evaluation of key performance figures, such as measurements of the test beds' runtimes and downtimes. These are certainly a major factor in test facility optimization. Many managers see the vision of fully automated test facilities with maximum utilization—also on a global scale—as a matter of the (near) future. The ability to handle the growing test complexity, combined with maximally utilized test beds, is a key factor of success for future test facilities.

5.1.3 Test Facility Processes

Test facility operation is characterized by diverse and alternating tasks. The superior quality standards and the growing economical demands necessitate consistent optimization of the work processes in test facility operation. Because of the high investment and operating costs for test bed technology, test operation accounts for a significant share of the total development costs. The quality standards, as well as the technical complexity and the desired cost transparency, require traceability of all processes throughout the test facility.

By interlinking and centralizing information in a controlled manner and ensuring the optimal provision of adequate information exactly where it is needed, it is possible to identify potential for economic success which should never be underestimated and very often remains untapped. In the powertrain development process, as in test facility operation, widely differing tools and systems are currently in use, each dedicated to a special purpose. Data exchange between these systems is often merely rudimentary. As a rule, identical data are managed in parallel across several systems. A central system that consolidates the entire process information and makes it commonly available is frequently non-existent—a circumstance which often produces duplication of work, unnecessarily long waiting times and unsatisfactory utilization of the test facility infrastructure.

Not least for quality and documentation purposes, the full range of process and process-accompanying information needs to be recorded. Data re-use for later utilization in simulation activities is not possible unless the complete data is recorded and the information is interconnected according to a logic that matches the process. There must be an efficiency-enhancing test facility management system in place that supports the standardization of tasks and makes the collected information, along with all the contained expertise, available for use. Central tasks also include quotation and order management, as well as utilization scheduling activities that cover all of the test facility's resources. Integrated acquisition of operating data and utilization times for each of the resources must ensure that the current state is transparent at all times.

Fundamental to the consistent utilization of the data available at the test facility (or "test factory") is a central filing system, i.e. a "data repository," that serves as a container for storing every piece of information. This information represents the test factory's entire knowledge. Building on this foundation, individual applications permit the actual use of the information pool. Applying this concept to the three basic phases of the test factory allows us to divide the tasks into logical groups according to their functions (Fig. 5.2):

Workflow management with task scheduling, monitoring and utilization optimization
Resource management with test equipment management
Data and information management.

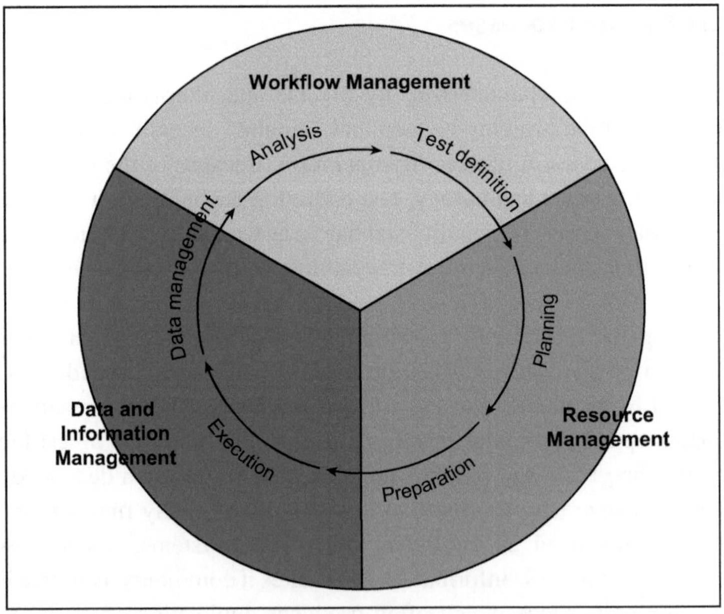

Fig. 5.2 Test facility management system process, based on the example Test Facility Status Monitoring of AVL's TestFactory Management Suite™ (TFMS) system

5.2 Workflow Management

An example of how the requirements mentioned above could be implemented is described below. It is based on the AVL Testfactory management concept, published in [2]. The model combines all of the processes needed to perform the testing tasks with a modular product approach that provides the tools to support workflow and management processes in test facilities (Fig. 5.3).

5.2.1 Task Scheduling in the Test Facility

Central to the Testfactory Management concept is the actual testing task. Fulfilling this task requires a range of service functions such as calibration data management, test factory scheduler (TFS), test order management (TOM), test equipment management (TEM), unit-under-test management (UUT), electronic test facility logbook (ELO), etc. Beside the classic service functions, the test order manager (TOM) is needed to steer and control the processes. On the one hand, the TOM integrates the service features and the related data and, on the other, TOM also ensures compliance with the defined standards (e.g. work instructions). The processes managed with the TOM undergo continuous optimization in line with the current state of knowledge. Test facilities control their resources in accordance with the test requirements and schedule the test bed capacities accordingly. Test order information is linked to the scheduled and actual data, which leads to an optimized development process. Smart and state-of-the-art workflow and scheduling systems permit linking these data.

The development test facility, in particular, faces unscheduled changes on a regular basis. Ongoing tests are interrupted, units under test or measuring devices are in need of repair, or employees fall sick. Every single change can affect the data configuration if, for example, a unit-under-test conversion changes the measuring tasks at short notice and makes it necessary to adapt the measuring setup's configuration.

In this extremely dynamic environment, a test facility management system makes sure that no information is lost, and that unnecessary meetings or idle times are avoided. Figure 5.4 shows how a consistent data management system, which links together all test facility data (e.g. test equipment, units under test or test orders), optimally supports planning activities in the test facility by their ability to automatically alert users to any scheduling conflicts in a graphical user interface at the moment they occur.

5.2.2 Utilization Optimization

The acquisition of key data that reflect operations performance is crucial if the operation of a modern test facility is to be efficient. Such data are often called key performance

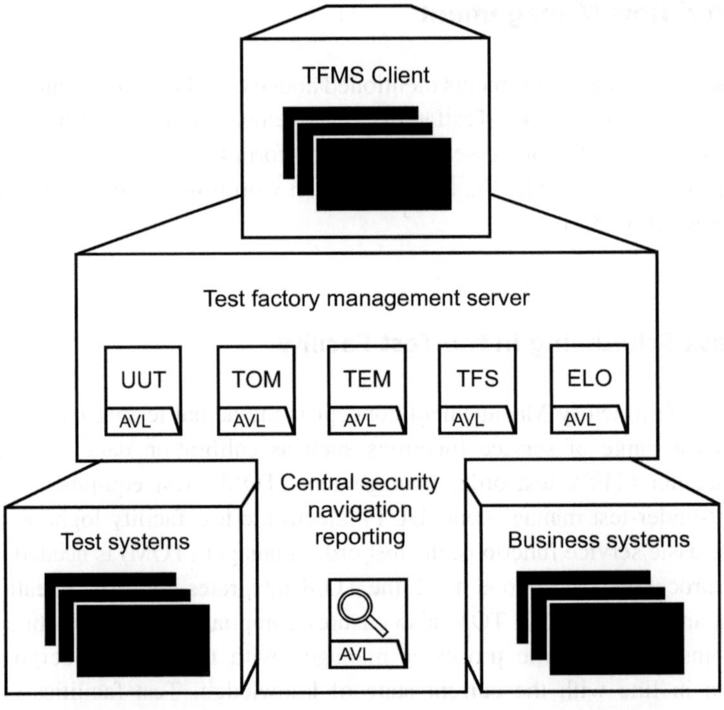

Fig. 5.3 Example of an implemented test facility management system

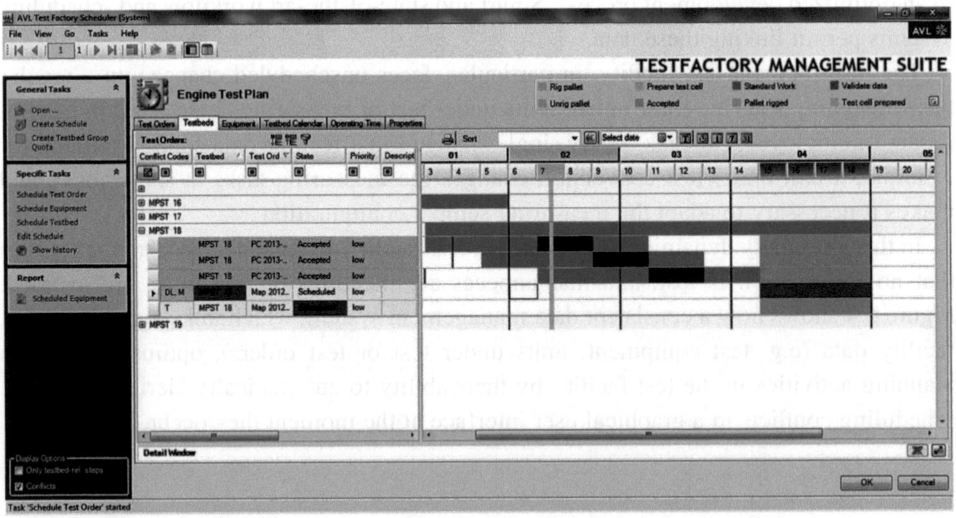

Fig. 5.4 Test facility scheduling (based on the example of AVL TFMS)

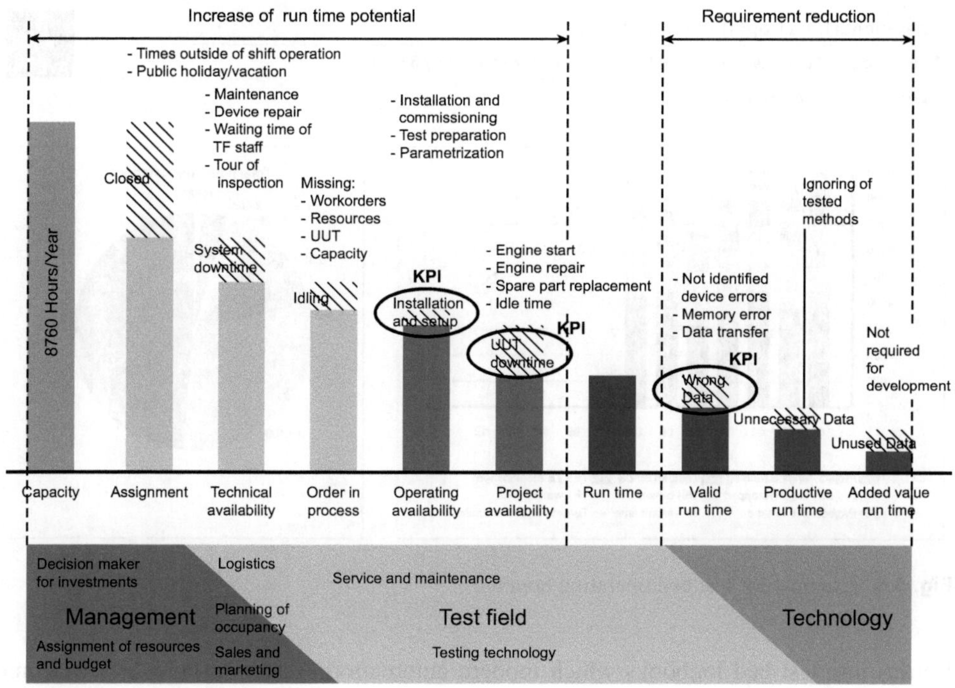

Fig. 5.5 Cost saving potential

indicators (KPI). Without such indicators, the sheer complexity of today's test facility environments would make it impossible to localize the parameters requiring fine-tuning.

The following questions indicate essential aspects of the analysis:

Which key performance indicators are needed to optimize operation?
What and how must be measured to determine the key performance indicators?

The overview shown in Fig. 5.5 compares the overall capacity of a test facility with the actually utilizable net availability. If a test bed is not operational, for example due to mechanical modifications, its availability will decrease, which is measured, for example, by a key performance indicator, in this case KPI #1.

Identification of the relevant key performance indicators and measurement of the necessary characteristic values allows optimizing the efficiency of test facilities by specific measures. To this end, it is necessary to acquire the data needed to generate the characteristic values. Examples are:

Operating times acquisition (e.g. test bed runtimes, testing times for units under test, downtimes of test equipment, time until the next calibration, etc.) (Fig. 5.6).

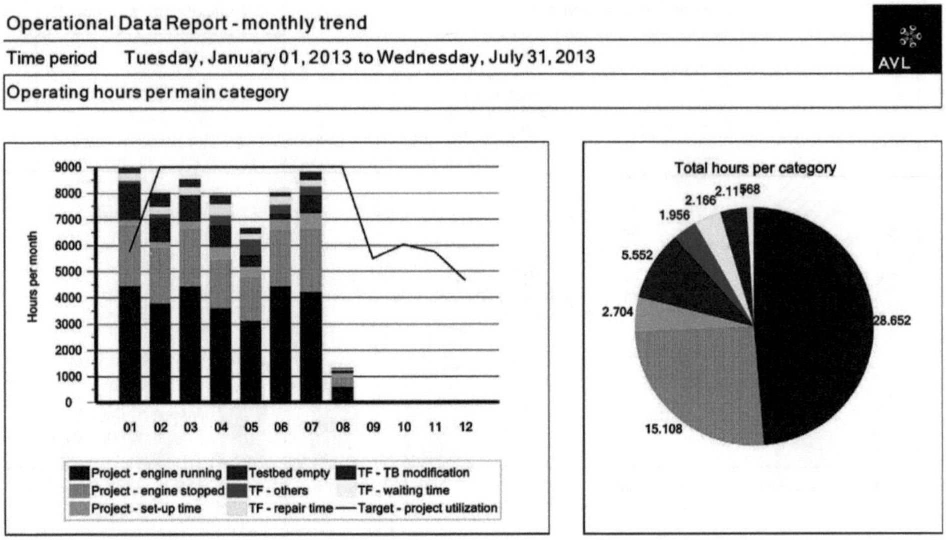

Fig. 5.6 Example for test bed operating hours

Entries into test-bed logbooks which modern automation systems provide in electronic form for easy post-processing

Entries relating to the used measuring systems, actuators, control and simulation systems as well as automation systems

Manual entries made by the test bed operator

Assignment of tags that assign the messages to their sources and indicate the criticality of the message, the effects of the message, the time of the messages and similar information.

For this purpose, data extracted from other management systems, e.g. SAP or PLM systems, are often linked.

Calibration databases that store the values and structures of control unit parameters in all variants and development stages

Databases for simulation model parameters (see Sect. 5.4.3)

Requirement databases, etc.

The KPI values are usually determined from the data sources online or in overnight sessions. They are visualized in data evaluation programs (see Sect. 4.7). It is common practice to view them in bar diagrams (Fig. 5.7), as these allow conclusions to be drawn on efficiency enhancement in the test facility:

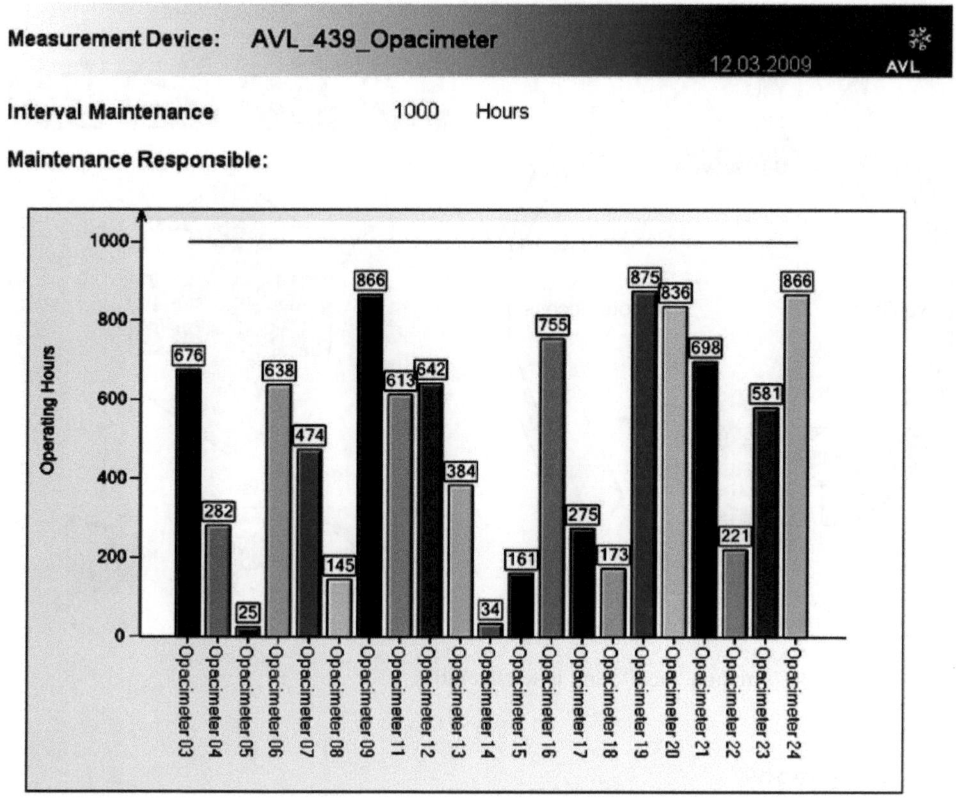

Fig. 5.7 Example for equipment use

Current data in trend chart
History
Classifications

5.2.2.1 Motivation

Modern test facilities are increasingly rated according to efficiency criteria, and face similar challenges as production operations, where investment-intensive machinery requires maximum capacity utilization. The requirement of fast response times in the event of unscheduled test interruptions calls for solutions (as shown in Fig. 5.8) that, firstly, display an overview of activities of all test beds within the test facility and, secondly, allow timely error diagnosis. Particularly where test beds and systems are operated unmanned, remote monitoring systems are of major significance.

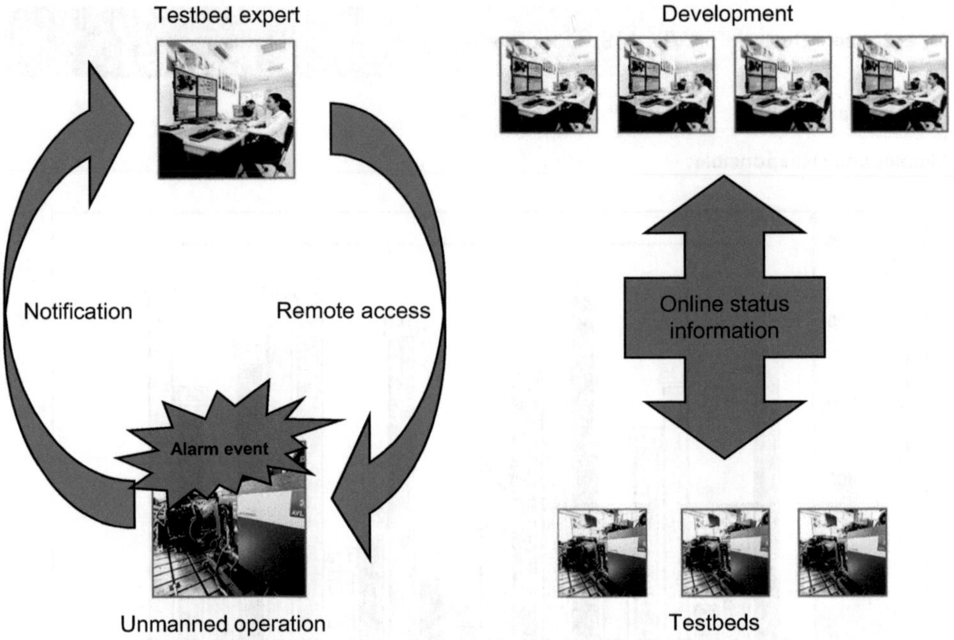

Fig. 5.8 Purpose of remote monitoring in test facilities

5.2.2.2 Solutions

In line with the specific responsibilities in the test facility, the remote monitoring process necessitates views from several perspectives.

Perspective 1: Development Engineer The responsible engineer wants to keep track of the test bed status. He or she must also be able to support the test bed operator from his/her work or home office over the Internet connection. For remote monitoring, test bed suppliers offer web-based solutions, allowing the different roles respective views of the test facility (see Fig. 5.9).

Perspective 2: Test Bed Operator Test bed operators would like to see the data of every test bed to get a detailed status overview of each of the test runs on one or several test beds.

Perspective 3: Test Facility Administrator Test facility administrators need an overview of the entire test facility. They must be able to determine the status of all test beds very efficiently and identify issues in a timely manner by navigating to a particular test bed and checking the details.

Safety-related functions are a central aspect when accessing test beds remotely because the personal safety requirements are very high. There must be no possibility of starting the

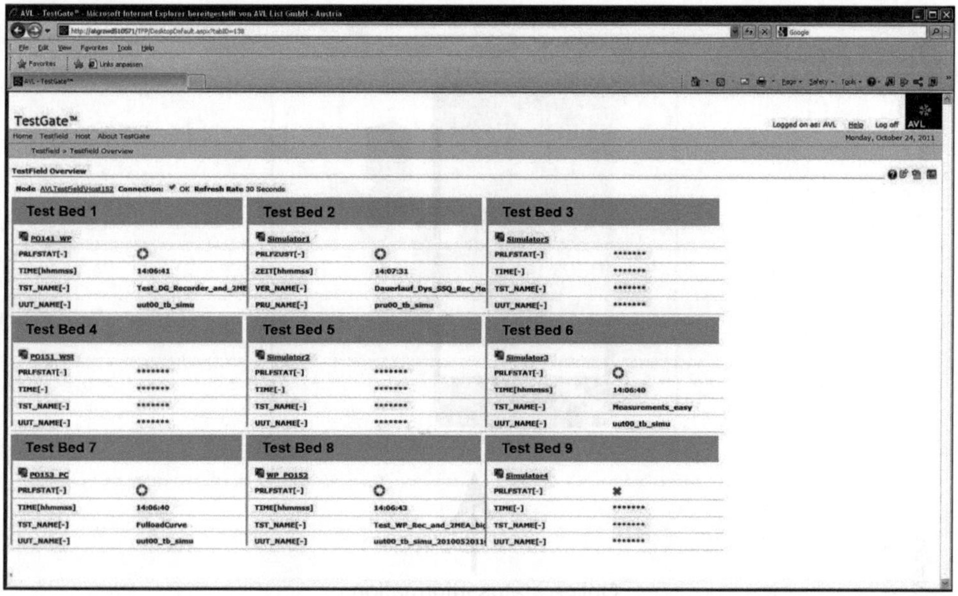

Fig. 5.9 Example of a test facility status overview (AVL TestGate™)

engine from a remote workstation while there are still people inside the test cell. For this reason, remote access systems are generally divided into two categories:

Remote monitoring solutions
Remote control systems

In the case of remote monitoring solutions, the software must ensure that users are unable to make any modifications to the test bed (e.g. start or operate an engine when people are in close proximity). Remote control systems must be integrated into the automation system and the test cell infrastructure. Special hardware measures, such as key switches on the test bed operating panel or door switches on the test cell door, can be taken to block any remote activities involving the unit under test.

5.2.2.3 Visualization on Mobile Terminals

Mobile devices, e.g. cell phones or tablets, have led to new applications in the automotive development sector, e.g. allowing the retrieval of test bed status information at any time. The high level of test bed automation with, in part, unmanned operation requires very flexible remote monitoring solutions, or remote control solutions that can be used on a variety of mobile platforms (Fig. 5.10).

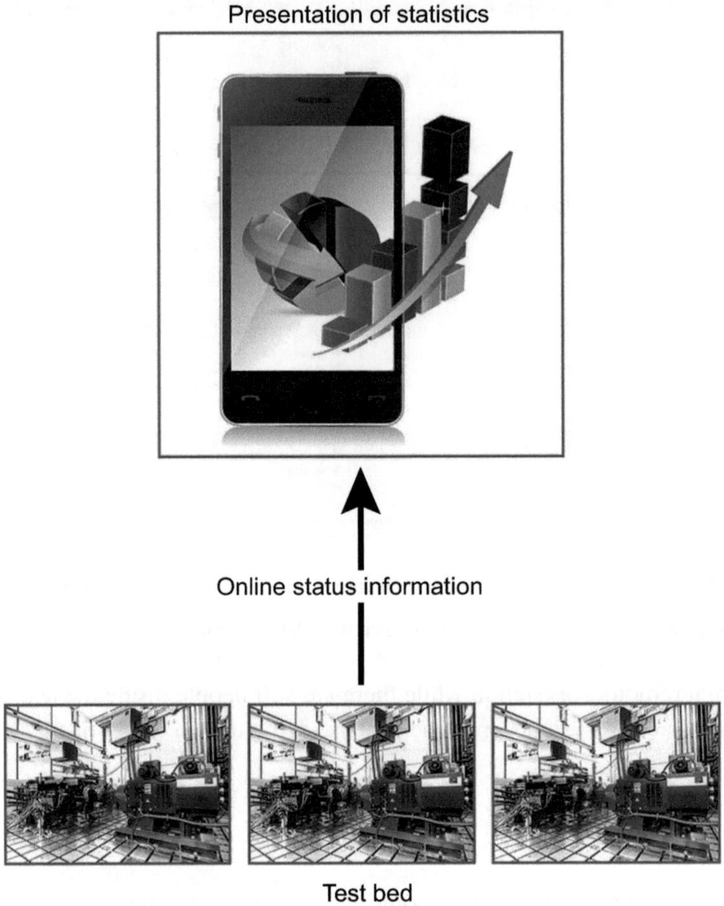

Fig. 5.10 Test bed status screen

5.3 Resource Management

5.3.1 Test Equipment Management Requirements

Managing test equipment, for example measuring devices, I/O modules, sensors or pallets, across their entire life cycle is one of a test engineer's main tasks. Central database software allows performing this task at a central location and provides support by applying a standardized methodology.

The measuring equipment management system serves as a central inventory and stores all of the relevant information about testing devices and their specific attributes, their current location, availability, calibration and maintenance states. Accompanying documents, such as images of the equipment or calibration instructions, can be added. The system includes report templates for various situations (e.g. use location of test equipment, calibration state, completed or pending maintenance activities) that allow report generation at the push of a button. Other features include the possibility to trace measurement validity with regard to a valid calibration whenever required.

The test equipment management system should be able to meet the following requirements:

Provide a central tool to manage and display all test equipment and measuring devices along their entire life cycle
Support test equipment monitoring in accordance with DIN EN ISO 9000
Provide status-controlled measuring equipment management including automatic notification of maintenance and calibration tasks and traceability of calibration and maintenance states
Manage calibration standards and calibration protocols including document versioning
Support the scheduling process in the test facility by providing information on the equipment's availability and state of calibration

5.3.2 Application Example

Figure 5.11 shows the example of a test equipment management system. The information is shown in a tree structure with hierarchically arranged detail information. So-called equipment types are used to define and manage not only individual pieces of test equipment but also complex combinations of test equipment in the form of test bed setups or pallet configurations.

The test facility administrator fills the structure with the available test equipment. In day-to-day test facility operation, this measuring equipment management system assists users in the following tasks:

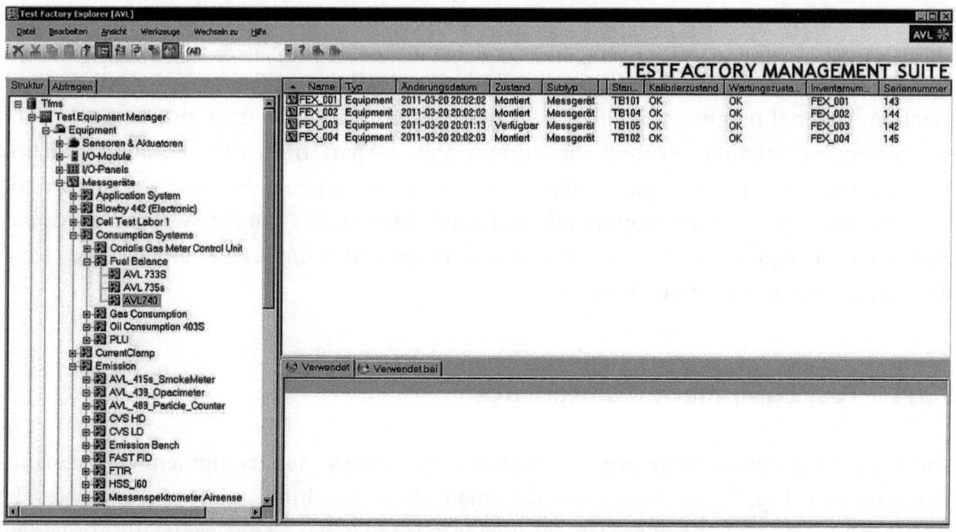

Fig. 5.11 Example of a test equipment structure (AVL's Test Equipment Manager)

Management of all physical test equipment based on equipment instances (individual pieces of test equipment)
Management and documentation of test equipment maintenance intervals and calibration details
Definition and management of test bed and pallet configurations
Generation of graphical or tabular reports
Use documentation for test equipment

All data are usually managed in an ASAM ODS database (see Sect. 4.1.2) on the host computer and are therefore accessible to all employees for test bed applications or software applications across the entire test facility. For the purpose of traceability, a full history is stored in the database for each piece of equipment. This allows tracing the use of each measuring device and their maintenance and calibration state for every point in time.

5.3.3 Test Equipment Data

For every physical test-equipment configuration an instance is created by using base types. These include:

Sensors and actuators
I/O modules
I/O panels
Measuring devices
Pallets
Test beds
Other (customer-specific configurations)

The test-equipment configuration description typically comprises serial number, type, function, channel number, manufacturer, quality and calibration data, maintenance data, the current installation site and the current state. Apart from that, detailed function descriptions, manuals and images may be added as attachments. For every configuration a full history is persistently stored in the database. This means that the system records any changes to the equipment's current location (storage/test bed/etc.) and any changes to its state (available/in use/defective/etc.).

5.3.4 Test Equipment Maintenance

The measuring device management system also supports test equipment maintenance activities (see Fig. 5.12). Users are informed about pending maintenance work. The system generates predefined reports for this purpose that can be electronically forwarded to the "owner" or the related "user group."

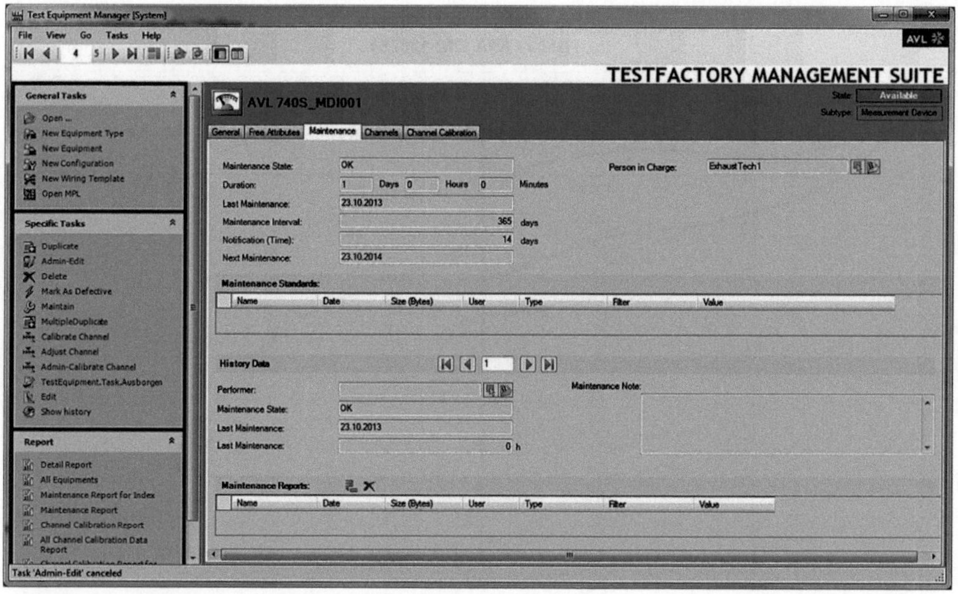

Fig. 5.12 Example for maintenance information

5.3.5 Sensor Calibration Data

To facilitate calibration jobs on the sensors, the calibration data management system must be made accessible across the entire test facility. This is independent of whether the calibration is carried out in the test room or elsewhere. Strictly speaking, one should use the term "adjustment," however in test bed environments the term "calibration" has become established and will therefore be used on the following pages (see also Sect. 3.4.5).

Different versions of calibration runs are stored persistently in the database and serve as a reference for future tasks. All measuring points are traceable back to the sensor and its calibration state. One of the relevant standards in this respect is ISO TS 16949.

Calibration information may contain a calibration interval, a calibration reference and a calibration state. The system calculates the next calibration date and shows when the last calibration took place. With certain test equipment, a scaling table is kept with up to 100 entries, containing the raw value and the physical value.

As with test equipment maintenance, the system automatically alerts users to any pending calibration jobs. After the calibration, the user is required to make the relevant changes to the calibration protocol (name of the person who performed the calibration, new calibration state). For an example of the graphical part of a calibration protocol, see Fig. 5.13.

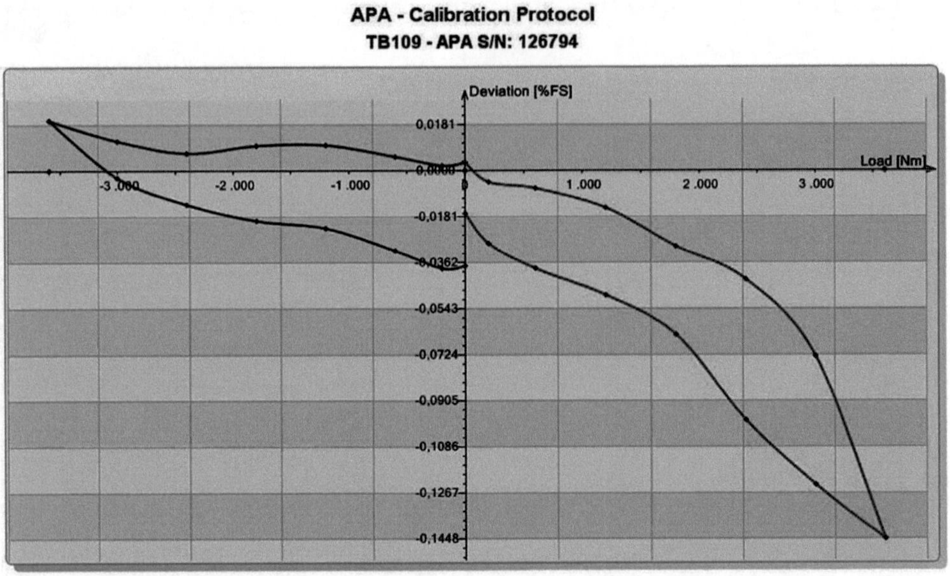

Fig. 5.13 Example of the graphical part of a calibration protocol

5.4 Data and Information Management

5.4.1 Result Data Management

A current U.S. study [3] on the topic of global data use has revealed that the volume of data is likely to increase eightfold over the next 5 years. However, the number of staff employed to manage such data is expected to remain basically unchanged.

It is safe to assume that the same can be expected for test facilities. As the complexity in product development further increases, so will the costs and demands on the test beds and methods, along with the necessary quantities of data (see Fig. 5.14). The same trend is apparent in the growing requirements in terms of measurement certainty and measurement speed. On top of that, the number of measuring devices and measurement channels is steadily increasing too. This is reflected in ever-larger quantities of data.

With development departments often being legally required to store measurement data for more than 10 years, this poses a tremendous challenge not only for the IT infrastructure, but even more so for the entire data management methodology. The question is how to manage these quantities of information so that it is easily retrieved when needed.

Organizations with several locations are in need of information pertaining to all resources involved in the development process. With the number of the measuring systems in use and the global utilization of test equipment increasing steadily, this task is becoming more and more complex. Already today, more than 15,000 different testing devices for test facilities need to be managed. Having access to information about their

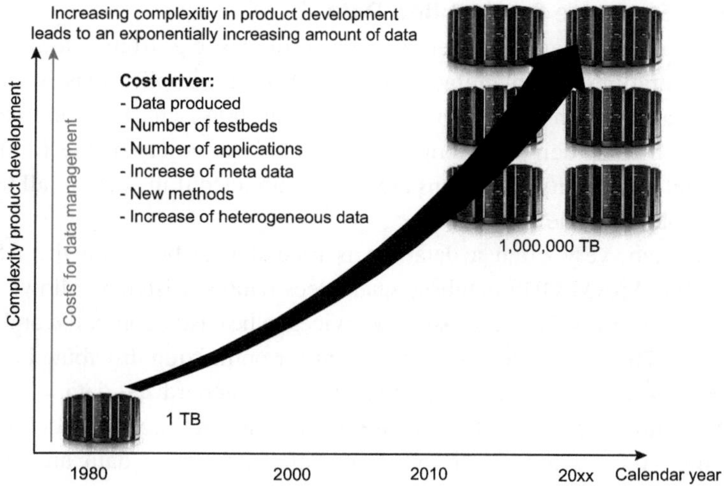

Fig. 5.14 Growing complexity in product development

Fig. 5.15 Data management on the test bed

current location and state including maintenance and calibration details is essential for efficient utilization across all sites. For that reason, such data have to be interpretable, analyzable and available anywhere around the globe (Fig. 5.15).

5.4.1.1 Filing Structure for Test Bed Data

Many of the results originate from heterogeneous sources, e.g. from automation systems of different generations and manufacturers. Test beds for applications, such as engine development, transmission testing, endurance testing, noise and vibration tests, control unit tests or crash tests, deliver incompatible results in a variety of file formats. In this context, efficient test facility operations are only guaranteed if the data of all test beds are easily comparable with one another.

For data storage, very often a database is used that is based on the ASAM ODS standard [1]. The ASAM ODS database guarantees data consistency, allowing the measuring results of all test beds and measuring devices to be easily compared and exchanged across locations. This is an essential requirement for optimizing distributed development work. Various analysis tools are able to process these standardized data.

As Fig. 5.16 shows, the ASAM ODS base model is a meta-model for defining test-bed and measuring-data storage. In simple terms, the measuring data are organized as follows:

Structure data that allow assigning the measuring data to the test context

Measuring matrices that contain the actual measuring values (raw data or calculation data).

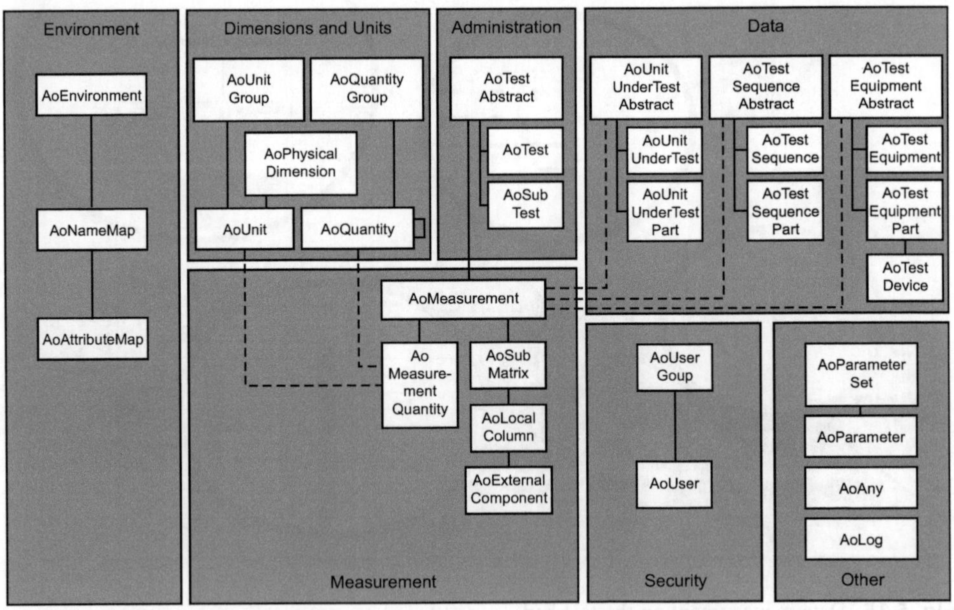

Fig. 5.16 ASAM ODS base model

Furthermore, the ASAM ODS model also defines the data structures for storing test run, unit-under-test and test equipment information. Another section of the model defines the security parameters that are needed to protect the measured data from unauthorized access.

To ensure the measuring data's comparability and exchangeability, all quantities, units and physical dimensions are defined centrally and independently of the measured data. All measuring data refer to a central quantities catalog. This ensures the comparability of the measuring quantities. Accordingly, "maximum torque", for example, has the same name everywhere along with a compatible unit, fully independent of the test bed or the evaluation system.

5.4.2 Calibration Data Management

5.4.2.1 Calibration Process

Depending on the control unit (engine, transmission, battery, ABS control, etc.) there are currently close to 60,000 parameters that require determining and tuning either on the test bed, in the vehicle or through simulation. The optimized values are then passed on to the relevant control unit (during calibration) and ultimately determine the vehicle's behavior. The parameters are thus crucial not only to emissions and drivability, but also to the vehicle's active safety.

The intensified trend toward standardization and modularization is causing automotive manufacturers to increasingly use the same drive components for different vehicle applications. As every vehicle requires specific adjustments to the control unit parameters, this is producing a continuous upward trend in the number of component variants in need of calibration. Based on a defined process, the parameters are applied, checked and released. This raises the demands on calibration quality drastically because possible calibration errors could lead to major recalls if the effects of error propagation concerned all or the majority of calibration variants. Accurate data flow control throughout the control unit calibration process is therefore essential to successful and error-free calibration project completion.

At the beginning of a control unit calibration project, the parameters are first preset based on prior knowledge and historical data (pre-calibration). The intended goal is to achieve a data status which, for example in an ECU calibration, allows starting the engine and running initial measurements at steady-state measuring points. The individual parameters are then managed in calibration databases from which the responsible calibration engineers export data with which they can continue their work. Next, these data are tested on the test bed and, where required, modified (re-calibrated) to achieve better results. The modified parameters are once again saved in the database as modification packages together with their complete history and the reasons for the changes.

After the fine-tuning process on the test bed or in the vehicle on the road, the optimized calibration data are merged, verified and released. To obtain the best possible results, it

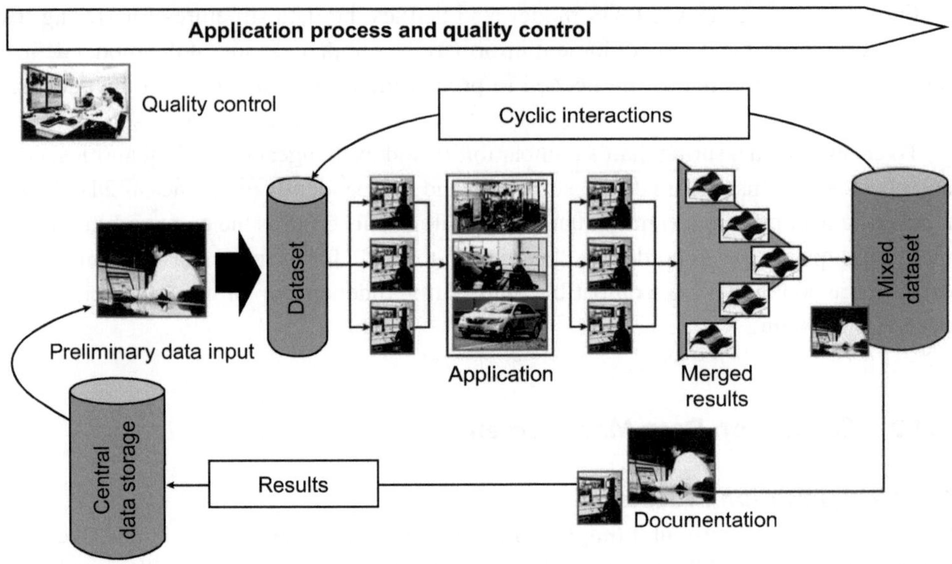

Fig. 5.17 Control unit calibration process

must be ensured upon calibration data release that all values were processed, and that every single one is the optimal solution. Particularly in the case of parameters accessed by multiple calibration engineers focusing on different objectives—e.g. cold-start behavior versus emissions at engine start—it is essential to establish clearly defined responsibilities for every parameter, so that calibration errors and significant differences in the variants become apparent in good time. The released calibration data serve as a basis for the next test-bed or in-vehicle optimizations.

As illustrated in Fig. 5.17, calibration data usually undergo several optimization rounds throughout the calibration process—from basic calibration, emissions calibration and consumption determination to drivability development and on-board diagnosis (OBD). Throughout the calibration process, progress and result quality are monitored and documented for reference in future calibration projects.

5.4.2.2 Calibration Database

Calibration database systems or calibration data management (CDM) systems are basically capable of representing the complete calibration process and consolidating all of the required data management work steps. Solutions like these are generally able to significantly improve and simplify all work steps relating to the handling and distribution of calibration data. As a result, the individual calibration engineers are much better able to focus on their actual work, which has a positive effect on overall efficiency in project execution.

The functionalities in the calibration data management system are very extensive and start with process support for the involved engineers. Each team can work from different locations on one or multiple projects. Automatically generated notifications and warnings sent by e-mail facilitate the workflow. Clearly defined responsibilities, which are saved in the system, enable global operation, ruling out any redundancies and conflicts between the calibration engineers' different work packages well in advance.

One of the most essential features of calibration databases is the ability to merge calibration results, which different people have created with different kinds of calibration systems, to create consistent overall calibration data snapshots. These data snapshots are usually created every 1 or 2 weeks, and the system makes them available to all of the involved engineers. As very old calibration data snapshots are rarely needed for tests, this ensures a high level of data security and data consistency. Thanks to reports and clearly defined versioning processes, calibration databases offer the possibility to trace every work step. Project managers, in particular, benefit a great deal from reports because they can check the project progress any time they like and immediately initiate corrective measures.

As there are several systems on the market that allow modifying, creating and combining calibration data, the ASAM consortium [1] has already standardized a large number of the data formats used in this environment (see also Sect. 4.1.2). The most common data formats supported by almost all systems in this environment are listed below:

A2L (ASAM MCD-2 MC, see also Sect. 4.1.2): The data format A2L is the description file for measuring and calibration quantities in an engine control unit (ECU). Within the A2L file, all characteristic quantities in the ECU are defined and described which external systems are able to access. The A2L file additionally stores the address of the individual parameters in the control unit microcontroller's hex dump.

HEX/S19: These are the most widespread formats used for ECU memory maps. The hex file contains several memory segments. The most important ones are:

- the operating system
- the control software that maps the engine control logic,
- and the calibration section containing the characteristic quantities to be calibrated (calibration data).

DCM: This is one of the most common exchange formats for calibration data. Its contents are used to distribute changes in the calibration parameters from the test bed or from the vehicle. Essentially, this format is a human-readable text format.

CDF (ASAM CDF): This is one of the most common exchange formats for calibration data. Its contents are used to distribute changes in the calibration parameters from the test bed or from the vehicle. This file format is an exchange format based on XML.

With the help of these formats it is possible to build a calibration database and trace every change made by the involved calibration engineers. Without this standardization, it would hardly be possible to ensure that all calibration parameters are correctly stored and that they behave identically in real-world vehicle operation.

A suitable interface to other calibration programs, the eCDM interface enables operation across calibration applications. The eCDM interface offers ways to exchange description files or calibration parameters with other calibration programs and is already available from a number of manufacturers.

5.4.3 Model Management

As shown in the previous sections, the significance of test-bed simulation in powertrain development is continuously on the rise. Accordingly, reliable simulation models are crucial for efficient and dependable test execution. These consist of two parts:

simulation model structure description, and
parameter values for each of the parts of the simulation structure (simulation model parameters).

Being split into two parts makes it possible to manage multiple variants of a unit-under-test component with the same simulation structure. This structure is very similar to that described in the previous section: The parameter values in this section correspond to the description file in A2L format mentioned in the previous section; the simulation model structure description in this section corresponds to the HEX/S19 file in the other section.

Likewise, the workflow needed to manage simulation data requires a similar approach to that for calibration states in control unit software, as described in the previous section.

We therefore distinguish between the actual model and the parameters. This facilitates comparability and the full traceability of changes. The model itself can be stored as an attachment. The model parameters are typically subject to many changes (when creating different variants of one and the same component described by the simulation model), meaning that lower data volumes are generated if the parameters are managed separately from the simulation model files.

For test facility operation, constant access to information about the reliability and relevance of the simulation models and parameters is extremely important. State attributes in the database for each of the model components can provide this capability. Meaningful states are:

In progress
Completed but not yet checked
Model checked but not yet calibrated
Model calibrated but not yet checked
Model calibrated and checked

The individual model creation steps are stored in a workflow. To begin with, experts at one of the component development departments develop a model (state "In progress"). Once the model is completed (state "Completed but not yet checked") it has to be checked and released through a quality assurance system (state "Model checked but not yet calibrated"). The model is now capable of describing many variations of a component, provided that the relevant parameters have been defined.

Defining simulation model parameters for a specific component instance usually requires measurements on a component test bed. Often, time-consuming identification procedures are needed. Once the simulation parameters are found, they are stored in the model database under the status "Model calibrated but not yet checked". A final quality assurance step, in which the simulation results are compared with measuring results from corresponding component test beds, leads to the status "Model calibrated and checked".

Only after ensuring that simulation results and measured values from corresponding component test beds are consistent can a model be used for XiL or HiL test-bed testing.

Simulation models or simulation parameters from different simulation domains are kept in a central database which we call a model management database. This is where simulation engineers can store their changes or download a new model version.

A structure customizable by the user with appropriate access rights and defined responsibilities ensures good traceability in the model management database.

5.4.4 Name Management in the Test Facility

All measured, calculated or simulated quantities are normally addressable by their names, which are also used in parameterizations, display screens or evaluations. To enable reuse of these parameterizations or evaluations in international organizations at different test beds or in different tests, standardized names for quantities are required. These names are often referred to as "normnames".

The total of these names is called a namespace. To operate test facilities efficiently, several namespaces are needed.

Namespace for equipment on the test bed
Namespace for the unit under test
Namespace for standardized tests
Global namespace for an entire organization

During runtime, a measurement, simulation or calculation quantity gets assigned one name from each of the namespaces. This process is also referred to as name mapping. One and the same test, which uses the names from the test facility's namespace for standardized tests in its test scripts, can be executed in various different test bed configurations. An example is a control unit function test which is executable on an MiL, SiL, HiL and powertrain test bed and requires the engine speed as measuring quantity. In German-speaking countries, this quantity in the namespace "unit under test on the test bed" is assigned the designation "Drehzahl" from the namespace of the German organization. On test beds in the U.K. the name "Engine Revolutions" was introduced. To run a test in a global test facility, both national names are mapped to the designation "Engine Speed" in the namespace "Test" during the test preparation phase. Separation of the "Test" namespace from the (local) test facility namespaces also allows easy reuse of unit-under-test setups on different test beds.

In many existing test bed automation systems, namespaces are not separated adequately. Instead, one namespace is used for multiple types of parameter units such as test runs and test bed configurations. This leads to a loss of flexibility and extra work when changes are needed because name changes in test runs often have to be carried out synchronously in multiple test bed description parameter files too. The lower complexity during commissioning due to the lower number of names to be defined leads to higher maintenance requirements during test facility operation. The optimum can only be found in specific applications, for example it depends on whether or not test descriptions are to be executed on differing test bed configurations.

The quantities typically have other additional attributes which are also managed in a central database together with the names. Examples of such attributes are:

SI unit
Quantity value range
Alternative display names that can be selected in the test bed visualization system
Additional display units (are often taken from a database that conforms to ASAM ODS)
Quantity description (e.g. by measuring site and physical quantity)

The namespaces are usually managed in a database. The ASAM ODS standard is often used for this purpose (see Sect. 4.1.2).

5.4.5 Result Data Warehouse

Apart from ASAM ODS databases, further de-facto standard formats for storing data from particular application areas have become established over the years. Unfortunately, this trend continues, so we have to assume that the number of data formats will not stop increasing any time soon. For these specific data formats, appropriate evaluation tools are on the market. Test engineers wishing to evaluate different data are therefore often forced to use several different tools with differing operating patterns. However, users would obviously like to be able to look for relevant information in all of the stored data using a uniform post-processing concept.

The solution to this issue is to consistently separate navigation from technical data evaluation. By doing so, it becomes possible to offer data navigation tools that operate across evaluation systems and formats. Such tools allow searching the data without having

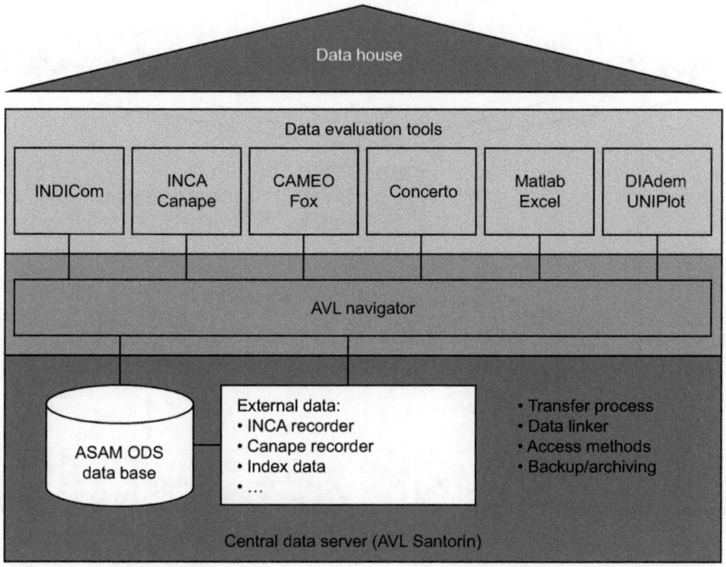

Fig. 5.18 Example of navigation integration

to convert them into a standard format first. Test-bed measured data are often saved to a database directly in ASAM ODS format. Other test results remain in their original format and are linked to the test bed data in the database. During import of the data (whether in ASAM ODS or a special file format), the data warehouse system extracts meta-data from the measured data files and creates searchable attributes in the database together with links to the bulk of the measured data in their native format. Figure 5.18 shows an example of such a concept which is called a data warehouse.

When searching for results, the extracted meta information is used for navigation. This allows quick search and navigation procedures. Special data access drivers convert the bulk data from their native format to the data warehouse exchange format. Evaluation tools can read this exchange format and therefore have access to all data formats.

New data formats are easily integrated into this framework and are then immediately available for navigation to all users. A further advantage of this method is that the navigation system is standardized across all kinds of application areas. All users use the same system to find their data and always work with the same software tool.

5.5 Data Management in Distributed Test Facilities

Globally positioned OEMs operate test beds at different locations, which they merge to shared virtual test facilities. Development tasks at one location can so be tested on test beds at another location to obtain a global test-bed utilization optimum. Keep in mind, though, that it is still necessary to leave the responsibilities with regard to setup,

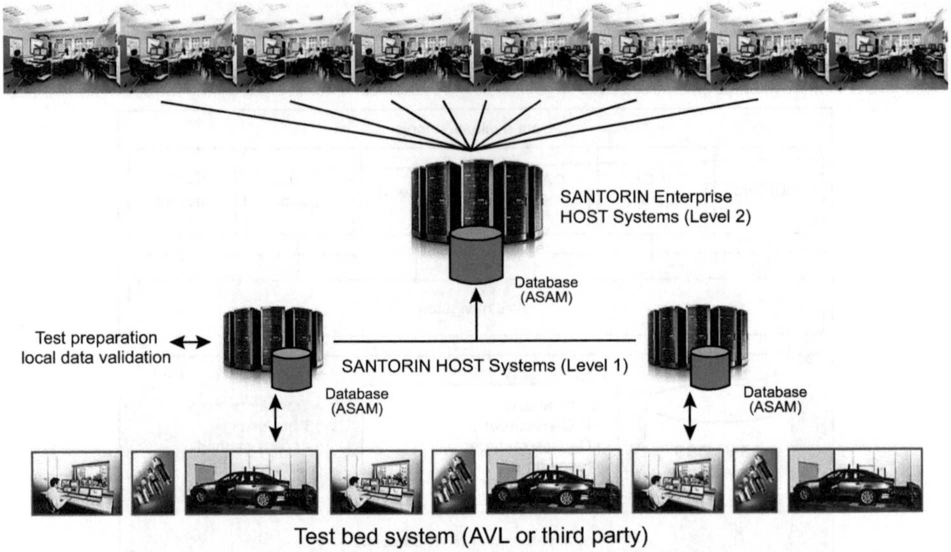

Fig. 5.19 Measured data management levels for test facilities

modification and maintenance with the relevant location itself in order to ensure efficient and effective test bed work.

One of the key factors in a virtual test facility distributed across different locations is a measured data management system which provides access to a central evaluation database for all systems and users as described in the previous chapter. One approach that has proved successful in practice is a three-level evaluation structure:

Level 1: Local test bed level
Level 2: Local server level (HOST Systems Level 1)
Level 3: Global server level (HOST Systems Level 2)—data warehouse

Level 1 and 2 are implemented for every location to provide each of them with a secure central evaluation platform. Level 3 is typically provided globally and, in most cases, represents a central measuring data server for all development locations (Fig. 5.19).

The measured data on the lower level are replicated to the next level. During this process, the measured data might also be filtered or restructured. After measurement completion, the Level 1 test bed systems send their data to the local HOST system (Level 1). On this local level, the data are typically filtered and processed. After their release, they are then forwarded to Level 3, i.e. HOST System Level 2. With the data being available across systems and locations, development engineers draw their data from HOST System Level 2.

References

1. ASAM e.V., ASAM Standards, 2013. [Online]. http://www.asam.net/nc/home/asam-standards.html. Accessed 28 Apr 2014
2. AVL List GmbH, AVL Testfactory Management Suite™—TFMS Test Facility Management, 2014. [Online]. https://www.avl.com/avl-testfactory-management-suite-tfms. Accessed 5 May 2014
3. J. Gantz, D. Reinsel, Extracting value from chaos. IDC iView Study, 2011

Index

M. Paulweber, K. Lebert, *Powertrain Instrumentation and Test Systems*, Powertrain,
DOI 10.1007/978-3-319-32135-6

MIX
Papier aus verantwortungsvollen Quellen
Paper from responsible sources
FSC® C105338

If you have any concerns about our products,
you can contact us on
ProductSafety@springernature.com

In case Publisher is established outside the EU,
the EU authorized representative is:
Springer Nature Customer Service Center GmbH
Europaplatz 3, 69115 Heidelberg, Germany

Printed by Libri Plureos GmbH
in Hamburg, Germany